Modern Oncology
An A–Z of Key Topics

To my family,
those who are gone, and those who remain:
with love and gratitude

Modern Oncology
An A–Z of Key Topics

Alastair Munro
BSc FRCP (Edin) FRCR
Professor of Radiation Oncology
Ninewells Hospital and Medical School
Dundee, UK

LONDON • SAN FRANCISCO

Greenwich Medical Media Limited
137 Euston Road
London
NW1 2AA

870 Market Street, Ste 720,
San Francisco,
CA 94102

ISBN 1 900 151 11 1

First Published 2001

A catalogue record for this book is available from the British Library.

Typeset by Charon Tec Pvt. Ltd, Chennai, India

Printed in the UK by the Alden Group, Oxford

Distributed by Plymbridge Distributors Ltd and
in the USA by Jamco Distribution

Visit our website at: **www.greenwich-medical.co.uk**

Contents

Preface

Disraeli famously compared a political career to the ascent of a greasy pole. His simile is equally apt as a description of any attempt to keep abreast of developments in the investigation and management of cancer. Just when you think that you have fully grasped a subject, new information, often from a previously unrelated area, erodes your understanding. Back down the pole you go. Change is rapid and involves a broad range of disciplines: from physics to genetics; from biochemistry to psychology. Advances occur within specialist enclaves, each with its own jargon and neologisms; new meanings are attached to old expressions. The language that we use can act as a mechanism for isolation when what is needed is integration and communication. Just because the cellular basis of cancer is largely founded in disordered communication, there is no reason to echo this in our approach to the understanding and management of the condition.

A glossary provides a simple method for restoring some sense of order to our communications. It is more than just a phrase-book for oncological travellers. There is the space to define and discuss a concept but brevity and the alphabetical organisation ensure ease of use. The juxtaposition of disparate concepts is often a helpful stimulus to creative thought, whether in science, medicine or the visual arts: a glossary provides ample opportunity for such fruitful juxtapositions.

This book should prove useful, not just to those who wish a rapid introduction to some of the concepts underpinning modern oncology but also to established scientists and clinicians who wish to explore areas outwith their immediate expertise. Those who have no experience of oncology will find it useful as a means of rapidly accumulating the new and various knowledge that is required to understand an area of investigation that cuts across so many disciplinary boundaries. Instead of having to consult a series of weighty tomes, the information is in one place, is expressed concisely, and the alphabetical organisation means that the pursuit of cross-references is easy.

Progress against cancer will occur when we are able to cut through the artificial barricades that separate the disciplines involved in studying and managing cancer. This book is an idiosyncratic attempt to help dismantle those barricades.

A. M.
Dundee
July 2001

Acknowledgements

I would like to thank all those patients, trainees and colleagues who, wittingly or unwittingly, have contributed to this book. They have provided the stimulus and the knowledge, I have provided the chaos, idiosyncrasies and errors. My thanks also to all those who contribute to the scientific literature: editors, authors, publishers, secretaries, reviewers. If it were not for their dedication and, largely unremunerated, efforts then I would have had nothing about which to write. Science is one of the last bastions of altruism, long may it so remain.

Abscissa

The x (horizontal)-axis of a graph.

Absolute Rate Difference

Absolute rate difference is the difference, in a comparative study, between the event rate in the treatment arm and the event rate in the control arm. Survival with adjuvant chemotherapy for breast cancer is 46.6% at 10 years; the corresponding figure for controls is 39.8% (data from EBCTG1 breast overview):

- Absolute rate difference = 46.6 − 39.8% = 6.8%.
- **Relative risk (p. 264)** reduction is 6.8/39.8% = 17%.
- **Number needed to treat (p. 216)** is 1/0.068 = 15.

Absolute Risk

Absolute risk simply indicates for an individual or group the probability of a specified event (usually adverse) occurring within a specified time interval. It is related to the concepts of relative risk and absolute rate difference. See also **Relative Risk (p. 264)**, **Absolute Rate Difference** (above).

Absorbed Dose

The energy deposited per unit mass of material by ionising radiation is the absorbed dose. The unit is the Gray (Gy), where 1 Gy = 1 joule kg^{-1} (J kg^{-1}). An older measure, the rad (radiation absorbed dose, 100 ergs g^{-1}) is equivalent to 1/100 of a Gray, i.e. 1 Gy = 100 rad, 1 rad = 1 cGy (centigray). Defined in this way, Gy is independent of the material irradiated or the type of radiation. Note, however, that the biological effect of any given dose of radiation may well depend on these two factors (see also **Relative Biological Effectiveness (p. 264)**, **Quality Factor (p. 253)**). It is sometimes difficult, particularly when dealing with indirectly ionising radiation, to measure dose directly — in these circumstances it is often useful to use **kerma (p. 170)** as the measure of radiation quantity.

Absorption Rate Constant

Absorption rate constant is a term used in pharmacology to describe the kinetics of drug absorption. It is defined as:

$$\frac{\text{rate of drug absorption}}{\text{amount of drug remaining to be absorbed}}.$$

Accelerated Fractionation

This term is only loosely definable. It implies that, keeping total dose and dose per fraction constant, a course of **fractionated radiotherapy (p. 131)** is given in an overall time which is less than standard. The problem is that there is no agreement on what constitutes a 'standard' time. In some departments, 6 weeks is standard, in others it is 3–4 weeks. A useful operational definition would be to define as accelerated any radical treatment given in <3 weeks. The problem here is that a department used to a 6–7-week schedule would define as accelerated a treatment given over 5 weeks; this 'accelerated' treatment would still take longer than the 4-week schedules used in many departments. All things are relative, accelerated fractionation is no exception.

Pure acceleration, by the conventional definition, would involve giving 60 Gy in 30 fractions in 3 weeks: this is not tolerated because the rate of dose-accumulation exceeds the tolerance of the acutely responding tissues. In practice, a combination of acceleration and **hyperfractionation (p. 156)** is used, e.g. 70 Gy at 2 Gy $fraction^{-1}$ day^{-1} in 7 weeks might be comparable with 72 Gy at 1.6 Gy (t.i.d.) in 5 weeks.

The biological rationale behind accelerated fractionation is that, by keeping the overall treatment time as short as possible, it will minimise the opportunities for clonogenic tumour cells to repopulate during treatment.

Accelerated Repopulation

The term 'accelerated repopulation' was introduced into radiobiology to describe an apparent increase in the rate of cellular proliferation in response to treatment with a **cytotoxic (p. 84)** agent such as radiotherapy. In squamous carcinomas of the head and neck, for example, the doubling time before treatment may be ~40–60 days: by the end of a course of radiotherapy the doubling time may be as short as 5–8 days. Although initially described in tumours, the phenomenon can also apply to normal tissues. There are several mechanisms whereby the proliferative rate of a population of cells might increase in response to an external stimulus such as radiation: the **growth fraction (p. 144)** could increase; the **cell loss factor** could be reduced; the duration of the **cell cycle (p. 52)** could be shortened. The clonogen doubling time will, with accelerated repopulation, approach the minimum **potential doubling time** (T_{pot}) **(p. 241)** for that particular population of cells: whether there is more to accelerated repopulation than simply, by reducing cell loss to zero, an unmasking of the latent T_{pot}, is debatable.

Another controversy concerns the timing of any accelerated repopulation: does it start with the first radiation treatment or is there an initial lag period ('hockey stick' relationship) before it begins. The lag has been estimated at ~20 days.

The practical consequence of accelerated repopulation is that overall treatment times should be kept as short as possible. If a lag period of 20 days exists then some schedules such as **CHART (p. 62)** may avoid the problem altogether since the treatment will have finished before accelerated repopulation has been established. With more protracted treatments, over ≥4 weeks, accelerated repopulation means that each extra day of treatment corresponds to a loss of effective dose of ~0.4–0.7 Gy. Unanticipated breaks in treatment will result in a loss of tumour control unless some compensatory strategy is used. Such strategies include treating twice a day on days subsequent to missed treatment, thus keeping overall treatment time the same; treating during weekends, again to keep overall time constant; and adding extra fractions at the end of treatment to compensate for missed days, the extra dose required being calculated by formulas based on the BED equation.

Accreditation

With the development of clinical standards and other benchmarks of performance, the process of accreditation is becoming increasingly important. Accreditation may be defined as a process based on a system of external **peer review (p. 229)** using written standards designed to assess the quality of an activity, service or organisation. It is an excellent concept.

Accreditation, as a process rather than as a concept, suffers from all the disadvantages and pitfalls associated with peer review. It has other additional problems associated with it. It could, potentially, be used as an instrument of fiscal control. It could be hijacked by special-interest groups and be used to push through changes that suit a minority, rather than the majority. Minorities are sometimes, but not invariably, right. A set of values imposed by a minority can never, by definition, be democratic. There is no doubt that accreditation has the potential to raise the standards of those organisations or systems that perform poorly; it also has the potential to reduce excellence to a lowest common denominator. Accreditation relies heavily on assessment of process, rather than outcome. The underlying assumption is that better process will produce better outcome.

Accrual

Accrual is a jargon term used to describe the recruitment of patients to a clinical study. The accrual rate that can reasonably be anticipated is a critical component of study design. Not all patients attending will be eligible for the study, not all eligible patients will enter the study and not all entered patients will complete the study: some horses are unfit for entry into the Grand National, some will not line up at the starting tape and some fall at the fences.

Problems of accrual and attrition have to be anticipated when the study is designed. The ultimate statistical power of the study will depend on the event rate in evaluable patients. There is no point in embarking on a study that is never going to accrue a sufficient number of patients. The issues of sample size, **power (p. 241)** and statistical **significance** were successfully addressed by Jacob Bernoulli over 250 years ago, and yet an appreciable number of published studies on comparisons of treatment continue to accrue so few patients that they would be unable to prove anything: this is the problem of the Type II statistical error.

Accuracy

Accuracy reflects the extent to which a statement or measurement corresponds to the truth: it is possible to be accurate without being precise. It is accurate to say that John Lennon was nearly 6 feet tall; it is, however, less precise than stating that his height was 5 feet 10 inches.

The accuracy of a measurement has to do with how closely it corresponds to the true state of affairs. It can be distinguished from **precision (p. 243)** since precision is no guarantee of accuracy. The value 3.05 would be an imprecise estimate of π, but it is more accurate than an estimate of 5.100087253. This latter estimate is more precise, containing detail down to the ninth decimal place: it is, unfortunately, more inaccurate. An accurate, but imprecise, measurement is usually preferable to a precise, but inaccurate, one. The ideal measurement is, of course, both accurate and precise.

Acquiescence Response Set

The term used to describe the tendency of subjects to agree with whatever is put to them. This is a problem in, for example, questionnaire studies of **quality of life (p. 253)** and it is certainly a problem on ward rounds: 'we're feeling much better today, aren't we Mrs Smith?', to which the only reply is, 'there's something happening here and you don't know what it is — do you Dr Jones?'

Acquired Immunodeficiency Syndrome (AIDS)

This emotive term defines a time-point within the natural history of infection with the **human immunodeficiency virus (HIV) (p. 151)**. As with any externally defined event in the natural history of disease, it is a somewhat arbitrary imposition, but one which has pragmatic usefulness in defining a certain degree of immunosuppression. The criteria for defining the existence of AIDS in an individual who tests positive for HIV infection have evolved since the first descriptions of the clinical syndrome in the early 1980s. At that time, it was the coincidence of immunosuppression with *Pneumocystis carinii* pneumonia and Kaposi's sarcoma that led, in 1984, to the identification of HIV infection. The current definition (Centers for Disease Control, 1993) regards any of the following as

AIDS-defining diagnoses in an individual known to be HIV positive:

- Severe immunosuppression: as defined by a $CD4^+$ T-lymphocyte count of <200 cells μl^{-1} or a $CD4^+$ level $<14\%$.
- Candidiasis of bronchi, trachea or lungs.
- Candidiasis, oesophageal.
- **Cervical cancer, invasive.**
- Coccidioidomycosis, disseminated or extrapulmonary.
- Cryptococcosis, extrapulmonary.
- Cryptosporidiosis, chronic intestinal (>1 month).
- Cytomegalovirus disease (other than liver, spleen or nodes).
- Cytomegalovirus retinitis (with loss of vision).
- Encephalopathy, HIV-related.
- Herpes simplex: chronic ulcer(s) (>1 month); or bronchitis, pneumonitis or oesophagitis.
- Histoplasmosis, disseminated or extrapulmonary.
- Isosporiasis, chronic intestinal (>1 month).
- **Kaposi's sarcoma.**
- **Lymphoma, Burkitt's** (or equivalent term).
- **Lymphoma, immunoblastic** (or equivalent term).
- **Lymphoma, primary, of brain.**
- *Mycobacterium avium* complex or *M. kansasii*, disseminated or extrapulmonary.
- *Mycobacterium tuberculosis*, any site (pulmonary or extrapulmonary).
- *Mycobacterium*, other species or unidentified species, disseminated or extrapulmonary.
- *Pneumocystis carinii* pneumonia.
- Pneumonia, recurrent.
- Progressive multifocal leukoencephalopathy.
- *Salmonella* septicaemia, recurrent.
- Toxoplasmosis of brain.
- Wasting syndrome due to HIV.

The tumours associated with the diagnosis of AIDS are highlighted. The importance of malignant disease in AIDS should not be underestimated, nor should the global importance of HIV infection in the context of patients with cancer be ignored. In sub-Saharan Africa up to 30% of patients newly diagnosed with cervical cancer are HIV-positive.

The AIDS epidemic has another, equally important, impact on oncology. It can serve as an example of how research can offer at least partial solutions to clinical problems. The pace of research into HIV–AIDS has far outstripped anything that has occurred in cancer research and this has been reflected in the ability to devise and test clinically useful therapies. The lessons that can be learned include:

- Identification of specific molecular targets (**reverse transcriptase** (**p. 272**), protease) and development of specific therapies that act on these targets.
- Research driven by a specific focus, as opposed to diffuse curiosity.
- Ability to take potential therapies rapidly from the laboratory to the clinic — streamlined bureaucracy and licensing procedures.
- Ability to carry out rapidly randomised trials that ask, and answer, specific questions.
- The importance of consumerism and advocacy: vocal demands by pressure groups result in action by government and other funding bodies.
- Partnership between the pharmaceutical industry and research agencies.

Cancer research suffers from the disadvantage, compared with research into HIV–AIDS, that it has been around a while, a period long enough for disillusion and cynicism to have set in, and long enough for vested interests and orthodox dogmas to dictate what is, and what is not, studied. There was no rulebook for HIV–AIDS whereas cancer research has been governed over the years by a reluctance to depart from set principles — even though those principles have not proved particularly successful in delivering effective therapies. The argument has always been that of the First World War generals, 'one final push', a plea for yet more resource. There has been a conspicuous reluctance to question whether the assumptions made are, in fact, valid. Are the screening programmes for new drugs effective or are clinically useful drugs eliminated because they fail in a rodent model? A treatment that does not produce **complete responses** (**p. 70**) is defined as ineffective, even though it may be clinically useful in terms of providing a patient with a period of clinical stability. Novel mechanisms of action, e.g. inhibition of angiogenesis, will require clinical assessments that have not yet been devised: an **angiogenesis** (**p. 14**) inhibitor is unlikely to produce a complete response, but it might lower the incidence and slow the growth of metastases. Experience with HIV–AIDS should help re-educate oncologists in how to think rationally from basic principles rather than simply repeating the mistakes of the past.

Activity

The activity of a sample of a radioactive isotope is defined as the rate of decay:

$$\Delta N/\Delta t,$$

where ΔN is the change in the number of radioactive atoms and Δt is the time interval. Its unit is the **becquerel** (Bq) (**p. 26**), where

$$1\ \text{Bq} = 1\ \text{disintegration s}^{-1}.$$

It appears in the definition of the **transformation constant** (**p. 313**), and by extension, in the **half-life** (**p. 147**) equation.

Actuarial

An actuarial method combines estimates of probability into a tabular structure. These methods originate in accounting and in the calculation of insurance premiums, but their main role in oncology is in survival analysis. The essential feature of the method is that where a data set is incomplete and not all patients have been followed up until death (i.e. the data are censored), the information we have can be used to predict the future pattern of the **mortality rate** (**p. 207**) in the whole group. The most commonly used actuarial method for survival analysis is the **Kaplan–Meier Method** (**p. 170**). Where survival times are only known approximately then the **lifetable** (**p. 179**) method can be used.

Actuarial Assumption

Actuarial assumption is an assumption used in the construction of a **lifetable** (**p. 179**) — it is assumed that the distribution of censoring is uniform during a given time interval. For example, if 30 patients enter a 6-month interval and at the end of the period 20 patients are known to have survived and six are known to have died, then there are four censored survival times. We can, using the actuarial assumption, calculate the average number of patients at risk during the interval as:

$$30 - (4/2) = 28.$$

Actuarial Methods in Calculating Complication Rates

Only those patients who survive long enough are candidates for late complications of treatment. Crude estimates of complication rates may underestimate the incidence of late complications associated with a particular regimen or technique: patients who have already died from cancer cannot be assessed for treatment-related complications. Actuarial methods offer only a partial solution to this difficulty. If the last survivor develops a complication then this will exaggerate any complication rate calculated actuarially. Figure 1 shows an actuarial plot from a series of patients in which the crude complication rate is 29% (6/21).

The complication-free survival would be calculated actuarially as 50% (not 71%, i.e. 100 − 29%). The use of the **crude rate (p. 81)** would have led to underestimation of the likely complication rate.

Figure 2 has data for a study in which the last surviving patient suffers a complication — the crude complication rate is still 6/21 (29%).

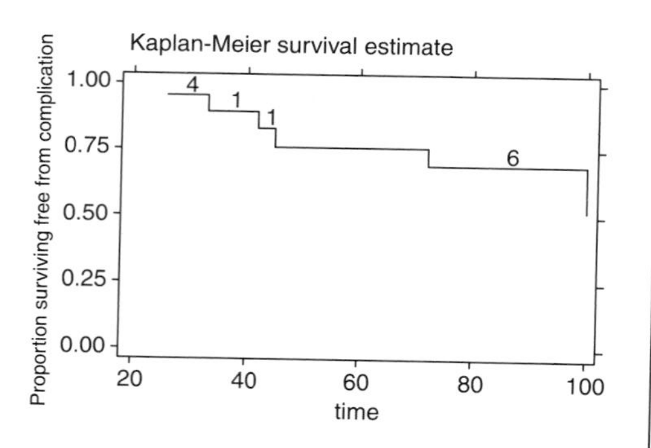

Figure 1.

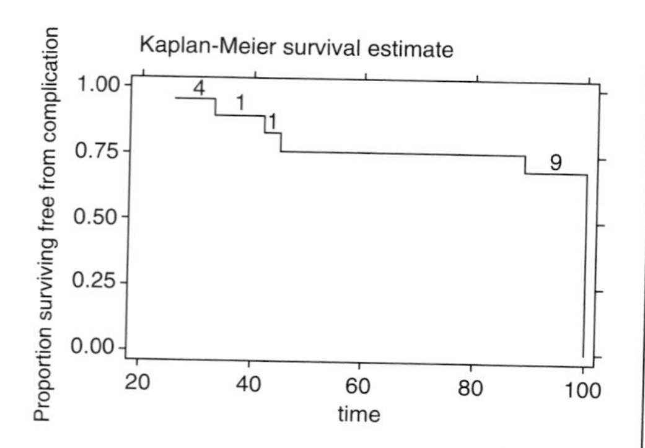

Figure 2.

The complication-free survival would be calculated actuarially as 0%. There is no easy solution to the problem of trying to use **censored data (p. 60)** to estimate the probability of events occurring later in follow-up: the important thing is to be aware of the pitfalls associated both with crude rates and with actuarial calculations.

Actuarial Lifetable

An actuarial lifetable is used to analyse time-dependent data, such as survival after treatment, when the exact dates of events are not known. Events may occur between follow-up visits: at the earlier visit they are recorded as absent but are recorded as present at the next visit. The precise date of onset cannot, however, be determined. The actuarial method simply defines the time of onset as the midpoint of the interval used to construct the table. A similar assumption (the actuarial assumption) is made concerning the timing of censored observations. Intervals commonly used for lifetables based on clinical studies are 3 months to 1 year.

Acute (Early) Side-Effects of Cancer Treatment

The side-effects of cancer treatment can, for convenience, be divided into early and late: in the jargon, acute and late morbidity. Acute effects are those that occur at and around the time of treatment. **Late effects (p. 174)** are those that arise at least 3 months after the completion of treatment. The distinction is not always clear cut — early effects may fail to resolve and lead directly to late effects, so-called **consequential late effects (p. 73)**.

The acute effects mainly affect those normal tissues that divide rapidly: the skin and its appendages (such as the hair follicles), the lining of the gastrointestinal tract, the bone marrow and the germinal cells of the **testis** (**p. 306**). Radiation will affect only those normal tissues within the irradiated area (patients who have radiotherapy to the chest do not go bald as a result of radiotherapy). Chemotherapy can potentially affect all rapidly dividing normal tissues, though the precise spectrum of toxicity varies from drug to drug.

Normal tissue	Symptoms and signs of cytotoxic damage
Oral mucosa	sore mouth, mucositis, pain on swallowing
Oesophageal lining	pain and difficulty on swallowing
Laryngeal lining	stridor, hoarse voice
Lining of stomach and small bowel	nausea, vomiting, diarrhoea
Lining of large bowel	diarrhoea, mucus discharge, bleeding, tenesmus, itch
Lining of bladder	frequency, dysuria, cystitis
Skin	itch, burning sensation, erythema, desquamation, hair loss
Germinal epithelium of testis	azoospermia, infertility
Bone marrow	leucopenia, thrombo-cytopenia, anaemia

Repair of acute damage to dividing tissues is an energy-dependent process. This may, in part, explain why fatigue is such an important, but often unrecognised, acute effect of cancer treatment.

Adaptation Response

Adaptation response is the process where cells, which appear to be in stable arrest in terms of the **cell cycle** (**p. 52**), can re-enter the cell cycle and resume growth and division. The process mainly depends upon environmental factors.

Address Labels

Address labels are specific amino acid sequences on newly synthesised proteins that ensure that the protein is directed to the appropriate site within the cell for processing. The **signal recognition particle** (**p. 286**) is one such destination; another, for which the PACS-1 sequence is the label, is the trans-Golgi network.

Adduct

A complex that forms when a chemical, such as a drug, binds to a biological molecule, such as DNA or protein. The formation of adducts between drugs and DNA is one of the mechanisms of action of **cytotoxic drugs** (**p. 84**).

ADEPT (Antibody-Directed Enzyme Prodrug Therapy) = ADC (Antibody-Directed Catalysis)

An enzyme-linked antibody is directed against an antigen that is tumour-specific. The enzyme is then used to convert an inactive precursor (**prodrug**) (**p. 246**) to an active cytotoxic agent. Since this activation will only occur where there is enzyme, and since, thanks to the antibody, there is only enzyme where there is tumour, the approach offers a means of improving the **selective toxicity** (**p. 282**) of cancer therapy. Examples of this approach include:

- MAb BW 431/26 conjugated to alkaline **phosphatase** (**p. 233**), the antibody binds to CEA, the enzyme activates **etoposide** (**p. 119**) phosphate (prodrug) and the locally active cytotoxic agent is etoposide.
- MAb 323/A3 (antibody to a membrane glycoprotein found in carcinomas) conjugated to β-glucuronidase, epirubicin

glucuronide is given as prodrug, epirubicin is the active agent.

This approach is also sometimes called antibody-directed catalysis (ADC).

Adhesion Molecules

A group of diverse molecules essential for normal tissue function and that has important roles in cell growth and differentiation, embryological development, and wound repair. There are four main families:

- Integrins.
- Immunoglobulin superfamily, e.g. N-CAM, PECAM-1.
- **Cadherins** (**p. 39**).
- **Selectins** (**p. 281**).

The first three interact with proteins. Selectins and lectins are unusual in that they interact not with proteins but with carbohydrate molecules, e.g. on the surface of white blood cells or endothelial cells. Lectins are of plant origin but have found wide application as stimulators of proliferation (mitogens) in the culture of mammalian cells. They are also useful in blood grouping and in the immunocytochemistry of carbohydrates.

Disruption of the function of **adhesion molecules** (**p. 8**) is of critical importance to several aspects of the **malignant process** (**p. 185**): uncontrolled proliferation, local **invasion** (**p. 167**) and **metastasis** (**p. 198**) formation.

Adhesion molecules include **E-cadherin** (**p. 111**), **fibronectin** (**p. 128**), the integrins and **desmogleins** (**p. 94**).

Adjuvant Treatment

Adjuvant treatment is treatment given in addition to definitive treatment in an attempt to lower the risk of relapse. The concept is based on the following assumptions and observations:

- There has been abundant time during the natural history of a cancer for a malignant cells to spread beyond the clinically demonstrable tumour (micrometastatic disease).
- There may be microscopic local or disseminated disease remaining after apparently successful treatment for cancer.
- No imaging technique currently available can detect single cancer cells.
- Accurate tumour markers are available only for a few rare tumours (e.g. choriocarcinoma).
- Microscopic residual disease is undetectable using currently available technology.
- A single cancer cell persisting after initial treatment is sufficient to cause recurrence.
- Dispersed, small-volume tumours are more sensitive to treatment than bulky tumours.
- Treatment given earlier is more effective than treatment given later: cells are more actively proliferating; drug resistance is less likely.

It is implicit within the concept of adjuvant treatment that there will be three categories of patient:

- Patients who have no residual disease after their initial treatment and for whom adjuvant treatment can confer no benefit.
- Patients who have residual disease but whose disease is refractory to therapy and who are destined to recur in spite of adjuvant treatment.
- Patients who have residual disease and whose disease will be eradicated by adjuvant treatment.

ADME

ADME is an acronym that describes the key processes by which the body handles an administered drug or chemical:

- Absorption.
- Distribution.
- Metabolism.
- Excretion.

Pharmacokinetics (**p. 232**) is the study of the time-course of these processes. The analogous processes at the cellular level are:

- Uptake.
- Intracellular distribution.
- Biotransformation.
- Efflux.

Adenosine Diphosphoribosyl Transferase (ADPRT)

Adenosine diphosphoribosyl transferase (ADPRT) is an enzyme concerned with repair of **DNA damage** (**p. 98**). It is bound to **chromatin** (**p. 63**) and is inhibited by **nicotinamide** (**p. 214**).

Ageing

The incidence of most forms of cancer rises with age. This is particularly true for cancers of the large bowel and prostate. This suggests that, for some reason, the development of cancer could be regarded as part of the ageing process. There is probably no one simple explanation for this. Many different factors are undoubtedly involved and these could include diminished immune competence leading to relative failure of immune surveillance; increased accumulation of mutations in nuclear DNA over time leading to increased risk of acquiring malignant genotype; age-related changes in the expression of tumour suppressor genes; mutations in mitochondrial DNA causing increased mitochondrial production of carcinogenic-**free radicals** (**p. 132**); and increased duration of exposure to environmental carcinogens. The change in the age distribution of the population in the developed world, with an increasing proportion of older persons, has implications for oncology. We can expect to see an increased number of older people with cancer, in both absolute and relative terms.

Age-Standardised Rate

Age-standardised rate (ASR) is a summary measure of a rate that a population would have if it had a standard age structure. Standardisation is necessary when comparing several populations that differ with respect to age because age has such a powerful influence on the risk of cancer. This is particularly true when cancer incidence in the developed world is compared with the incidence in the developing world. The most frequently used standard population is the World standard population. The calculated incidence or **mortality rate** (**p. 207**) is then called World Standardised incidence or mortality rate. It is also expressed per 100 000 population.

A-Kinase (Cyclic AMP-Dependent Protein Kinase)

The enzyme through which the intracellular messenger, cyclic AMP, acts to control cellular functions. It transfers phosphate groups from ATP to threonine and serine residues on target proteins. It activates phosphorylase kinase and glycogen **phosphorylase** (**p. 234**), with important effects on energy metabolism (release of glucose from glycogen). A-kinase also activates genes via the **CREB binding protein** (**p. 80**).

ALARA

The ALARA concept (As Low As Reasonably Achievable) is used to underpin modern guidelines and regulations concerning radiation protection. The concept involves conceding the principle that some degree of radiation exposure is, for each individual, unavoidable. Background radiation, from rocks and cosmic rays, affects us all. Any exposure as a result of medical or industrial uses of radiation is simply an addition to an inevitable exposure. Against this background it is both unnecessary and impossible to achieve zero exposure. The exercise becomes one of damage limitation: to increase exposure no more than is absolutely necessary. The goal is simply to minimise, using reasonable precautions and procedures, the dose of radiation received by any individual as a result of medical intervention or occupational exposure.

Alkylation, Alkylating Agent

Alkylation is the process where an alkyl group replaces a hydrogen atom in a compound. An alkyl group is simply an alkane that has lost a hydrogen atom. An alkane is an organic molecule consisting of chains, either branched or unbranched, of carbon and hydrogen atoms where all of the bonds between carbon atoms are single bonds.

An alkylating agent is a substance, typically a cytotoxic drug, that replaces the hydrogen in the hydrogen bonds between the complementary DNA base pairs with an alkyl group. This binds the complementary **DNA strands** (**p. 101**) irreversibly and interferes with cell division by preventing DNA **replication** (**p. 267**). **Cyclophosphamide** (**p. 83**), busulphan, melphalan and chlorambucil are typical examples of alkylating agents used in the chemotherapy of malignant disease.

Allelic Heterogeneity

This occurs when a disease may be caused by more than one type of mutation within a specific gene.

All or None Phenomenon

A term used in grading evidence dealing with treatment or prevention of illness. It occurs when before the treatment became available all patients died, but after introducing the treatment some patients survive. Alternatively, it would occur if before introducing the new treatment some patients died, but after its introduction none died. In the NHS R&D classification the all or none phenomenon is categorised as Grade A level 1c.

α/β Ratio

This radiobiological term, derived from the linear–quadratic model of **radiation-induced cell killing** (**p. 256**), describes the curviness of the shoulder of the cell survival curve. The α component of the ratio, which is directly proportional to dose, indicates single-target single-hit killing and, as such, gives a measure of the intrinsic **radiosensitivity** (**p. 260**). The β component, which is proportional to $(\text{dose})^2$, gives an indication of the killing due to multiple hits on several targets. The α/β ratio is that dose of radiation at which single- and multiple-hit killing are equal. Units are: α (Gy^{-1}); β (Gy^{-2}); α/β ratio (Gy).

Tumours, and acute radiation effects such as skin erythema and mucositis, have fairly high α/β ratios (typically 10 Gy) compared with those for late radiation effects, such as subcutaneous fibrosis and **radiation myelopathy** (**p. 256**), (typically 3 Gy): the initial slope of the cell survival curve is curvier for late responding tissues than for tumours and acutely responding tissues. Reducing fraction size will, therefore, selectively tend to spare late responding tissues compared with tumours or acutely responding tissues. These considerations underlie the use of **hyperfractionation (p. 156)** as a means of improving the **therapeutic ratio (p. 307)** in clinical radiotherapy.

It is difficult to measure α/β ratios directly for tissues *in vivo*. One indirect method that can be used to provide an estimate of ratio is to use the **intercept (p. 163)** to slope ratio of the reciprocal iso-effect plot. The basic linear–quadratic equation can be rearranged to give:

$$\log_e S/n \cdot d = \alpha + \beta \cdot d,$$

where S is the number of surviving clonogens, n the number of fractions and d the dose per fraction. It is reasonable to assume that, for any given iso-effect, i.e. equal degree of damage, the surviving cell number will be equal. Total dose $= n \cdot d$, and, if S is held constant as in an iso-effect plot, then the above equation is that of a straight line: when a reciprocal dose is plotted against dose per fraction, the slope of the line is β and the intercept is α. The intercept to slope ratio is, therefore, the same as the α/β ratio. This statement is only true if repair is complete between fractions, if repair is incomplete then the data will follow a curved line that approximates to the expected line, but which is concave downwards.

Figure 1 is an example of a reciprocal iso-effect plot showing points corresponding to different combinations of dose per fraction and total dose at which the biological effect of the radiation is equal, e.g. the endpoint could be moist desquamation of the irradiated skin:

- Slope of the line $= \beta/\log_e S = 0.000872\ \text{Gy}^{-2}$.
- Intercept on the y-axis $= \alpha/\log_e S = 0.011\ \text{Gy}^{-1}$.

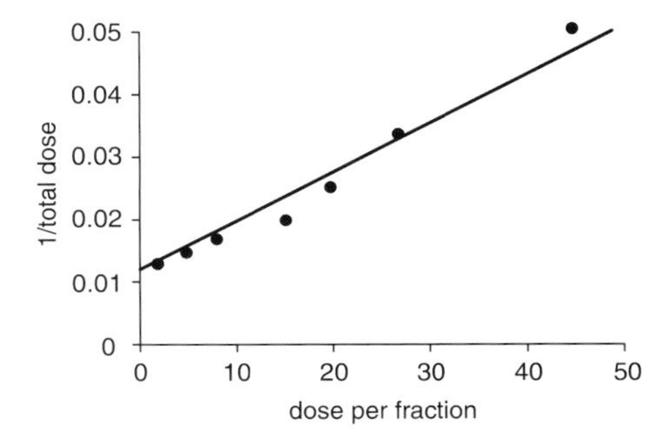

Figure 1 — Reciprocal iso-effect plot.

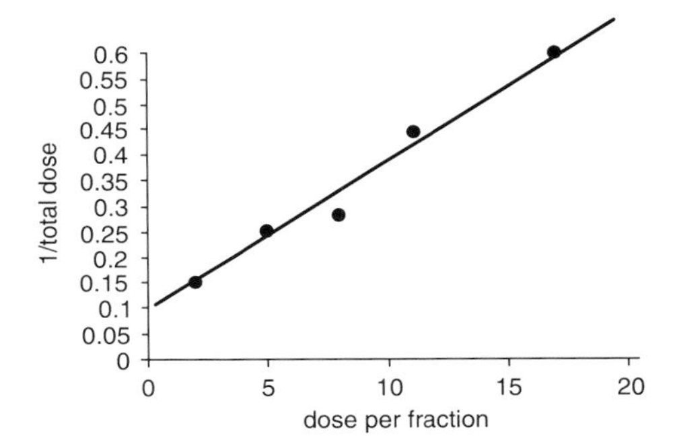

Figure 2 — Reciprocal iso-effect plot.

- Ratio of the intercept to the slope $= 0.011\ \text{Gy}^{-1}/0.000872\ \text{Gy}^{-2} = 12.6\ \text{Gy}$ (an α/β ratio characteristic of an acute response to radiation).

Figure 2 shows a reciprocal iso-effect plot using data characteristic of a late effect:

- Slope of the line $= \beta/\log_e S = 0.0333\ \text{Gy}^{-2}$.
- Intercept on the y-axis $= \alpha/\log_e S = 0.1\ \text{Gy}^{-1}$.
- Ratio of the intercept to the slope $= 0.1\ \text{Gy}^{-1}/0.0333\ \text{Gy}^{-2} = 3\ \text{Gy}$ (an α/β ratio characteristic of a late response to radiation).

Figure 3 shows a reciprocal **isodose (p. 169)** plot showing data for both early and late effects. The late effect has a greater slope and, therefore, a lower intercept to slope ratio. This indicates that **late effects (p. 174)**

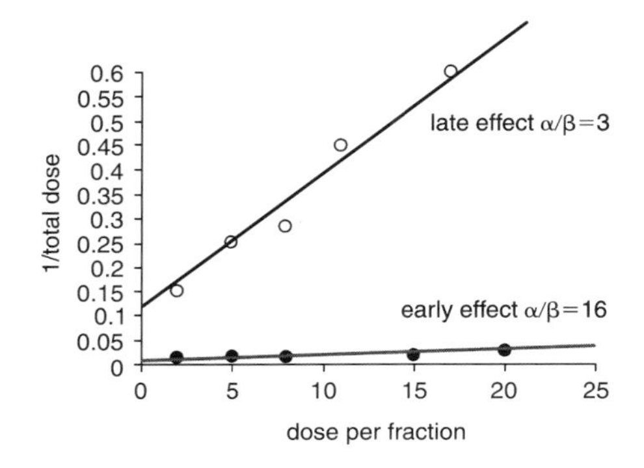

Figure 3 — Reciprocal iso-effect plot.

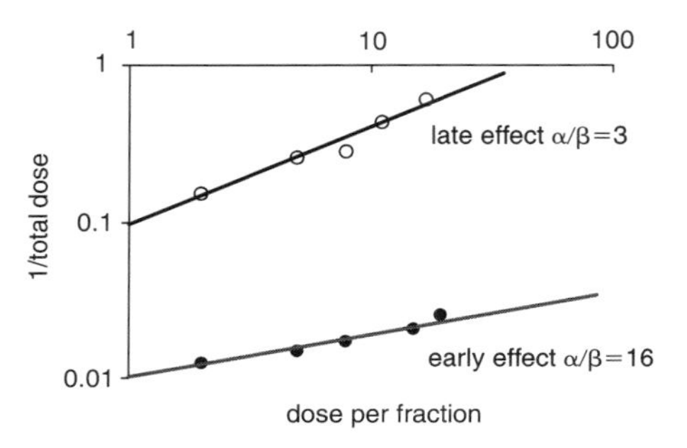

Figure 4 — Reciprocal iso-effect plot.

are, for any given dose, going to be more dependent than acute effects upon dose per fraction.

By using log scales the data from figure 3 can be more clearly displayed. The difference in slope for early, as opposed to late, effects is still obvious.

If two radiotherapeutic schedules are known to produce identical effects, then the BED equation (E/α = total dose × (1 + [dose per fraction/{α/β}])) can be used to calculate the α/β ratio. If the schedules are iso-effective then E/α for each schedule will be equal, so that:

$$\text{Total dose}_1 \times \left[1 + \frac{\text{dose per fraction}_1}{\alpha/\beta}\right] = \text{total dose}_2 \times \left[1 + \frac{\text{dose per fraction}_2}{\alpha/\beta}\right]$$

or, in symbols:

$$D_1 \times \left[1 + \frac{d_1}{\alpha/\beta}\right] = D_2 \times \left[1 + \frac{d_2}{\alpha/\beta}\right]$$

This can be rearranged to solve for α/β:

$$\alpha/\beta = (D_1 \cdot d_1 - D_2 \cdot d_2)/(D_2 - D_1).$$

If a regimen of 55 Gy in 20 fractions produces Grade 3 telangiectasia in 15% of treated patients and if the same rate of Grade 3 telangiectasia is produced by 62 Gy in 30 fractions then the regimens are, at least approximately, clinically iso-effective. The α/β ratio for this late effect can be calculated, using the above equation, as 3.3 Gy. Similarly, if 55 Gy in 20 fractions controls 50% of T3 bladder tumours and if the same local control probability can be obtained using 58.5 Gy in 30 fractions then, in terms of killing of tumour cells, the regimens can be considered iso-effective. The α/β ratio is 10.6 Gy.

Amifostine

This analogue of cysteamine has been in use for many years as a **radioprotector** (**p. 259**): formerly it was known as WR 2721. It also has a role in protecting against the renal and neurotoxicity associated with *cis*-platinum treatment. It may also protect against chemotherapy-induced myelosuppression. It is administered immediately before exposure to drugs or radiation. It has a distribution half-time of 60 s and the **half-life** (**p. 147**) for elimination is ~8 min.

It is converted, by membrane-bound alkaline phosphatase, to WR 1065. WR 1065 acts as a scavenger of free radicals and can also bind directly to the activated carbonium ions of alkylating agents, thereby protecting DNA bases from damage. Since normal tissues have more membrane-bound alkaline phosphatase than most tumours, and since the lower pH in tumours lowers the activity of any alkaline phosphatase present, amifostine preferentially protects normal tissues rather than tumours. Its protective action against ionising radiation is synergistic with that provided by compounds, such as glutathione, which are rich in thiol groups. The sulphhydryl groups in amifostine protect the renal tubules against the toxic effects of *cis*-platinum.

Its main, proven, benefit in clinical practice is protection of the function of the salivary glands in patients treated with radical irradiation for head and neck cancer. Its main side-effects are nausea, giddiness and hypotension.

Analysis of Variance (ANOVA)

The variance of a series of measurements made on a population is an indication of how spread out the data are. When there is little variance the results will cluster around the mean or median; when there is considerable variance then, when plotted, the values will spread across a wide range of values. There could be a variety of explanations for this dispersion: the variance in the weights of a population could be determined by numerous factors including ethnic origin, dietary fat intake, amount of exercise, daily calorie intake, etc. The analysis of variance explores just how much such factors might account for the observed variance — what proportion of the variance is explained by values of key variables.

In a one-way analysis of variance only one factor that might account for the spread of the data is studied. In two-way ANOVA two variables are studied, etc. The analysis is based on a comparison of the squared differences of the means both between and within groups. If there is as much variation within a group as there is between groups then it is unlikely that much of the observed variance seen in the overall population can be explained by belonging to, or not belonging to, that particular group. The ratio of the between group variance to the within group variance is expressed as F, which can be calculated and compared, for the appropriate **degrees of freedom** (**p. 91**), in a table. The statistical significance of any observed F can, therefore, be assessed. **Regression** (**p. 264**) coefficients can be calculated to indicate which values for each factor have the most effect. The higher the regression coefficient the greater the effect.

Below are some data on an experimental tumour. At day 0 the tumours were all 5 mm in diameter. The animals were treated with three different therapies and the tumour size at 15 days was the outcome measure used.

Rx type	Tumour diameter (mm) at 15 days
1	12
1	10
1	8
1	10
2	30
2	25
2	26
1	10
1	11
3	28
3	42
3	35
3	41
3	45
1	15
2	35

F for the data is 55.04, which corresponds to $p<0.00001$. The type of treatment has a highly significant effect upon outcome. But which treatment is the most effective?

The regression coefficients for the treatments are: treatment 3 is dropped from the model

because it is the least effective; treatment 2 has a coefficient of −9.2 (CI −15.8 to −2.6); treatment 1 has a coefficient of −27.3 (CI −33 to −21.5). The minus signs indicate that tumour size is less when these factors are present and that the highest negative value is for treatment 1, indicating that treatment 1 is clearly the most effective of the three.

The **Kruskal–Wallis test** (**p. 172**) is a non-parametric one-way analysis of variance.

Anaphase

The phase of **mitosis** (**p. 201**) immediately after **metaphase** (**p. 198**). It lasts ~5 min and begins when the chromosomes start to migrate towards the mitotic poles. It is divided into phases: A and B. During anaphase A, the chromosomes move to the poles; during anaphase B, the spindle poles themselves move apart. The rate of movement during anaphase A is about twice as fast as that during anaphase B.

Anastrozole

This drug, used in the treatment of breast cancer, is an orally active aromatase inhibitor. Aromatase is the enzyme that catalyses the last step in oestrogen biosynthesis. Anastrozole is a competitive inhibitor of the enzyme and, because it has a non-steroidal structure, it lacks steroid-like side-effects. It does not inhibit adrenal function and, therefore, unlike other aromatase inhibitors such as aminoglutethimide, steroid replacement therapy is not required. Most of the oestrogen synthesis in post-menopausal women takes place in the peripheral tissues and, therefore, aromatase inhibitors have their main effect on the peripheral production of oestrogen. The main toxicities of anastrozole are weakness, fatigue and gastrointestinal disturbances.

Androgens/Anti-androgens

Androgens of relevance to cancer treatment include testosterone proprionate, fluoxymesterone, testosterone enathate, testolactone, methyltestosterone and danazol. Androgens have been used in the treatment of metastatic breast cancer and in the supportive management of patients with impaired bone marrow function — infiltration or hypoplasia. The main problem is that they cause virilisation and can also cause liver damage, peliosis hepatis.

Anti-androgens are used primarily in the management of prostate cancer and include **nilutamide** (**p. 214**), cyproterone acetate, finasteride, **flutamide** (**p. 131**) and bicalutamide. The concept of **total androgen blockade** (**TAB**) (**p. 310**) in which an anti-androgen is combined with a **luteinising-releasing hormone** (**LHRH**) (**p. 178**) agonist is popular but expensive. There is no good evidence that combination therapy with TAB offers any advantage over monotherapy.

Aneuploidy

A cell that contains the wrong number of chromosomes is said to be aneuploid. Aneuploidy arises as the result of chromosomes being lost from, or gained by, a cell.

Angiogenesis

The process by which blood vessels are formed. It is becoming increasingly important in oncology because of an appreciation that the formation of new blood vessels is an essential part of tumour development. For many tumours, a mass of clonogenic cells alone is insufficient to sustain development into an invasive metastasising neoplasm — the tumour has to attract and/or form new blood vessels to develop fully. Since pathways distinct from those of normal vascular formation may control tumour neovascularisation, the tumour vasculature offers a potential target specific to tumours. Since the pattern of vasculature within a tumour may, to an extent, reflect the genetic heritage of that tumour, quantitative and qualitative analysis of the pattern of blood vessels within a tumour may provide prognostic information. In general, the more tumour vessels there are the higher the likelihood of metastatic disease, and the worse the prognosis.

Angiogenesis can be assessed using a variety of methods: microvessel area as an absolute value (or percentage) per unit volume or area of tumour, scoring systems using staining for Factor VIII antigen or other vascular markers such as CD34 (**p. 52**), von Willebrand's Factor or *Ulex*, molecular expression of angiogenic cytokines or their receptors, e.g. VEGR, expression of proteinases used to degrade the extracellular matrix ahead of sprouting capillaries.

A mass of proliferating cells can grow to ~0.4 mm in diameter and thereafter its growth is limited unless it can obtain a vascular supply. The formation of new blood vessels is a complex process controlled by a network consisting of stimulators and inhibitors. Angiogenesis in normal tissues, e.g. in wound healing, is under physiological control. Tumour angiogenesis is outwith normal control and is largely driven by the demands of the proliferating malignant cells. To invade and metastasise, a tumour has to have developed its own blood supply. In the absence of angiogenesis the tumour fails to grow and clonogens are lost as a result of necrosis and apoptosis. Tumour angiogenesis is not a passive event, something that happens simply because a tumour has achieved a certain size. An active process, the angiogenic switch, has been postulated as the controlling mechanism for tumour angiogenesis. Histological examination of *in situ* carcinomas of the breast and cervix suggests that only those *in situ* lesions that show capillary formation ('angiogenic CIS') progress to invasive cancer. The angiogenic switch is turned on at a fairly early stage in neoplastic development and expression of the genes that turn it on is probably essential for the full expression of the malignant phenotype. The mechanism that controls the angiogenic switch depends upon the relative balance between promoters and inhibitors of angiogenesis. There are many examples of each, but the most important promoters are **vascular endothelial growth factor** (**VEGF**) (**p. 324**) and the **fibroblast growth factors** (**p. 128**), both acidic, aFGF, and basic, bFGF. The key inhibitors include thrombospondin-1; a 16 kD fragment of the prolactin molecule; Platelet Factor 4; and angiostatin (a recently categorised fragment of plasminogen).

Recent experimental data show that continuous administration of low-dose **cyclophosphamide** (**p. 83**) can inhibit division of endothelial cells lining new blood vessels and thereby arrest growth of cyclophosphamide-resistant tumours. This philosophy is very different from that which underpins the concept of high-dose therapy.

Angiogenic Switch

The acquisition by a tumour of the ability to recruit new blood vessels from the surrounding vasculature involves an event, or series of events, termed the angiogenic switch. This ability is crucial, not only just for tumour growth and nutrition, but also for the ability to invade and metastasise. **Fibroblast growth factor-binding**

protein (**FGF-BP**) (**p. 128**) is a 17 kD protein that may displace bound FGF from extracellular proteoglycans. When bound to proteoglycans, FGF cannot stimulate angiogenesis, when FGF-BP is present the FGF is freed and angiogenesis occurs. FGF-BP could thereby function as a major component of the angiogenic switch. In some tumour cell lines there is evidence for increased expression of FGF-BP, and also some showing that inhibiting FGF-BP production retards tumour growth. Only very small relative reductions in FGF-BP production are required to produce measurable growth restraint, the implication being that FGF-BP might offer a tempting therapeutic target for cancer treatment.

Angiopoietins

The angiopoietins are a family of molecules that act as growth factors and are concerned in the development of blood vessels. So far, five members have been identified. Ang1, unlike members of the vascular endothelial growth factor (VEGF) family, decreases vascular permeability and increases vascular diameter.

Ann Arbor System

The Ann Arbor System is a system for the clinical staging of Hodgkin's disease. It was introduced in 1971 as a modification of a previous system, the Rye Classification. It has subsequently been modified as the **Cotswold Staging** (**p. 78**).

The Ann Arbor System is as follows:

- Stage I: involvement of a single lymphoid region or structure (spleen, Waldeyer's ring).
- Stage II: involvement of two or more lymph node regions or lymphoid structures on the same side of the diaphragm.
- Stage III: involvement of lymph node regions or lymphoid structures on both sides of the diaphragm.
- Stage IV: involvement of liver or bone marrow.

Extra nodal disease, arising as direct local extension from nodal involvement, is designated E. Disease arising at a single extranodal site (e.g. localised Hodgkin's disease of the parotid) is designated IE.

Each stage is divided according to absence (A) or presence (B) of certain cardinal systemic symptoms. These comprise any one of:

- unexplained weight loss amounting to >10% of body weight over the 6 months before diagnosis;
- unexplained persistent or recurrent fever >38°C during the previous month; or
- recurrent drenching night sweats during the previous month.

Anoikis

Anoikis, or having to with homelessness, is a type of **apoptosis** (**p. 19**) that occurs in cells, e.g. endothelial cells, that are prevented from adhering to and spreading on an appropriate matrix.

Anthracyclines

Anthracyclines are a class of cytotoxic drug whose members include **daunorubicin** (**p. 88**), **doxorubicin** (**p. 107**) and mitozantrone. Anthracyclines are **intercalating agents** (**p. 163**). Their most important toxicity is myocardial damage, which is mediated by the formation of superoxide **free radicals (p. 132)** that damage the mitochondria by peroxidating the lipid in the inner membrane. The superoxide radicals are more readily formed in the presence of iron, an effect that may be mitigated by the prior administration of **dexrazoxane (p. 95)**.

Anti-angiogenesis

The problem is to define a target specific to newly formed, as opposed to established, blood vessels. Integrin $\alpha\nu\beta3$ may provide such a target. Its activity can be inhibited by a combination of tumour necrosis factor (TNF) and interferon, and this combination, by disrupting tumour microvasculature, can be clinically useful. LM609 is a monoclonal antibody to integrin $\alpha\nu\beta3$, which is available in a humanised form (Vitaxin), currently being tested in clinical trials.

Anticipation

A phenomenon, sometimes observed in inherited diseases, in which the disease manifests in a more severe form, and at an earlier age, in successive generations. Anticipation has been described in a family with **Cowden syndrome (p. 79)**, which has been observed over four generations.

Anti-idiotype Network

An anti-idiotypic antibody is directed against unique features, usually the variable region, of another antibody. A region recognised as antigenic is known as the idiotope. The totality of all the regions recognised as antigenic is the sum of the idiotopes: the idiotype. If an antibody is raised against a tumour-associated antigen and if an antibody is, in turn, formed against the variable region of the first antibody, then the second antibody will express features similar to the antigen recognised by the first antibody. We have entered into the looking glass world: the mirror image of a mirror image is the original image. In this way a human antibody can express an antigen identical to a tumour-associated antigen: this is an anti-idiotype network. Some of the successes observed in monoclonal antibody therapy may be due to the formation of anti-idiotype networks: effectively, patients produce an endogenous tumour vaccine. The price paid may be **HAMA (p. 147)**, the human anti-mouse response, with the attendant risks of serum sickness like syndromes and anaphylaxis.

Anti-metabolites

The anti-metabolites are a group of cytotoxic drugs that subvert normal cellular synthetic pathways. Often they act as false substrates for enzymes involved in the synthesis of **purines (p. 252)** and **pyrimidines (p. 252)**. They may inhibit enzyme action through competition with the normal substrate or they may, by substituting in the finished product, interfere with DNA replication, hence an alternative term, 'counterfeit incorporation mechanism'. Examples include methotrexate, cytosine arabinoside and **5-fluorouracil (p. 130)**.

Antisense Oligonucleotides

These are short, chemically modified sequences of single-stranded DNA that are complementary to the premessenger RNA and mRNA sequences of a target gene. By combining with these sequences, the antisense oligonucleotides will inhibit gene expression.

Antoxidants

A group of substances and mechanisms that protect cells and tissues against damage induced by free radicals. Endogenous examples include **superoxide dismutase**

(**p. 300**), α-tocopherol, reduced glutathione, urate and ascorbate. Diet is an important source of antoxidants: vitamins C and E, carotenoids, flavenoids and other plant-derived phenolics. These dietary factors, and their pharmaceutical derivatives, might, by minimising the damage caused to DNA by **free radicals** (**p. 132**), be useful in the prevention of cancer.

(Apaf) Apoptosis Activating Factors

Apoptosis activating factors (Apaf) are required for **apoptosis**. Apaf-1 is the mammalian homologue of CED-4; Apaf-2 is **cytochrome c** (**p. 84**); while Apaf-3 is a 45 kD protein about which little is known. The three Apaf, together with dATP, are sufficient to activate **caspases** (**p. 47**) — the main effector proteins in the process of apoptosis.

APC Gene

Located on the long arm of chromosome 5 (5q21), the *APC* gene is mutated in familial polyposis (FAP) and **Gardner's syndrome** (**p. 135**). Functionally, it acts as a tumour suppressor gene since loss of both alleles seems essential for full expression of its carcinogenic potential. The *APC* gene product normally binds to catenin, suggesting that it is involved in cell adhesion and/or communication. Interference with the interaction between catenin and the *APC* gene product may cause loss of contact inhibition of cell division and, consequently, hyperplasia. This hypothesis is consistent with the observation that loss of activity of the *APC* gene product is an early stage in the multistep process that leads to carcinomas of the large bowel. An alternative explanation is that there is a loss of cell–cell adhesion that permits the abnormal cells to detach themselves, proliferate and, ultimately, to invade and metastasise. Most (>90%) of *APC* mutations associated with bowel cancer involve the production of an *APC* protein which is shorter than normal. Acquired mutations of *APC* (somatic mutations, as opposed to the **germline mutations** (**p. 141**) of *APC* involved in FAP) are common in non-familial cases of bowel cancer. The vast majority of mutations occur at a site designated MCR (mutation cluster region). This site occupies <10% of the coding sequence for the gene. Location of the mutations within the gene may also control the number of polyps formed, mutations between codons 1250 and 1464 tend to be associated with large numbers of polyps.

There is an interaction that may be clinically relevant between *APC* and *Ptgs2*, a gene that controls cycloxygenase 2 (COX2). The mouse with mutant *APC* requires wild-type *Ptgs2* to express fully its ability to form bowel polyps. This suggests that inhibitors of COX2, which include many non-steroidal anti-inflammatory drugs (NSAID), may have a potential role as chemopreventative agents in the control of bowel cancer.

Apodictic

A useful term describing the degree of certainty exhibited by certain molecular biologists about the importance of **molecular biology** (**p. 203**) and the rapid impact that it will have on the understanding and practical management of human diseases. Such individuals are usually **reductionists** (**p. 263**) or **determinists** (**p. 94**) by nature and fail to appreciate that human beings are somewhat more complex than cells growing in culture. A few days' exposure to clinical medicine can be a useful cure for apodicticism; unfortunately, relapses are common.

Apoptosis

Apoptosis is a distinct mechanism of cell death that may be exploited in cancer therapy. Apoptosis, literally dropping off, is a mechanism of intermitotic cell death. Intermitotic in this context means that the cell does not wait until the next cell division before apoptosis occurs. Dead means dead: the cell breaks up and the fragments are absorbed. We are not concerned here with the subtleties of reproductive integrity. It affects non-cycling cells and does not evoke an inflammatory response.

One definition of apoptosis is a process of active **cell death (p. 55)** in the absence of immune surveillance. Apoptosis has many physiological roles: the involution of the prostate after androgen deprivation; the disappearance of the webs between the fingers and toes during embryological development. Its relevance to cancer is that apoptosis occurs in many tumours and is a potential therapeutic target. Indeed, many cancer treatments thought only to act on cell replication can stimulate apoptosis. Probably the best example is the effect of radiation upon the long-lived recirculating T-lymphocytes. It was long recognised that the cells died an intermitotic death after irradiation — now it is realised that apoptosis is the mechanism. **Cytotoxic drugs (p. 84)** can also induce apoptosis: examples include the platinum analogues and **topoisomerase (p. 310)** inhibitors.

An essential feature of apoptosis is that it is under genetic control, hence another term for it is **programmed cell death (p. 55)** — when your (pre-programmed) time is up then its time to die. Stimulation of the **Fas-receptor (p. 126)** (Apo-1, **Fas-ligand**) **(p. 126)** on the surface of a cell will cause that cell to undergo apoptosis — this is the mechanism by which certain T-lymphocytes are cytotoxic. Conversely, expression of the ***bcl*** **(p. 26)** gene, or increased levels of the *bcl* gene product, can protect cells against apoptosis. *p53* and other tumour suppressor genes can induce apoptosis in cells that have suffered genetic damage — an evolutionary mechanism whereby the organism can rid itself of potentially harmful mutations. There are at least five pairs of agonist–antagonist interactions involved in the control of apoptosis and it is the overall balance that controls whether apoptosis occurs. Agonists include ***Bax*** **(p. 26)**, *Bak*, *Bcl-Xs*, *Bad* and *Bvd*. Antagonists include *Bcl2*, *Bcl-XL*, *Bcl-w* and *Mcl*-1. Other genes are also influential: *Ras*, *Raf*-1, ***caspase*** **(p. 47)**, *ced*-4. A major problem in research on apoptosis is to disentangle the causes of the process from its effects.

Changes in **mitochondrial membranes (p. 200)** appear to be important in governing the all-or-nothing nature of apoptosis. These changes particularly affect the permeability transition (PT) pores, also known as the mitochondrial megachannel. One of the major constituents of this channel is the peripheral benzodiazepine receptor. The changes destroy the mitochondrial transmembrane potential and thereby totally disrupt mitochondrial function. The effects on the PT pores can produce a self-amplifying cascade that, once started, ends in the total destruction of the cell.

The process is rapid: from initiation to completion takes <4 h, and most of this time is taken up by phagocytosis of the fragmented cell. The cellular components condense and the endoplasmic reticulum swells. Membranes remain intact and the cell breaks itself up into number of membrane-bound packages, apoptotic bodies. The fact that the cellular constituents are sequestered in this way may explain the lack of

immunological or inflammatory response. In addition, the packaging process leads to the expression of antigens that signal that apoptosis is occurring and recruit appropriate phagocytic mechanisms.

Apoptosis is a mechanism by which radiation can kill cells without necessarily causing any damage to DNA. It is proving difficult to unravel the mechanisms by which radiation can induce apoptosis. Wild-type p53 may be essential for some, but not necessarily all, radiation-induced apoptosis: there is considerable variation between cell lines in terms of susceptibility to radiation-induced apoptosis and this variation does not entirely correlate with the presence or absence of wild-type p53. Radiation can induce apoptosis in certain glioblastoma cell lines by mechanisms that appear completely independent of p53.

Apoptotic Index

The percentage of the number of apoptotic cells divided by the total number of cells counted: counts should be obtained from regions that do not show obvious necrosis. The apoptotic cells can be identified using fluorescence activated cell sorting (**FACS**) (**p. 129**) or by TUNEL (**p. 317**) staining.

Apparent Volume of Distribution

This pharmacological term is calculated as:

$$\frac{\text{amount of drug in body}}{\text{plasma drug concentration}}.$$

It is an estimate of the total fluid volume required to contain the drug at the same concentration as in plasma. It tells nothing, however, about where the drug has actually gone. Acidic drugs such as warfarin are protein bound and have small volumes of distributions. Basic drugs, in contrast, are widely distributed and concentrated in the tissues: their apparent volume of distribution may be >70 litres.

ARCON (Accelerated Radiotherapy with Carbogen and Nicotinamide)

This is a complex method of radio-therapeutic treatment devised to counteract the problem of hypoxic, but viable, cells within tumours imposing a limit to the therapeutic effectiveness of radiotherapy. Patients are treated with **nicotinamide** (**p. 214**), which, in part at least, acts as a vasodilator and may improve the oxygen supply to tumours. Radiotherapy is accelerated, given over a shorter overall time, so that proliferation during treatment is minimised. In an attempt further to improve tumour oxygenation, during each treatment with radiotherapy patients breathe carbogen (95% oxygen, 5% carbon dioxide) at atmospheric pressure. **Carbogen** (**p. 46**) is used because breathing pure oxygen would produce vasoconstriction. The technique is clinically feasible and has produced encouraging results in the treatment of advanced laryngeal cancers: it is currently undergoing testing in randomised controlled trials.

Area Under the Curve (AUC)

The calculation of area under the curve represents an attempt to reduce a complex time-dependent relationship to a single number. The concept is most easily understood when applied to measurement of drug concentration over time, but it does have other applications, e.g. in studies of **quality of life** (**p. 253**). For a pharmacological study the area under the curve is the integral of the function that describes the time-course of the concentration of the drug from time 0 to infinity. It is a means of conveniently describing total exposure to the drug in units of concentration × time (mM h or mg ml^{-1} h). **The Calvert Formula (p. 39)** is used for calculating **carboplatin (p. 46)** dosage, the dose (mg) is chosen according to the desired AUC (e.g. 4 for low dose therapy, 6 for more intensive treatment).

The approach has its limitations. In pharmacology, the effect of a drug may not solely be related to AUC: threshold effects may also be important. Peak levels in plasma may be higher with bolus administration as opposed to continuous infusion. These peak levels may be critical for either therapeutic effect or toxicity. The changing toxicity of 5-fluororuracil is a practical example of this: the dose-limiting toxicity for bolus administration is myelosuppression, when the drug is infused continuously the predominant toxicities are diarrhoea and hand–foot syndrome. Similar problems apply when AUC is applied to quality of life research. A short period of intense discomfort may have the same AUC as a longer period of much milder distress — we are still left with the question: which is worse, a bang on the thumb with a hammer or backache for a month? AUC can most easily be calculated by the **trapezoidal approximation (p. 314)**: AUC for value plotted against time is estimated as the sum of the measurements:

$$(t_2 - t_1)(y_1 + y_2)/2,$$

where $t_2 - t_1$ is the interval between the measurements and $y_1 + y_2$ are the values at the two time points. More elaborate methods are available for those who wish to apply a little more mathematical sophistication.

Arithmetic Fallacy

A problem that arises when attempting to combine results from different studies. It is best indicated by an example. We wish to know whether **tamoxifen (p. 304)** is of benefit to women with early breast cancer who are <50 years of age. There are several randomised studies that address the issue, albeit indirectly:

Study	Control Rx	Experimental Rx
A	radical mastectomy	radical mastectomy + tamoxifen
B	local excision + XRT	local excision + XRT + tamoxifen
C	any surgery + CMF chemotherapy	any surgery + CMF chemotherapy + tamoxifen
D	local excision alone	local excision + tamoxifen
E	radical mastectomy + XRT + CMF	radical mastectomy + XRT + CMF + tamoxifen

Since the only difference between arms, for all these studies, is that the experimental group received tamoxifen and the control group did not, it might seem reasonable to pool data from all these comparisons and to use it as a measure of the effect of tamoxifen. This would completely ignore the possibility that, for example, tamoxifen might interact with other therapies rendering them less effective. This is, in fact, biologically plausible. If tamoxifen were to increase the probability of a cycling cell entering G_0, then this would antagonise the effects of those therapies, chemotherapy and radiotherapy, which depend on the presence of proliferating cells.

The fallacy arises from the assumption that, if in comparing groups of studies, the only difference between them is a single factor, call it factor X, then the studies can assess the impact of factor X. The fallacy arises because this assumption ignores the potential for interaction between factor X and other interventions.

Study (%)	Control Rx	Survival (C) (%)	Experimental Rx	Survival (E) (%)	Difference
A	radical mastectomy	30	radical mastectomy + tamoxifen	40	10
B	local excision + XRT	23 (local excision) + 7 (XRT) = 30	local excision + XRT + tamoxifen	23 (local excision) + 2* (XRT) + 10 (tamoxifen) = 35 *2 = 7 − 5	5
C	any surgery + CMF chemotherapy	28 (surgery) + 8 (chemotherapy) = 35	any surgery + CMF chemotherapy + tamoxifen	28 (surgery) + 3 (chemotherapy) + 10 (tamoxifen) = 41	6
D	local excision alone	23	local excision + tamoxifen	23 (local excision) + 10 (tamoxifen) = 33	10
E	radical mastectomy + XRT + CMF	30 (radical mastectomy) + 7 (XRT) + 8 (chemotherapy) = 45	radical mastectomy + XRT + CMF + tamoxifen	30 (radical mastectomy) + 2 (XRT) + 3 (chemotherapy) + 10 (tamoxifen) = 45	0

Assume that the following facts represent the true effects of each **adjuvant therapy** (**p. 8**) and their interactions:

- Tamoxifen increases survival by 10%.
- Tamoxifen reduces the survival benefit from chemotherapy or radiotherapy by 5%.
- Radiotherapy improves survival by 7%.
- Chemotherapy improves survival by 8%.

Trials A and D give the true measure of the effect of tamoxifen, with the survival difference measured as 10%. The other trials all underestimate the potential benefit of tamoxifen because of the **confounding** (**p. 72**) effects introduced by interaction with other therapies.

This is only one example of an arithmetic fallacy: other fallacies arise when, often unwittingly, numbers are divided by zero. Another fallacy, which intrigued Leibniz, arises when, in the series $1, -1, 1, -1, 1, \ldots$, the terms are grouped differently: sometimes you obtain the answer zero and sometimes the answer is 1, but the numbers have not changed.

Arsenic Trioxide (As_2O_3)

As_2O_3 can induce apoptosis in leukaemic cells from patients with acute promyelocytic leukaemia (APL). It appears specifically to interact with the PML moiety of the PML/RAR α-**fusion protein** (**p. 133**) produced as a result of the t(15 : 17) translocation characteristic of APL. The fusion protein appears to be protective against apoptosis and As_2O_3 abolishes this protection.

As_2O_3 can be used in combination with **retinoic acid** (**p. 271**) and conventional chemotherapy in the treatment of APL.

Ascertainment Bias

A form of bias that can particularly afflict studies and trials in cancer. The harder we look for something, the more likely we are to find it. If treated patients have regular follow-up CT scans and untreated patients do not, then the incidence of recurrent disease may appear to be higher in the treated group — simply because the detection of subclinical disease in this group is going to be more efficient. The low incidence of cancer in the developing world, relative to the developed world, may be because only the more gross cancers are diagnosed in countries with less effective provision of healthcare. The use of screening and easy access to diagnostic services in the developed world will mean that not only are cancers likely to be diagnosed earlier, but also that death from an undiagnosed and unsuspected cancer is likely to be less common than in the developing world.

Askin Tumour

Extraskeletal Ewing's sarcoma presenting as an intrathoracic mass is sometimes described as an Askin tumour.

Asparaginase

This enzyme, which hydrolyses asparagine to aspartic acid, is used in the treatment of leukaemia. The malignant cells lack asparagine synthetase and, therefore, unlike the normal cells, they cannot manufacture asparagine from aspartic acid and glutamine. Asparagine is, therefore, a non-essential amino acid for normal cells but, for tumour cells, it is essential. Resistance to asparaginase treatment will develop if the malignant cells can acquire the ability to synthesise asparagine synthetase. It is a toxic drug that often causes hypersensitivity. It has been combined with **polyethylene glycol** (**PEG**) (**p. 230**) to produce a less toxic derivative: pegaspargase.

Ataxia-Telangiectasia (AT)

This condition is inherited as an autosomal recessive. The genetic abnormality involves the *ATM* gene at 11q22.3. It is distinct genetically form other, clinically similar, conditions such as the Nijmegen breakage syndrome and Berlin breakage syndrome. Most of the mutations in *ATM* involve truncation.

Clinically, these children present with cerebellar ataxia at ~12 months of age. Progressive spinal muscle atrophy develops later. Immunoglobulin levels are decreased. Telangiectasia develops by ~7 years of age. The tumours associated with AT are lymphomas (usually B-cell), leukaemias and CNS tumours. Survival >40 years of age is rare. Extreme **radiosensitivity** (**p. 260**) is a feature of the syndrome and AT heterozygotes may account for some of the unexpectedly severe radiation reactions encountered in apparently normal individuals. Over 5% of women with breast cancer may be heterozygous at the AT locus.

ATM

The gene that is mutated in ataxia-telangiectasia. It codes for a large molecule, homologous to *rad*3, *mec*1 and *tel*1, genes responsible for mediating the response to **DNA damage** (**p. 98**), particularly to double-strand breaks. A common feature is **kinase** (**p. 171**) activity similar to PI-kinase, towards the 3′-end. The *ATM* gene product is an important factor in signalling to *p53* that damage to DNA has occurred. In common with the gene for p53, *ATM* can be regarded as one of the guardians of the genome. Heterozygotes, individuals with only one mutated gene, are at increased risk of breast cancer.

Atomic Mass Unit (amu)

This is used in particle physics as a measure of mass. 1 amu is defined as 1/12 of the mass of ^{12}carbon, i.e. $^{12}C = 12.0000$ amu.

Atomic Nomenclature

The number of protons in the nucleus, or the number of electrons orbiting in an electrically neutral atom, defines the atomic number (Z). The atomic number is unique to each element: hydrogen = 1, helium = 2, carbon = 6, etc. The number of nucleons

(protons and neutrons) in the nucleus defines the mass number (A) of the atom. The number of neutrons in the nucleus is given by $A - Z = N$ (neutron number).

Atoms are described conventionally in the form $^{A}_{Z}X$. Elements can exist in different forms: the defining Z stays the same but A can differ. These forms, where Z is the same but A changes, are called isotopes. The chemical symbol stands proxy for Z and, therefore, isotopes are usually written as, for hydrogen isotopes, ^{1}H (native hydrogen), ^{2}H (deuterium), ^{3}H (tritium) or, for carbon, ^{12}C, ^{14}C, etc.

Atoms are normally electrically neutral, the negative charge on an electron exactly balances the positive charge on a proton, and there are equal numbers of electrons and protons. An ion is an atom that has an unbalanced electrical charge, if electrons are lost then the ion is positive, if extra electrons are acquired then the ion is negative.

Atzpodien Regimen

The atzpodien regimen is a combination of interferon-α (IFα) **interleukin 2 (IL-2) (p. 164)** and **5-fluorouracil (5FU) (p. 130)**. It has been used, with some success, in the treatment of metastatic renal cancer. The contribution of the 5FU is uncertain. Other studies suggest that the combination of IFα and IL-2 is sufficient to produce therapeutic benefit.

Audit

Audit is something we should do, something we probably do and something we often do without even knowing it. Its derivation is from the old custom of hearing the accounts. Today, it is used to describe a structured, systematic attempt to assess to what extent an individual or organisation has been successful in achieving predetermined objectives. Audit can deal with process, are the right things being done correctly? Or it can deal with outcomes: are the right things happening, regardless of how these results may have been achieved?

The problem of audit has been parsed by the World Health Organisation as follows:

- Identify the problem.
- Set priorities.
- Determine methods to be used.
- Set criteria and standards.
- Compare performance with standards.
- Design and implement action to deal with any shortfalls between standard and performance.
- Re-evaluate the process again.

The final three items are, in the jargon, involved with 'closing the audit loop'. Closing the loop means that it is not sufficient to identify the problems and deficiencies it is necessary to do something about them and then to check that what has been done is having the desired effect. One of the problems with audit at present is that it is too often used simply to criticise. Since remedial action often costs money, and since resources are usually limited, the loop is not always closed. This leads to erosion of morale, among the criticised, and erosion of confidence, among those who will have to use a service that has been shown to fail and for which there is no immediate prospect of affordable improvement.

Medical audit assesses the performance of individual doctors. Clinical audit assesses the performance of the system as a whole.

Audit is often treated and defined as if it were an isolated process, in fact there is considerable overlap with quality assurance (concerned with ensuring that things are done properly) and clinical research (assessing which interventions produce the best results). A popular question, for those who would speculate about angels and the heads of pins, is what is the difference between research and audit? Sometimes, in truth, there is no difference. One possible difference is that research is about finding new ways to do things and that audit is about assessing whether the old way of doing things is being implemented correctly.

Audit should be subjected to same methodological scrutiny as any other clinical investigation: prospective audit according to a protocol defined in advance of the study is more likely to approximate the truth than a retrospective canter through a sifted data set performed with the intention of confirming pre-existing prejudices.

There is nothing new about audit even though, for some, it is a bandwagon onto which they have but recently leapt. E. A. Codman, a surgeon at Massachusetts General Hospital, began in 1900 to record cases and to assess the long-term outcomes of treatment. In 1914, he proposed that this 'end results system' become part of the institution's policy — a proposal rejected by the hospital's trustees. His publication on the subject was presciently entitled 'The product of a hospital'. He resigned in protest: it was said at the time, 'There is nothing difficult about the system except the human nature part'.

Autocrine Stimulation

This describes cellular self-stimulation. The cell secretes a substance that acts on that cell and any similar cells in the vicinity. The process can serve to reinforce commitment to a particular line of differentiation. Critical concentrations of autocrine factors may be achieved within a group of cells that could not be attained by a solitary cell. This mechanism can drive localised areas of differentiation within an undifferentiated tissue.

Avidin–Biotin

Avidin is a protein found in egg white. It is positively charged and highly glycosylated, and it binds avidly to certain **lectins** (**p. 177**) expressed by tumours, hence its name. Biotin (vitamin B_{26}) is a coenzyme in carboxylation reactions and can link covalently to both protein, e.g. horseradish peroxidase, and nucleic acids. Streptavidin (from bacteria) and avidin bind tightly to biotin. The biotin–avidin system is useful for increasing the sensitivity of systems designed to detect particular sequences of amino acids or nucleotides. Biotinylation is a process whereby incorporation of biotin-dUTP into DNA can be used as a label that can be demonstrated using appropriate binding with avidin. Enzyme histochemistry or chemoluminescence can demonstrate where biotin–avidin label, and hence the DNA of interest, is located. This is a way of investigating DNA without using radioactivity.

Avidin forms a bridge between a tumour and labelled biotin. The advantage of this system is that unlabelled avidin can be administered first and allowed to localise to the tumour. Any unbound avidin is excreted. The labelled biotin can be administered later. When radioisotopes are used to label biotin the exposure of normal tissues to radioactivity can be minimised. This approach, known as avidin pretargeting, can improve the tumour to a non-tumour ratio of administered radioactivity. The technique can be used for intracellular localisation of **tumour-associated antigens** (**p. 315**) and has implications for both diagnosis and therapy.

Bannayan–Zonana Syndrome

This is a clinical syndrome in which hamartomatous polyps of the bowel are associated with melanin freckling of the penis. Inheritance is as autosomal-dominant. It is similar to the **Peutz–Jegher syndrome (p. 232)**, but the candidate gene maps to a completely different chromosome: 10q23–q24. This is the location of the ***PTEN* (p. 251)** gene and Bannayan–Zonana syndrome may be a variant of the group of disorders known as the **PTEN (p. 251)** MATCHS syndrome.

Base–Base Mismatch Repair System (syn. DNA Mismatch Repair Genes)

This system is, in evolutionary terms, highly conserved. It is responsible for detecting and repairing incorrect base pairings on the opposite strands of the DNA helix. It can also identify and eliminate abnormal single-stranded loops of DNA that arise when inappropriate insertions of up to 4 bp have occurred within a helical strand of DNA. The system in humans concerns the genes *hMSH2*, *hMSH3*, GTBP, *hMLH1*, *hPMS1* and *hPMS2*. In mouse, the *MutS* and *MutL* genes are involved. The human genes and their abnormalities are critical components in the adenoma ⇒ polyp ⇒ cancer sequence for colorectal cancer. **Germline mutations (p. 141)** of these genes are frequently found in HNPCC kindreds.

Bax

The *bax* gene promotes apoptosis by binding to bcl-2 and thereby preventing bcl-2 from inhibiting the process of programmed cell death. The gene is on chromosome 19q13.3–q13.4.

BCL2 (B-Cell Lymphoma Gene)

This gene, located at 18q21.3, produces a protein that protects cells from apoptosis. This effect may be mediated via the protein kinase RAF1 in mitochondria. In many follicular B-cell lymphomas, there is a characteristic translocation t(14;18)(q32;q21) that results in persistent uncontrolled expression of *BCL2*. This in turn diminishes apoptosis in the affected cells and so there is an accumulation of abnormal lymphoid cells. Patients with lymphomas with translocations involving *BCL2* respond less well to therapy than patients with normal *BCL2*.

Beams Eye View

A concept used in radiotherapy treatment planning in which the view 'seen' by a beam of radiation as it traverses the patient is used to dictate the shielding, beam intensity and field arrangement used to treat the patient. The concept is extremely useful in the planning of **conformal therapy (p. 71)**.

Becquerel

The becquerel (Bq) is the unit used to measure the activity of a radioactive isotope. It is defined as:

$$1\ \text{Bq} = 1\ \text{disintegration}\ \text{s}^{-1}.$$

Benchmarking Query

A process whereby the structure and performance of a system is evaluated using a set of widely accepted criteria. As such, benchmarking could be regarded as a specific form of **audit (p. 24)** in which judgements are made in relative, as opposed to absolute, terms. Benchmarking provides a set of standards deemed, through usage, to be acceptable. They may not, however, represent the best that can be achieved.

Bernoulli's Balls

A classic thought experiment (which *Blue Peter* graduates should perform without difficulty) and which forms the basis for statistical sampling and power calculations for clinical trials. Jacob Bernoulli formulated the **Law of Large Numbers (p. 176)**: increasing the size of the sample will increase the probability that the observed value will lie within a specified distance from the true value. The key concepts are probability and specified distance. The Law does not state that we will be more likely to hit on the true answer by increasing sample size, simply that the estimate will be more likely to be within a specified range around the true value.

Imagine an urn filled with 3000 white pebbles and 2000 black pebbles. This is the truth of the matter but we are not told this truth: we have, by sampling from the urn, to produce an estimate as close to the truth as possible: Bernoulli's expression for this was to be morally, as opposed to absolutely, certain. He stipulated that the estimate of the ratio of white to black had to be within 2% of the true ratio (3 : 2), i.e. white/black should lie within the range 1.47–1.53, and that the chance of achieving this accuracy had to be >1000/1001. He estimated that it would require 25 500 drawings from the urn to provide him with his 'moral certainty'.

In this one thought experiment, devised in the seventeenth century and finally published in 1713, we face many of the problems fundamental to statistics. We cannot know the truth for sure, we have to rely on sampling and estimation, we have to be sure not to find differences where there are none and not to miss differences when they exist.

Bernoulli set himself a very hard target: he wanted to have a very high probability (>99.9%) of achieving his 2% accuracy, hence his calculation of the necessary sample size of >25 000. In these more hard-pressed times where 'moral certainty' is, if it exists at all, a luxury, we are less demanding.

Bernoulli's experiment can be used to show, quite simply, that it is the number of events that drives the power of a study. Imagine that the white balls represent survivors following treatment A and the black balls survivors following treatment B. Assuming that, in a randomised study, equal numbers of patients were allocated to each treatment then the survival rate is 1.5 times higher with treatment A than with treatment B. If the survival rate for treatment A is 3% and that for B is 2%, then this would replicate the proportion used in the Bernoulli experiment. The rate difference is 1%. Statistically, the task we would now set is as follows. We wish to have a reasonable chance (conventionally 80%, 'power') of detecting this rate difference. Conversely, we wish to be reasonably certain that any difference observed could not have arisen by chance — that we have not taken a misleading sample and that, in reality, the number of white balls equals the number of black. Conventionally, this is defined as 'significance' and described as p. $p = 0.05$ indicates that, when there is truly no difference, five times in 100 experiments (1 in 20) the difference that we have observed might have arisen by chance. Using much stricter criteria, a power of 99%, $p = 0.001$, we can calculate the size of sample needed for the experiment. The answer is 823 patients in each treatment group. This would yield 24 survivors in group A (white) and 16 survivors in group B (black). If the survival rates were, again preserving Bernoulli's proportion, 30 and 20%, the sample size for each group (power 99%, $p = 0.001$) is 500.

Best Evidence Synthesis

A form of **systematic review (p. 302)** in which weak evidence is ignored and only

evidence of high quality is used to frame the conclusions. This selective approach is in contradistinction to the promiscuous, all-inclusive approach used in **meta-analysis (p. 195)**. Although this approach sounds eminently sensible, it suffers from the disadvantage that it ignores process–outcome dissociation: that sometimes sloppy work may be closer to the truth than work which is both diligent and comprehensive. This disadvantage is almost certainly more apparent than real. The main problem with best-evidence synthesis is framing the criteria used to define 'best', and what do we do with a large study that almost, but not quite, fulfils the criteria chosen?

Bias

Any process at any stage of inference that tends to produce results or conclusions that differ systematically from the truth. The elimination of bias, whether at source or at point of consumption, is what much of the so-called 'scientific approach' is all about. Rigor in **experimental design (p. 122)** and healthy scepticism in assessment of results and conclusions are necessary defences against bias. The taxonomy of bias is extensive and, for each category of bias, a specific defence is required. Bias in clinical research can occur when:

- The literature is reviewed.
- The study sample is selected.
- The intervention is executed.
- Measurements are made.
- Data is analysed.
- Data is interpreted.
- The results are disseminated (or not).

Bias in literature reviews

- Rhetoric: 'it is unethical not to …'.
- Nelson's syndrome:
  - ignore references that do not support your opinion
  - ignore data that erodes your claims.
- Publication bias.
- Fashion.

Bias in study sample

- Self-selection: volunteers will differ from conscripts.
- Iatro-tropism: certain doctors will attract certain, possibly untypical, patients.
- Filters: barriers imposed by income, postcode, etc. Not all individuals have equal access to healthcare.
- Diagnosis:
  - access
  - expertise
  - suspicion.
- Historical controls.
- **Stage shift (p. 294)**.
- Missing data.
- Naming.

Bias in executing studies

- Exclusions/withdrawals/eligibility.
- Intention-to-treat.
- Protocol assignment.
- Compliance and deviation.
- Supportive or ancillary care.

Bias in measurement

- Instruments faulty or insensitive.
- Rounding.
- Fear.
- Obsequiousness.

- Expectations.
- Surrogates.
- Observer (Heisenberg).

Bias in data analysis

- *Post-hoc* **significance testing** (**p. 286**).
- Data dredging.
- Collapsing and **confounding** (**p. 72**).
- Multiple peeking.
- Inappropriate deletion of outliers.

Bias in interpretation

- Confounding.
- Difference not actually due to intervention.
- Measurement problems.
- Choice of weapon (endpoint).
- Statistical and clinical significance are not the same.
- **Correlation** (**p. 75**) does not imply causation.
- Framing effects.
- Alternative explanations not considered.

Bibliometric Methods

These attempt to provide a means of assessing the quality of a scientist or scientific study. It is based on the simple premise that good work is published and extensively quoted, whereas bad work, even if published, is usually ignored. The problem is that bad work may extensively cited — as an example of how not to do things. Mediocre work may be excessively cited, and self-citation is always advantageous. The assumption that quantity equates to quality is dubious in the street market ('never mind the quality, feel the width'), it is equally suspect in assessing scientific endeavour. Gregor Mendel was never cited in his own lifetime, he practised excellent science in total isolation and obscurity. It is ironic that the Catholic Church in the nineteenth century proved itself, in this instance, to have been a more effective sponsor of good basic research than many modern universities. How would Mendel sell himself today?

Bioavailability

This term is used to define the extent to which a drug is absorbed into the systemic circulation. It defines the rate and degree to which the active agent gains access to its site of action. It can be expressed as:

$$\frac{\text{amount of drug absorbed}}{\text{drug dose administered}}.$$

Bioequivalence

A term used in pharmacology to indicate that chemical equivalents of a drug when given to the same person at the same dose will produce identical levels of drug in the blood and tissues. It differs from **chemical equivalence** (**p. 62**) in that chemical equivalence only implies that equal amounts of active agent are present, it does not allow for the fact that different binders and other constituents of the formulation may affect absorption, metabolism and excretion. Bioequivalence is a necessary, but not a sufficient, cause of **therapeutic equivalence** (**p. 307**).

Biologically Effective Dose (BED)

The biological effect of radiation will depend not only on physical dose, but also on factors such as type of tissue and patient's condition. It will also depend on the amount (volume) of tissue irradiated, the number of treatment sessions (or fractions) into which the total dose is divided and on the overall time taken to deliver treatment. Thus, a basic minimum for describing a dose of radiation would include: total dose, number of fractions, overall treatment time and size of radiation fields used. Typical radical schedules would be 5000–5500 cGy in 20 fractions over 4 weeks or 6000–6500 cGy in 30–35 fractions over 6–7 weeks.

The concept of biologically effective dose (BED) represents an attempt to produce a single number that will take into account total dose, fraction size and number of fractions. An extension of the concept also incorporates consideration of overall treatment time. The calculation of BED is based on a linear–quadratic model of **radiation-induced cell killing (p. 256)**.

The linear–quadratic equation can be modified to produce the expression:

$$E/\alpha = \begin{matrix}\text{total}\\\text{dose}\end{matrix} \times \left[1 + \frac{\text{dose per fraction}}{\alpha/\beta}\right],$$

where E/α defines a dose that produces a given level of effect and is, therefore, termed the BED, sometimes known as the biologically equivalent dose. This is identical to ETD, the extrapolated tolerance dose, as originally defined by Barendsen in 1982. The BED concept is extremely useful in calculating the changes in radiotherapy dose prescription necessary when changing from one fractionation schedule to another. It has superseded other methods, such as the Ellis NSD formula, based on less accurate assumptions concerning the relationship between radiation dose, fraction size and biological effect. The term:

$(1 + [\text{dose per fraction}/\{\alpha/\beta\}])$

is defined as **relative effectiveness (p. 264)**. The BED equation can be described in words:

> BED is the product of total dose and **relative effectiveness (p. 264)**.

This version of the equation does not take overall treatment time into account — it simply deals with the biological effects of altering total dose and dose per fraction. A tumour or normal tissue is a dynamic, not a static, population. Cells will proliferate during a course of radiation treatment; indeed, they may proliferate more rapidly as a result of the disruption caused by the treatment. Proliferation during treatment will, whatever its rate, increase the number of cells surviving after treatment. The longer the overall treatment time the more important this effect will be. The original BED equation can be modified to take account of proliferation by introducing a term that includes components describing overall duration of treatment, exponential growth, doubling time and the initial slope of the **cell survival curve (p. 57)** (α).

The modified equation is:

$$\begin{aligned}\text{BED} &= E/\alpha\\ &= \begin{matrix}\text{total}\\\text{dose}\end{matrix} \times \left[1 + \frac{\text{dose per fraction}}{\alpha/\beta}\right]\\ &\quad - \left[\frac{\log_e 2 \times t}{\alpha \times T_{pot}}\right],\end{aligned}$$

where t is the overall treatment time, T_{pot} is the potential doubling time and $\log_e 2 = 0.693$.

The problem with the use of the BED equation incorporating overall time is that it introduces a two parameters with uncertain values. T_{pot} varies widely between tumours and its estimation involves some guesswork, the value to use for α causes even more problems. There are indirect methods available for estimating the α/β ratio, but to measure α precisely requires a detailed cell survival curve — something that cannot be obtained *in vivo* for human tissues. A value of 5 days is usually chosen for T_{pot} and α is typically in the range 0.2–0.4 Gy^{-1}.

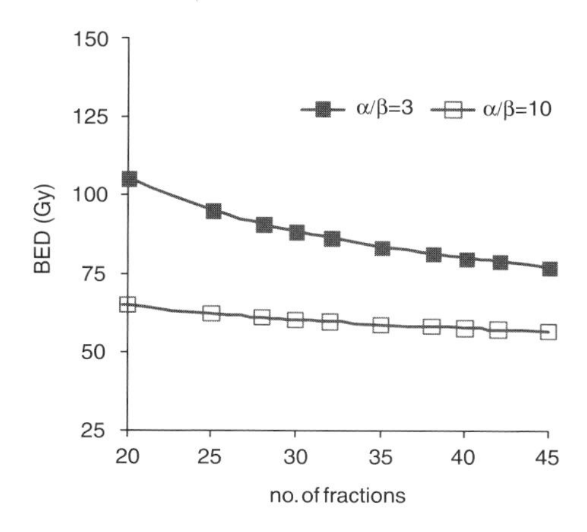

Figure 1 — BED equation showing the relationship between the change in the number of fractions and BED according to the α/β ratio. The dose and overall treatment time have been held constant at 55 Gy and 35 days respectively. T_{pot} for α/β = 3 has been set at 1000 days and at 5 days for α/β = 10. The lower the α/β ratio, the steeper the increase in BED as the number of fractions decreases (or as dose per fraction increases). An α/β ratio = 3 would be typical of a late radiation effect, values of ≥10 are typical for tumours and for the acute effects of radiation.

As T_{pot} approaches infinity, BED approximates to those obtained with the original formula, without a component dependent on overall treatment time. A smaller α will be associated with lesser biological effect — since the divisor in the time-dependent term is $\alpha \times T_{pot}$, a smaller α has the same effect as shortening T_{pot}, there will be a greater number of cells surviving at the end of the course of treatment.

The BED equation can be easily entered into a spreadsheet and the consequences of altering the various parameters can be illustrated graphically. Tables of biologically equivalent doses, based on the BED equation have also been produced.

Biomedical Research

Like **molecular medicine (p. 207)** this is another term that can be employed to mean

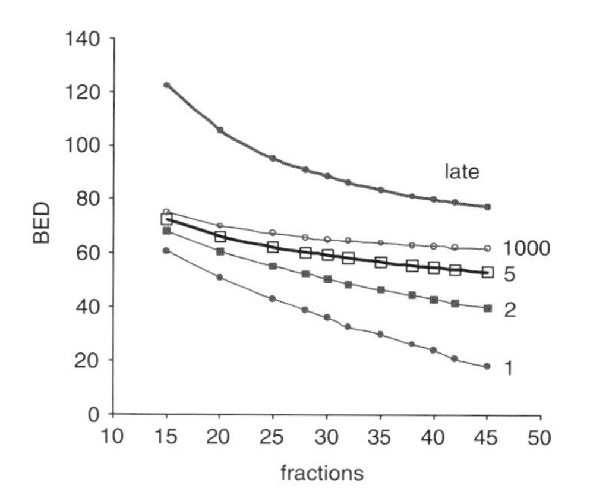

Figure 2 — BED equation (with time component) showing the effect of altering T_{pot} on BED. The dose is constant at 55 Gy, and treatment is at five fractions per week (e.g. 15 fractions are equivalent to 21 days treatment duration). Late effects use an α/β ratio = 3 Gy; the curves for T_{pot} = 1, 2, 5 or 1000 days use an α/β ratio = 10 Gy. α is constant at 0.3 Gy^{-1}. The biological effect of radiation decreases as treatment is prolonged and the decrease depends on the value chosen for T_{pot}.

whatever the user wants it to mean. It simply implies that the research links biology and medicine: it could, therefore, include zoology, botany, pharmacology, anatomy biochemistry and molecular biology. As a catchall it has some publicity value, however, it has little intrinsic meaning.

Bioreductive Agents

The essential characteristic of a bioreductive drug is that, through chemical reduction (electron gain), a relatively inactive compound is transformed into a highly toxic moiety. The process of electron gain is reversed in the presence of oxygen: bioreductive drugs should, therefore, be selectively more toxic to hypoxic, as opposed to well-oxygenated, cells. **Mitomycin C (p. 200)** was one of the first bioreductive drugs to be used in clinical practice. Porfiromycin is a more recently developed bioreductive drug that has been used in conjunction with radiotherapy in the treatment of head and neck cancer. Another clinically useful bioreductive drug is Tirapazamine. The toxicity of **Tirapazamine (p. 309)** is highly specific for hypoxic cells (×50–200). It has minimal myelosuppressive marrow effect, its main toxicity is gastrointestinal but it can cause muscle cramps. It is currently in clinical trials in combination with *cis*-platinum for the treatment of non-small cell lung cancer.

Classification of bioreductive drugs

- Dual function: 2-nitroimidazoles with clinically active group on side-chain, e.g. RSU 1069 and RB 6145, a **prodrug (p. 246)** for RSU 1069.
- Quinones: Mitomycin C, Porfiromycin.
- Indoloquinones: EO9.
- Benzotriazine-di-*N*-oxides: SR4233.

Bisphosphonates

This class of drugs is used in the management of hypercalcaemia and the skeletal complications associated with malignancy. The bisphosphonates may also have a role to play in the prevention and treatment of osteoporosis. They act by inhibiting the ability of osteoclasts to destroy bone and to release calcium from the bone into the blood. Bisphosphonates form complexes by binding, through their double phosphonate groups, to the hydroxyapatite of the bone matrix. This effectively coats the bone surfaces and insulates them from the attacks of osteoclasts. Examples of bisphosphonates include clodronate, pamidronate, etidronate and alendronate.

Black Literature

This term describes data, usually on file with a drug company or other commercial operator, not in the public domain. It is an important hidden source of bias in systematic reviews: if all negative studies are concealed, then only the positive can be included in the review.

Bleed and Run Research (syn. Helicopter Science)

A group of scientists from the developed world descend on a 'primitive' community with a particularly high **prevalence (p. 244)** of an interesting disorder. Samples are grabbed from the bemused population who never see the scientists again or hear of the research results. Bleed and stay research is similar, except there is a continuing commitment to the community, e.g. Nancy Wexler and Lake Maracaibo (Huntington's chorea).

Bleomycin

Bleomycin is an antitumour antibiotic with a broad spectrum of clinical activity: from lymphomas to squamous cancers of the head and neck. It is obtained by fermentation from *Streptomyces verticillis*. It is a complex mixture of active agents and is usually supplied, and prescribed, in units rather than milligrams. It causes both single- and double-strand breaks in DNA. It is **cell cycle (p. 52)** phase-specific, most active during G_2 and M. It is given parenterally and may also be given intrapleurally or intraperitoneally for the palliative treatment of effusions. Its main toxicity is pulmonary fibrosis. This problem is, to an extent, dose-related with a significant incidence above a cumulated dose of 400 U per m^2. Idiosyncratic responses do occur, with severe fibrosis after <100 U. Clinically, the syndrome is of shortness of breath, dry cough and interstitial infiltrates on CXR. Lung function tests show a fall in Tco with some restrictive deficit. One of the earliest detectable changes may be interstitial fibrosis on high resolution CT of the chest. Skin pigmentation, often prominent at sites of minor trauma, e.g. scratch marks, is a common problem. Bleomycin causes little myelosuppression and, for this reason, is an extremely useful drug in combination chemotherapy.

Block Randomisation

This is also known as restricted randomisation or permuted block randomisation. It is a procedure designed to ensure that randomised groups remain as closely balanced in size as possible. It is particularly useful for small-scale clinical studies in which the play of chance could lead to significant imbalances in the size of allocated groups. In a trial of 16 patients, it is entirely possible that unblocked randomisation could yield a group of five patients allocated to one treatment and 11 to the other. Blocked randomisation would ensure that, in a study of this size, the groups were about equal in size. It should be distinguished from stratified randomisation, which is designed to ensure that the groups are as balanced as possible in terms of known prognostic factors.

Blocked randomisation works as follows. We wish to have equal numbers of patients allocated to treatments A and B. With a block size of four, we have six possible sequences with 2 As and 2 Bs in them:

1. AABB.
2. ABBA.
3. ABAB.
4. BBAA.
5. BABA.
6. BAAB.

We have a sequence of random numbers as follows: 1, 1, 3, 5, 3, 2, 4, 6, 3, 1, 4, 3, 5, 5, 6, 1. With unblocked randomisation, we might allocate odd numbers to treatment A and even numbers to B: this would give 11 patients allocated to treatment A and five to B. If we let each number now code for one of the six blocks, we have the sequence:

AABB AABB ABAB BABA.

The blocks have been chosen at random but the groups balance: eight allocated to treatment A and eight to B.

Bloom Syndrome

A recessively inherited syndrome that involves a mutation of the *BLM* gene. It mainly affects Ashkenazi Jews. It is characterised by low birth weight, decreased growth, sun-sensitive skin, deranged pigmentation, telangiectasia and an increased incidence of malignancy. The tumours include leukaemia, **Wilms' tumour** (**p. 328**), oral cancers and gastrointestinal cancers.

The defect is in *BLM*, which maps to 15q26.1. Although DNA **ligase 1** (**p. 180**) activity is abnormal in cells from patients with Bloom syndrome, this is probably not the primary defect. *BLM* belongs to a family of genes, *RecQ*, many of which have helicase activity. It is likely that the *BLM* gene product is an enzyme that interacts with **topoisomerase** (**p. 310**).

Chromosomal instability is a feature of the syndrome with the majority of cells showing high rates of sister chromatid exchange (SCE).

Body Mass Index (BMI)

This is a measure of bodily habitus and is usually calculated according to a formula described by **Quetelet** (**p. 255**):

$$\text{BMI} = \text{weight}/(\text{height})^2.$$

It is used in epidemiology, e.g. in studying the incidence of breast cancer, to see whether general physique has any influence on the risk of developing a disease. It can also be used in pharmacology for individualising doses of drug.

Bone Marrow Transplantation (BMT)

BMT involves the infusion, usually intravenously, of cells of marrow origin. Many of the cells will be stem cells and, once they have migrated to the appropriate niches within the haemopoietic microenvironment of the host, they can restore or supplement haemopoiesis. The procedure is useful when there is either endogenous, as in primary aplasias, or exogenous, as after high-dose chemotherapy, impairment of marrow function that is severe enough to be life threatening. The clinical indications for BMT have expanded widely. Originally used in leukaemia, BMT is now used for aplastic anaemia, lymphomas, thalassaemia, refractory solid tumours, osteopetrosis and sickle cell anaemia.

The broadest clinical experience with BMT is in leukaemias and lymphomas. The process of BMT can be divided into a sequence of stages:

- The conditioning regimen to kill malignant cells and prevent graft rejection.
- Marrow manipulation:
  - — to purge autologous marrow of malignant cells
  - — to purge allogeneic marrow of T-cells to prevent or ameliorate **graft versus host disease** (**GVHD**) (**p. 143**).
- Marrow re-infusion (transplantation).
- Post-transplant regimen:
  - — to support patient through cytopenia and minimise risks of infection and haemorrhage
  - — to prevent or ameliorate GVHD.

There are two main types of BMT: allograft, from another donor; and autologous, in which an individual is treated with their own stored or manipulated marrow.

- Allografts can be identical, from an identical twin; and HLA matched, either from a relative or from an unrelated donor identified via a registry; mismatched. The chance of an individual having an HLA matched relative is ~35%.
- Autologous marrow is given as an autograft, marrow previously harvested from the patient is re-infused with or without conditioning *in vitro*. Marrow for autografting may be cryopreserved for several years.

The marrow, whether allograft or autograft, can be treated *in vitro* to increase the chances of therapeutic success. An allograft may be treated with antibodies to T-cells to lower the risk of GVHD: T-cell-depleted marrow. Autologous marrow can be treated *in vitro* to purge any malignant cells that might be present. This can be accomplished either with **cytotoxic drugs** (**p. 84**) or using antibodies directed against antigens expressed preferentially by the tumour cells.

Bonferroni Correction

The Bonferroni correction is a method of adjusting the criterion for statistical significance when multiple tests have been performed. It is calculated as the desired ('true') significance level, usually $\alpha = 0.05$, divided by the number of tests performed. In an experiment involving 25 comparisons in which statistical significance has assessed, the p defined as 'significant' for each comparison would be calculated as $0.05/25 = 0.002$. By using this criterion, we could be reasonably certain that any differences defined as significant were truly significant at $p < 0.05$. Had we chosen 0.01 as the overall significance level, then each individual significant comparison would have had to have $p = 0.01/25 = 0.0004$.

Boron Neutron Capture Therapy

This therapeutic approach is based on the fact that if a boron isotope, ^{10}B, is exposed to low-energy neutrons, it will capture the neutrons and produce α-particles, which will then produce highly localised cellular damage. If the boron is selectively targeted to the tumour and then activated, then this should offer an effective means for improving the **therapeutic ratio** (**p. 307**) for external beam therapy. The concept is simple and elegant, but it has proved extremely difficult to put it into practice.

The first problem is to obtain sufficient selective targeting of a boron compound to yield useful target to normal tissue ratios. BSH (sodium mercaptoundeca-hydrododecaborate) does not reach all tumour cells, and uptake of *p*-borono-phenylalanine (BPA) may be even worse. New boron compounds conjugated to porphyrins or low-density lipoproteins are being developed. Attempts have also been made to exploit the selectivity of monoclonal antibodies for boron capture therapy. Progress is slow because there is little commercial incentive to develop compounds for such a small market. The second major problem is the availability of neutron beams suitable for capture therapy. These beams are usually produced in nuclear reactors in facilities not really suitable for patient care.

Bragg Peak

When ionised particles interact with matter, their path length is limited by two main factors: their physical mass and initial velocity. They decelerate rapidly towards the

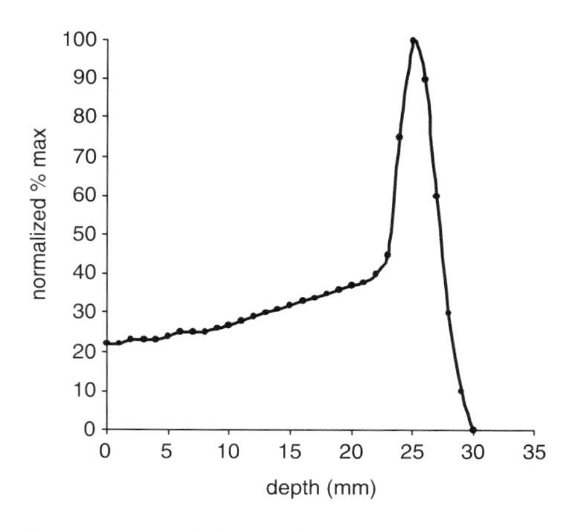

Figure 1 — Depth–dose profile for a proton beam of ~60 MeV traversing a Perspex phantom. The peak is, by itself, too narrow for therapeutic use and can either be modulated, by scattering or attenuation, or used in multiple-field techniques to produce a larger high-dose volume.

end of their path and this deceleration produces an increase in ionisation. The maximum deposition of energy, therefore, occurs over a limited range at the end of their path in tissue.

The narrow peak at depth, with limited superficial deposition of energy, means that charged particle beams are, in many ways, ideal for radiation treatment. There are, however, many practical problems, particularly that of production. Proton beam therapy requires a cyclotron; heavy-ion therapy requires vast accelerators. There are problems with collimating and focussing beams from equipment designed primarily for research into particle physics. However, the physics of the Bragg peak, to say nothing of the radiobiological advantages of densely ionising radiation (independence from the oxygen effect, cell killing less dependent on cell cycle phase), are too tempting to ignore. Considerable effort has been put into exploiting the Bragg peak, particularly proton beam therapy, and, world-wide, there are now over 15 clinical facilities.

Proton beams have been used mainly for treating CNS tumours and ocular melanomas.

Brassicas

Also known as the cruciferous vegetables; a group of green vegetables including broccoli, Brussel sprouts, cauliflower and cabbage that are rich in sulphoraphane. Sulphoraphane can induce enzymes that are protective against **oxidative stress (p. 221)** and a diet high in brassicas may, therefore, be protective against bowel cancer. It is a minor irony that these are the very vegetables that patients are told to avoid during pelvic radiotherapy for rectal cancer.

Brattleboro Rat

A strain of rat that has a naturally acquired single base-pair deletion that results in diabetes insipidus. The deletion is in the portion of the arginine-vasopressin (AVP) gene that codes for the carrier protein neurophysin. A stop codon is missing and so a mutated AVP precursor accumulates in the endoplasmic reticulum — the result is a deficiency of functional AVP. The Brattleboro rat provides a useful model for **gene therapy (p. 137)** designed to correct a single mutation.

*BRCA*1

Mutations in this gene are associated with a high incidence of breast cancer and ovarian cancer in women and of prostate cancer in men. Families that carry mutations in *BRCA*1, therefore, show a higher than expected incidence of these tumours (hereditary breast–ovarian cancer syndrome). Typically, these tumours develop at an early age, with the majority of breast cancer presenting at <60 years of age. The lifetime risks of developing cancer for women with *BRCA*1 mutations have been estimated at 80% for breast cancer and at 40% for ovarian cancer. Family studies suggest that 81% of breast–ovarian cancer families have *BRCA*1 mutations with only 14% associated with *BRCA*2 mutations. Ashkenazi Jews have a particularly high incidence of *BRCA*1 mutations, 1% of Askenazi women may carry *BRCA*1 mutations. *BRCA*1 is located at chromosome 17q21.

The *BRCA*1 gene product shows **sequence homology (p. 284)** to granins, a type of protein predominantly expressed in endocrine and neuroendocrine tissues. It probably functions as a tumour suppressor gene by influencing **transcription (p. 313)**. That it is a component of the RNA polymerase holoenzyme (polII) is consistent with this mode of action. *BRCA*1 causes increased expression of genes such as *GADD*45 and *p21*, which are involved in apoptosis and **cell cycle (p. 52)** arrest. It also appears to have a role in DNA repair and, like *BRCA*2, interacts with *RAD*51. Activation of *BRCA*1 in response to **DNA damage (p. 98)** is by **phosphorylation (p. 234)** and probably involves a protein kinase produced by expression of *Chk2*. *BRCA*1 may also be activated by the **ATM (p. 23)** gene product. The net effect of *BRCA*1 mutation is to increase genomic instability.

A **knockout mouse (p. 171)** missing both *BRCA*1 alleles fails to develop, probably because DNA damage accumulates and activates *p53*-mediated apoptosis.

*BRCA*2

Mutations in this gene, in common with *BRCA*1 mutations, are associated with a increased susceptibility to breast cancer. The risk of breast cancer in women with *BRCA*2 mutations has been estimated at 60% (95%, CI 26–79%) by 60 years of age. Families carrying *BRCA*2 mutations also have a high incidence of cancer of the male breast, the link with ovarian cancer is not as strong as it is for *BRCA*1 mutations: only 14% of breast–ovarian cancer families have *BRCA*2 mutations. In contrast, *BRCA*2 mutations are in 76% of families with high incidence of both male and female breast cancer. *BRCA*2 is on chromosome 13q12–q13, and it interacts with both *p53* and *RAD*51 and it is probably concerned with cell cycle control, effectively functioning as a tumour suppressor gene. Cells deficient in a normally functioning *BRCA*2 gene product may show increased sensitivity to radiation. *BRCA*2-deficient embryos are more radiosensitive than embryos with normal *BRCA*2. This may have implications for treatment in patients known to have breast cancer associated with mutations in *BRCA*2. The question of whether any increased sensitivity is confined to the malignant cells, or whether the normal tissues also show increased **radiosensitivity** (**p. 260**) is, of course, crucial.

Breakpoint Cluster Region (BCR)

This is an area on chromosome 22 that is conspicuously involved in the various translocations that can produce a **Philadelphia chromosome** (**p. 233**). Nearly all the translocations appear to involve the sub-band 22q11.21. The high frequency of breaks at this site has led to its designation as the BCR. This region, which is ~135 kb in length and contains 23 exons, codes for a 160 kD protein with kinase activity and with a C-terminal **domain** (**p. 102**) which has GTPase activity that can activate p21.

Breast Cancer — Hereditary Susceptibility

Only ~10% of all cases of breast cancer are associated with an inherited susceptibility, usually involving *BRCA*1 or *BRCA*2, other genes which may be involved include *BRCA*3, *p53* and *BWSCR*1*A*. In 1757, Le Dron described a 19-year-old nun with breast cancer and felt that to treat her would be a waste of time because 'her blood was corrupted by a cancerous ferment natural to her family'. Although inherited mutations are a relatively rare cause of breast cancer in general, the risks to individuals can be very high. The sister of a woman who developed bilateral breast cancer when <50 years of age faces a 28% lifetime risk of developing breast cancer. The **Li–Fraumeni syndrome** (**p. 179**), which involves the inheritance of mutant *p53*, is associated with breast cancer and soft tissue sarcomas. In Lynch syndrome, Type I breast cancer can be associated with cancer of the colon and endometrial cancer.

Brock Assay

The Brock assay is an *in vitro* method for assessing the survival of irradiated cells. The tumour is biopsied and a cell suspension prepared. The suspension is plated and then cultured for 24 h. The plated culture is irradiated and then incubated for 2 weeks. Those colonies, which are growing exponentially, are counted. The results are usually expressed as SF_2.

Bromocriptine

A dopamine agonist that, by inhibiting secretion of prolactin, is useful in the management of prolactin-secreting pituitary adenomas.

Bystander Effect

An important collateral advantage of gene therapy based on the targeted reprogramming of malignant cells to induce them to self-destruction. Treatment may kill

even those cells that have not incorporated the construct. When **suicide gene therapy (p. 299)** is attempted it is clear that some cells die, even though they have not actually incorporated the exogenous genetic material into their genome. The mechanism may involve the passage of a toxic metabolite, e.g. phosphorylated ganciclovir, via gap junctions, or it could involve apoptosis of the transfected cells with subsequent ingestion by adjacent non-transfected cells of vesicles containing toxic metabolites. Immunological mechanisms may also play a role since, in some systems, the bystander effect is less apparent when tumours are grown and treated in the immune-suppressed mouse. In immunocompetent animals, specific antitumour responses have been demonstrated following suicide gene therapy. This has implications, not only just for the local effects of gene therapy, but also for systemic effects.

Cadherins

The cadherins are a family of cell adhesion molecules involved in strong binding between cells, particularly epithelial cells. Interactions between cadherins and a small intracellular protein, catenin, are an important means of communication between the extracellular matrix and the interior of the cell. Loss of function or expression of cadherins by tumour cells can, through permitting escape of cells that would otherwise be firmly bound to their neighbours, be associated with **invasion** (**p. 167**) and **metastasis** (**p. 198**). Cadherins are named according to their predominant tissue of origin: N-cadherin is found in nerve, lens and muscle; epithelial cells express **E-cadherin** (**p. 111**). The gene for N-cadherin is on chromosome 18q.

Calman–Hine Report

The Calman–Hine report arose from the demonstration that there are geographical variations in the quality of care for patients with cancer. This is rightly perceived as unjust and unacceptable. Why should one woman have to attend three separate clinics before she is reassured that her breast lump is benign, while her friend, living two streets away, has triple assessment in a one-stop clinic?

The Calman–Hine report has recommended that the care of patients with cancer be reorganised as follows:

- Primary care is seen as the focus of care.
- Designated cancer units should be created in many district general hospitals: these should be of a size to support clinical teams with sufficient expertise and facilities to manage the commoner cancers.
- Designated cancer centres should provide expertise in the management of all cancers including less common cancers by referral from cancer units. They will provide specialist diagnostic and therapeutic techniques including radiotherapy.
- Centres or units using different methods of treatment should be expected to justify them on scientific or logistical grounds.
- High-quality cancer registration data enable health authorities, trust and clinicians to monitor and **audit** (**p. 24**) service performance.

Cancer centres and units are now trying, in the cruel absence of adequate funding, to reorganise themselves along the above lines.

Calmodulin

Calmodulin is concerned with intracellular binding of calcium. It is a 150 amino acid polypeptide that has four binding sites for calcium. Once calcium is bound, its conformation changes. Calcium can function as an intracellular messenger. A family of protein kinases, the Ca^{2+}/calmodulin-dependent protein kinases, is controlled by the Ca^{2+}/calmodulin system. These kinases control the phosphorylation of threonine and serine residues on target proteins. The Ca^{2+}/calmodulin system is very similar in concept to the cAMP/A-kinase system. The pathway is widely used. Cells that produce neurotransmitters are particularly rich in calmodulin.

Calvert's Formula

A formula used for calculating the dose (mg) of **Carboplatin** (**p. 46**) that will produce a specified area under the curve (AUC). The usual procedure is to choose an AUC (4 for low-dose therapy, 6 for intermediate dose, 8 for high-dose treatment). The calculation is:

$$\text{Dose (mg)} = \text{target AUC} \times [\text{GFR (ml min}^{-1}) + 25],$$

where GFR is glomerular filtration rate.

The following adaptation should be used for children:

$$\text{Dose (mg m}^{-2}) = \text{target AUC} \times [(0.93 \times \text{GFR}) + 15].$$

CAMPATH Antibodies

The CAMPATH 1 series of antibodies recognises the CD52 antigen expressed by T-cells and has been used to remove T-cells from allogeneic bone marrow used in transplantation. Treating the marrow *in vitro* with CAMPATH 1 reduces the incidence and severity of **graft versus host disease (p. 143)** in recipients. A variety of CAMPATH antibodies are in clinical use — CAMPATH-1M, a rat IgM; CAMPATH-1G, a rat IgG; CAMPATH-1H, a humanised IgG1 — and they have been used for treating graft rejection, chronic lymphatic leukaemia and autoimmune diseases.

Camptothecins

The camptothecins are a family of anti-cancer drugs derived from a Chinese tree, *Camptotheca acuminata*. These agents are plant alkaloids with the ability to inhibit topoisomerase I. Members include **irinotecan (p. 168)**, **topotecan (p. 310)**, rubitecan and 9-AC.

Cancer Control

This concept acknowledges that there is more to mitigating the impact of cancer on a community than simply treating established cases. Cancer control involves strategies for the prevention, early detection and optimal management of cancer. A coherent and feasible strategy for cancer control is essential if both morbidity and mortality rates from cancer are to be successfully reduced. It is equally essential that resources be allocated to the areas of greatest need: socio-economic gradients in cancer **mortality rate (p. 207)** often exceed, in quantitative terms, the marginal benefits achieved from well-accepted adjuvant treatments. These gradients are often due to a combination of potentially remediable factors: social behaviour, smoking, alcohol, diet; a delay in seeking medical attention; and inadequate treatment. The elimination of such differences should be a basic component of any cancer control programme.

Cancer Management

Cancer management is not just treatment. To minimise the impact of cancer on a population, it is necessary to do a lot more than simply treat those individuals unfortunate enough to develop the disease. A sequence of opportunities for intervention can be defined, starting with prevention and moving through treatment and beyond:

- **Primary prevention (p. 245)**: provide the community with advice about how each individual might lower his or her personal risk of developing cancer:
  - — avoid **tobacco use (p. 309)**
  - — eat plenty of fresh fruit and vegetables
  - — avoid excessive intake of alcohol.
- **Secondary prevention (p. 281)**: take measures to decrease the incidence of second primary cancers in patients successfully treated for their first tumour:
  - — retinoid therapy for patients who have had squamous carcinomas of the head and neck
  - — advise all patients treated for cancer to stop smoking.
- **Screening (p. 278)** the healthy population: identify patients with malignant disease before they become symptomatic and

thereby increase the chances of successful therapy:

— breast screening and cervical cancer screening have been accepted as worthwhile and national screening programmes exist in the UK

— screening for colorectal cancer is probably effective but as yet there is no national **screening programme** (**p. 278**)

— screening for prostate cancer is widely practised on an *ad hoc* basis: there is little evidence to support such activity

— screening for lung cancer can be effective in affluent motivated populations with ready access to cardiothoracic surgery, few areas in the UK fulfil these basic criteria.

- Early diagnosis of symptomatic disease: encourage patients and their family doctors to take seriously the early symptoms of cancer: haemoptysis (lung cancer); breast lump (breast cancer); rectal bleeding, change in bowel habit (colorectal cancer):
- staging: ensuring that, before treatment, the extent of the disease is known. This involves delineating the local disease precisely as well as actively seeking metastatic spread to distant sites and/or lymph nodes. The process involves clinical examination, imaging and other appropriate investigations. Various staging systems are used to summarise results the most widely used are the TNM and AJC systems. By using a common staging system results from different centres, and from different times, can be compared — at least approximately
- treatment: the most obvious, but not always the most important, aspect of cancer management
- rehabilitation and follow-up: a series of procedures aimed at restoring fully to the community an individual who has been unfortunate enough to require treatment for cancer. Follow-up, attending clinics regularly after treatment, is not just about early detection of recurrence or obtaining statistics on outcome, it is about reassurance and psychological support and demonstrating a continuing commitment of the treatment team to the wellbeing of the individual patient
- evaluation of treatment: classically, the evaluation of cancer treatment has been in terms of survival statistics — not a particularly demanding exercise intellectually, since it simply involves discriminating between the living and the dead and enumerating appropriately. Recently there has been an increasing appreciation that, particularly when treatment has been given with palliative intent, a more thoughtful approach is required. Evaluation of **quality of life** (**p. 253**) is but one example of the trend towards more elegant solutions to the problem of evaluating treatment
- **palliative** (**p. 226**) care: when curative treatment fails, or is impossible, then measures are required that will relieve the suffering and distress of patients, and their families. Palliative care is the portmanteau term used to summarise the necessary interventions. Historically, palliative care has been based upon, and associated with, the hospice movement. This is not necessarily desirable: for many patients palliative care should be based in the community, in their own homes. The healthcare system is lurching towards fulfilling this need but, in the UK at least, there is an unhealthy reliance upon charitable funding for something that should be provided routinely to those in need.

Cancer Statistics

In 1990, the year for which the most recent world-wide data are available, 4.28 million people developed cancer. The number of deaths from cancer was 2.95 million.

Rates per 100 000 (age-standardised rate, world population as reference) are:

	World	More developed	Less developed
Incidence, men	203	299	151
Incidence, women	154	208	122
Mortality rate, men	140	182	105
Mortality rate, women	90	116	77
Mortality rate: incidence ratio	0.64	0.58	0.66

Rates in men are consistently higher than those in women, and rates in the developed world are higher than those in less developed countries. The mortality rate : incidence ratio is higher in less-developed countries suggesting that survival may be better in the developed world. Some of this apparent improvement may be artefactual since ascertainment **bias** (**p. 28**) will dictate that, in the developing world, only the most severe cancers will be diagnosed. If the mortality rate : incidence ratio in the developing world could be brought down to the level in the developed world, then this could represent a saving of ~200 000 lives year^{-1}.

The most common cancers in men, world-wide, are, in descending order: lung, stomach, prostate, colon and rectum, and liver. The commonest tumours in women are: breast, cervix, colon and rectum, stomach, and lung.

In the more developed regions the common tumours in men are: lung, prostate, colon and rectum, stomach, and bladder.

In women, the common tumours are: breast, colon and rectum, lung, stomach, and cervix.

The pattern is very different in the less-developed world. For men it is: lung, stomach, liver, oesophagus, colon and rectum. In women it is: breast, cervix, stomach, colon and rectum, and lung (figures 1 and 2).

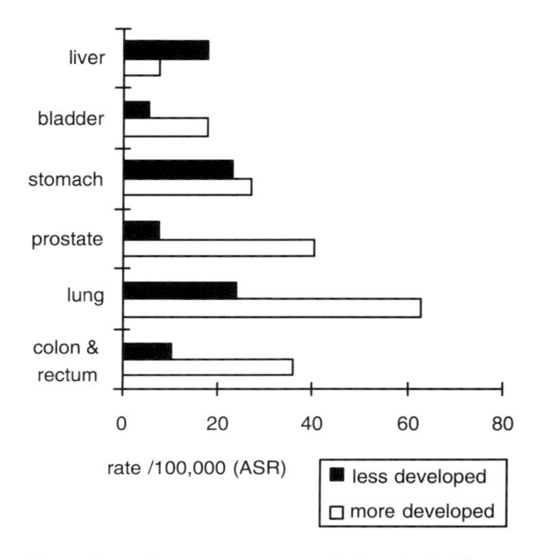

Figure 1 — Cancers in men : world data divided by level of economic development.

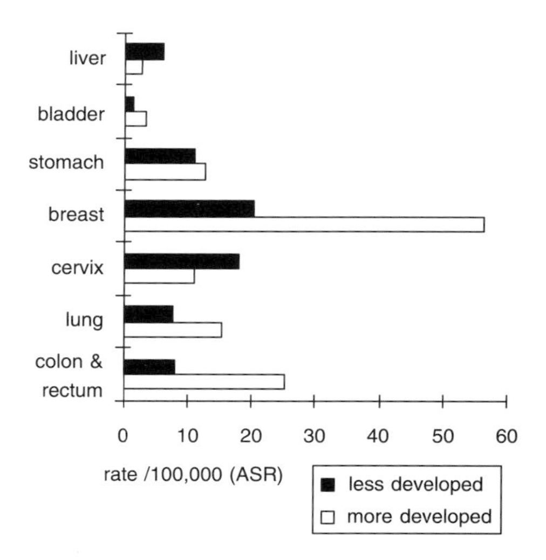

Figure 2 — Cancers in women: world data divided by level of economic development.

There are 1.35 million new cases of cancer, and ~900 000 deaths from cancer, in the EU each year. For the UK, the corresponding figures are 235 000 and 160 000. Of the ~64 000 deaths each year in Scotland, 15 500 (24%) are from cancer and 17 000 (26%) are due to ischaemic heart disease (figures 3–5).

Smoking causes between 25 and 40% of all cancers: on current figures this corresponds to at least 1 million new patients with smoking-induced cancer in the world each year. Up to 50 000 deaths in the UK and 250 000 deaths in the EU each year may be due to cancers caused by tobacco.

The world-wide budget for cancer drugs was £8 billion in 1997. Cancer drugs account for only 5.3% of total prescription costs, even though >30% of the population will develop cancer and 22% of deaths in the developed world are due to cancer.

Cancer statistics are not an end in themselves. They should be regarded as hypothesis-generating. Why is the 5-year survival for patients with colorectal cancer so much better in Switzerland than it is in Scotland? Are Swiss patients different? Are Swiss doctors better? Is the discrepancy real or is it because of **bias** (**p. 28**) in the selection and interpretation of the data? If it is real, does it simply reflect a different disease affecting a different population? (figure 6).

Why is the survival rate for patients with breast cancer in South East England affected by postcode, with lower long-term survival for patients living in socially deprived areas? The absolute magnitude of the difference,

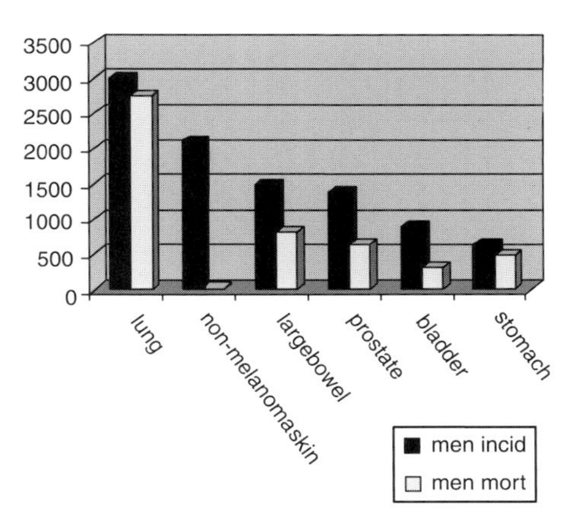

Figure 4 — Cancer incidence and mortality rate per annum in Scotland (1986–95).

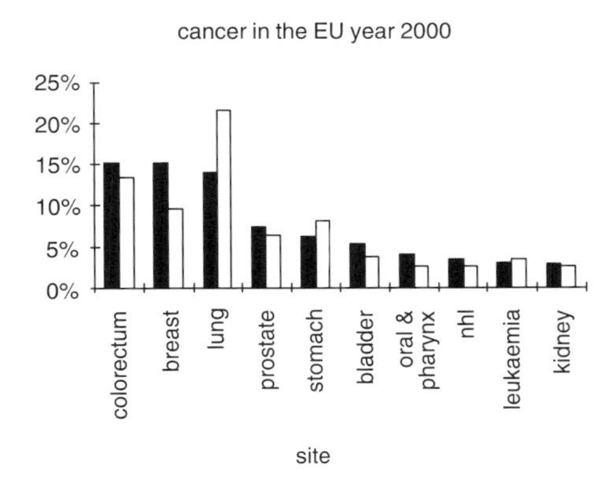

Figure 3 — Projected figures for the EU in 2000. Solid bars indicate new cases (% of all cancers); white bars indicate deaths (% of all cancer deaths).

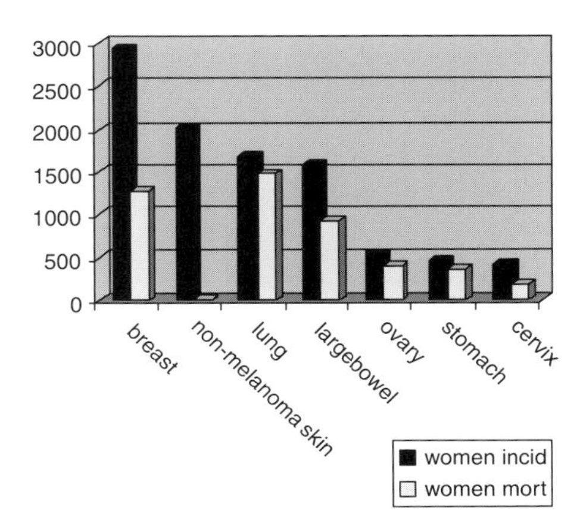

Figure 5 — Cancer incidence and mortality rate per annum in Scotland (1986–95).

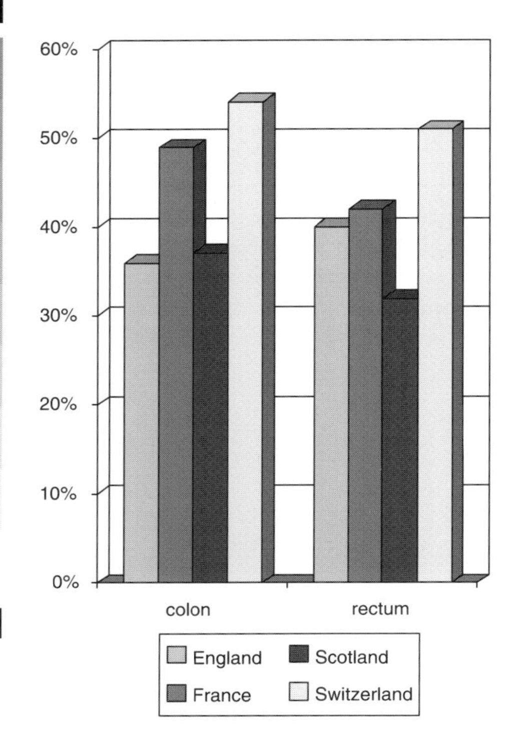

Figure 6 — Five-year survival rates for patients with colorectal cancer.

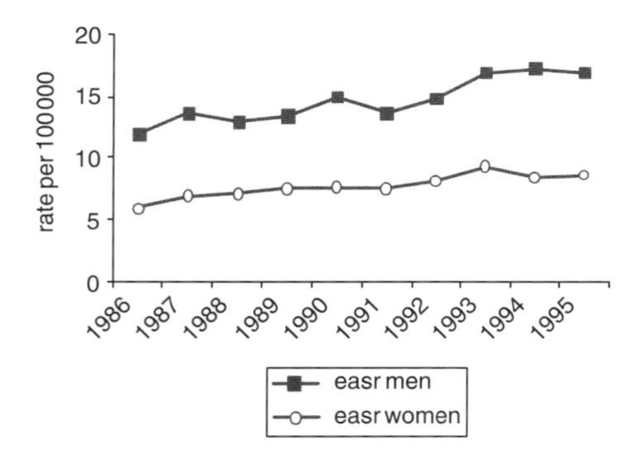

Figure 7 — Oesophageal cancer in Scotland (1986–95).

59 versus 48%, is such that, were it achieved by a new treatment for cancer, it would be hailed as a major breakthrough. Is it because poorer people have problems gaining access to high-quality healthcare? Do poorer

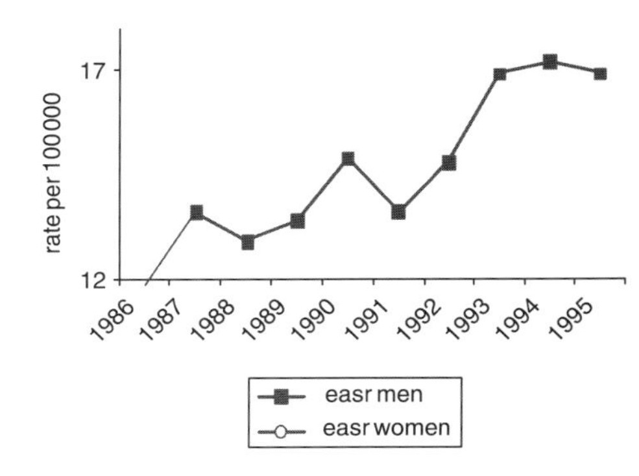

Figure 8 — Oesophageal cancer in Scotland (1986–95).

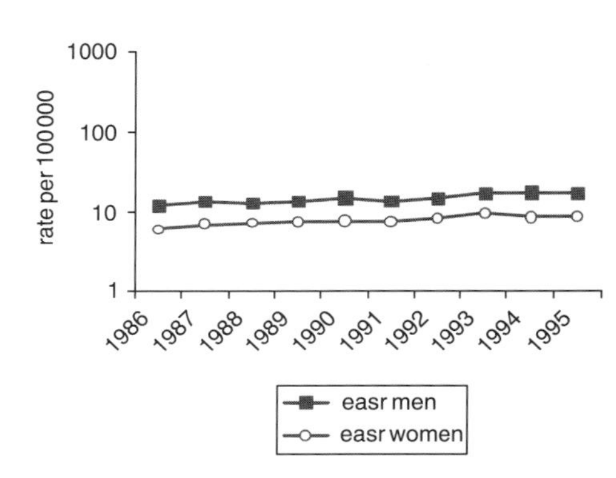

Figure 9 — Oesophageal cancer in Scotland (1986–95).

people neglect themselves and present with more advanced disease? Is the observed difference remediable, by social or medical intervention, or is it simply determined by a biology that is unalterable?

Secular trends in cancer incidence are also provocative. Is the recent increase in oesophageal cancer, particularly in adeno-carcinomas around the gastro-oesophageal junction, related to the increasing use of drugs to decrease gastric acid production (H2 antagonists, proton pump inhibitors)? (figure 7).

Cancer statistics can, like any other statistics, be made to lie. Differences can be exaggerated (figure 8) or made to disappear (figure 9).

The raw data are all the same: in these examples, as with modern politics, it is all a matter of scale, of presentation rather than content.

Cancer Vaccines

Vaccines have theoretical potential in both the prevention and treatment of cancer. One approach is to produce vaccines that stimulate the immune system in general. Such vaccines would not be specific to any particular tumour.

An alternative approach is to immunise the patient against an antigen specific to its particular tumour. One approach is to exploit **anti-idiotype networks** (**p. 17**) and the world through the looking glass. The **HAMA** (**p. 147**) response noted after therapy with murine monoclonal antibodies could possibly be exploited in this way. A more direct, and certainly conceptually more simple, approach is to try to immunise patients by injecting them with large amounts of tumour-specific antigen in a purified form. The antigen can be obtained by incorporating tumour-specific DNA into a plasmid. The DNA is expressed once the plasmid has been incorporated into a host cell. Direct intramuscular injection of DNA plasmids has been tried, but to obtain adequate amounts of antigen large amounts of DNA are required and the production of adequate amounts of DNA becomes the rate-limiting step. The **gene gun** (**p. 136**) is a method for circumventing this problem; the amounts of DNA required for successful antigen expression are very much less.

Other approaches include targeting with viral vectors or using **liposomes** (**p. 182**) for vaccine delivery.

Although a considerable amount of effort, over many decades, has been put into the development of cancer vaccines, the clinical results have been disappointing. The concept is simple and attractive, implementation is complex and requires an attention to tedious detail which is, ultimately, unappealing. It is one thing to design a plausible vaccine, it is quite another to define the schedule that produces the optimal benefit and then to demonstrate that the approach is clinically worthwhile. One problem is the need to define intermediate **endpoints** (**p. 114**) to permit more rapid evaluation of agents and schedules. It is dispiriting to have to wait several years to assess an approach simply because tumour prevention or eradication is the only available endpoint.

Cancer Statistics: National Comparisons

Comparison of cancer statistics between nations can provide interesting clues about aetiology, when incidence is compared, and about both incidence and management, when **mortality rate** (**p. 207**) is compared. Even within the so-called developed world, there are major differences. Here are some data on annual mortality rates for breast cancer:

Country	Period	Rate/100 000 women
Belgium	1986–89	26.5
England and Wales	1990	28.4
Germany	1990–91	22.1
Italy	1990	20.6
The Netherlands	1990–91	26.9
USA	1988–92	22.3

The annual mortality rate from breast cancer is 40% higher in England and Wales than in Italy, and 27% higher than in the USA.

Cancerlit

A database service provided by the National Cancer Institute that includes only those articles, published since 1963, considered of relevance to cancer research or treatment. It is a companion to the clinical information service **PDQ** (**Physicians Desk Quarterly**) (**p. 229**). The services are available over the World Wide Web or by subscription on a CD-ROM. Free access to **Cancerlit** (**p. 45**), with downloading and printing of references and abstracts, can be obtained via Cancerweb (www.graylab.ac.uk) or, more directly, via the NCI itself (www.cnetdb.nci.nih.gov/cancerlit.shtml) or via www.mds.qmw.ac.uk/lib/db/cancer.htm

Cannabinoids

A group of compounds isolated from marihuana. The compounds can act on a central neuroreceptor (the CB1 receptor) and produce analgesia. There is also a second receptor, CB2, and cannabinoids may also have a peripheral action. Cannabinoids have also been used as anti-emetics in patients receiving treatment for cancer; Nabilone, a cannabinoid derivative, had a brief vogue but was relatively inactive and produced dysphoric reactions, particularly in elderly patients.

Capecitabine

This is a **prodrug** (**p. 246**) that can be administered orally and which is ultimately converted to **5-fluorouracil** (**5FU**) (**p. 130**). It is absorbed intact and is then converted to 5′-deoxy-5-fluorocytidine (5′-DFCR) in the liver. The 5′-DFCR is then converted by cytidine deaminase to an intermediate metabolite, 5′-deoxy-5-fluorouridine (5′-DFUR). The conversion to 5FU occurs at sites where there is a high concentration of thymidine **phosphorylase** (**p. 234**) (dThdPase). Tumours, particularly in areas where there is hypoxia and/or active **angiogenesis** (**p. 14**), contain high levels of thymidine phosphorylase and, therefore, there is the potential for improving the therapeutic ratio. Phase III studies have shown that capecitabine is useful in the treatment of both breast and colorectal cancers. Synchronous administration of capecitabine and radiation might, in theory, improve the **therapeutic ratio** (**p. 307**). Since 5FU levels will, theoretically, at least, be higher in the tumour than in normal tissue, its **radiosensitising** (**p. 260**) effect might preferentially affect the tumour, with selective sparing of adjacent normal tissue. The main dose-limiting side-effects are diarrhoea, mucositis, **hand–foot syndrome** (**p. 148**) and myelosuppression.

Capping

A mechanism by which **telomeres** (**p. 306**), at the end of chromosomes, stabilise and protect the tip of the chromosome. Capping contributes to nuclear architecture by helping to ensure that chromosomes are correctly positioned. This is particularly important during **mitosis** (**p. 201**).

Carbogen

This is a gas used in **ARCON** (**p. 20**) therapy to improve the oxygenation of tumours during radiation treatment. It is a mixture of 95% oxygen and 5% carbon dioxide. Carbogen is used because breathing pure oxygen would produce vasoconstriction, which could impair, rather than improve, tumour oxygenation.

Carboplatin

Carboplatin is an analogue of **cisplatin** (**p. 66**), and is more myelotoxic, but less nephrotoxic, than the parent compound. It is activated within the cell by hydroxylation and, like cisplatin, forms intrastrand links, particularly involving guanine and adenine. It is given as a short intravenous infusion and because, unlike cisplatin, there is no need for extensive pre- and post-treatment hydration, it can safely be given as to outpatients. Carboplatin is one of the few

cytotoxic drugs (**p. 84**) for which dosage is routinely based on AUC (area under the curve) estimation. The AUC for carboplatin is usually calculated from the glomerular filtration rate using Calvert's formula:

$$\text{Dose (mg)} = \text{target AUC} \times [\text{GFR (ml min}^{-1}) + 25],$$

where GFR is the glomerular filtration rate. The dose is typically in the AUC range 4–7.

Carstairs' Index

An index of socio-economic deprivation that, through postcodes, can be linked to census data and thereby used in assessing the influences of social deprivation, and affluence, on health.

Case-Control Study

A study design in which 'cases' (individuals with the attribute of interest: a particular disease; response to a particular therapy) are compared with 'controls', individuals who lack the attribute but who, in every other important respect, appear to resemble the cases. This, mainly because results can often be achieved quickly, and usually within the time frame imposed by funding cycles, is a popular research design. It is, however, extremely susceptible to bias and case-control studies are probably best regarded as hypothesis-generating exercises since they are unlikely to produce an unbiased and robust conclusion. They may form a useful and essential preliminary to a prospective study: cohort study or randomised trial. They are particularly useful in the investigation of rare events and conditions — when a cohort study would be impossible.

There are important methodological issues to be considered in a case-control study:

- How to define and identify cases?
- How to define and identify controls?
- How to assess the extent to which the cases and the controls match each other?
- How many cases (or controls) to recruit to avoid the possibility of an under-powered study with the probability of a Type 2 statistical error?

The main problems with case-control studies are concerned with:

- **Confounding** (**p. 72**): how to be sure that any observed difference is due to the variable we have implicated rather than some other hidden factor?
- The identification of cases where selection bias could lead to difficulties (only severe cases included or self-referral of untypical patients).
- In assessing the controls, ascertainment **bias** (**p. 28**): if we have not tested them how do we know whether they have the condition of interest?
- In assessing history, recall bias: cases, to whom something unpleasant has happened, are more likely to have ruminated on possible causative factors than 'controls', who feel perfectly well and have no particular reason to think too much about the past.

There are many examples of conflicting results from case-control studies: this is scarcely surprising given the many opportunities for bias to affect their design and execution.

Caspases

A family of cysteine- and aspartyl-specific proteinases whose activation may be essential for some types of apoptosis. They act downstream of the *bcl-2*-dependent checkpoint. Caspase-1 is also known as ICE, i.e. interleukin 1-converting enzyme. Targets for the action of caspases include PARP (poly[ADP-ribose] polymerase), fodrin and

the presenilins. Inhibition of caspases may have therapeutic potential in the treatment of degenerative neurological diseases such as amyotrophic lateral sclerosis (ALS) and Alzheimer's. The classic mode of caspase activation is the binding of **Fas** (**p. 126**) to **Fas ligand** (**p. 126**).

Cause

Association and/or correlation does not indicate cause. Cause, and causation, implies that B exists as a result of the direct effects of A and not as the result of some confounding process or variable. It is often difficult confidently to attribute a causal relationship between two entities — in spite of all the excellent evidence some people still doubt that cigarette smoking can cause lung cancer.

There are some common-sense approaches to assessing whether A causes B. Sir Austin Bradford Hill (1965) was one of the first medical statisticians to summarise the epidemiological criteria to be fulfilled for an agent to be defined as the cause of a disease:

- A should precede B in time, John Stuart Mill's principle of uniform antecedence: only in the realms of quantum physics can events precede their cause.
- A should be reliably associated with B, they should be correlated. If B has a habit of occurring in the absence of A then it is unlikely that A consistently causes B. This is of course the classic objection to smoking as a cause of lung cancer: 'my Uncle Josh smoked 150 cigarettes a day all his life and he was shot to death at the age of 105, don't tell me that smoking causes cancer'.
- As A increases then so should the likelihood of B occurring. This is an example of dose–response and is a telling piece of evidence in favour of smoking as a cause of lung cancer. The more cigarettes a person smokes the more likely they are to get lung cancer.
- There should be a plausible mechanism whereby A can cause B.
- There should be no obvious factor that can cause, or be associated with, both A and B and thereby spuriously implicate A as a cause of B.
- Controlled experimental evidence shows that A can cause B.

Causes can be categorised as necessary and sufficient, all combinations of necessary and sufficient are possible.

a) A sufficient cause is one without which the event cannot occur.

b) A necessary cause is one that precedes an event, but other causes for the event are possible.

Pregnancy cannot occur without conception. Conception is, therefore, both a necessary and a sufficient cause of pregnancy.

Sexual intercourse may lead to pregnancy, but, perhaps mercifully, does not always do so. In these days of assisted conception, pregnancy can be initiated without intercourse; sexual intercourse is, therefore, neither necessary nor sufficient to cause pregnancy.

Hodgkin's disease is sufficient to cause lymphadenopathy, but it is not a necessary cause of lymph node enlargement — there are many other causes.

A bone marrow transplant is a necessary cause of **graft versus host disease** (**p. 143**) in a patient with leukaemia, but it is not a sufficient cause: other factors, in particular additional immunosuppression, must also be present.

Cause-Specific Survival

A type of survival analysis in which only deaths due to a specific cause, in oncological practice this is usually cancer, are considered. All other deaths are censored. In studies on patients with cancer, the approach can be used to eliminate confusion caused by deaths from incidental conditions, such as heart disease or strokes, but may mislead. If a cancer treatment itself is associated with increased **mortality rate** (**p. 207**), e.g. an increase in the rate of fatal myocardial infarction, then the use of cause-specific survival will overestimate the overall benefits of treatment.

Causes of Cancer

There is no one cause for cancer, any more than there is any one cure. There has been a strenuous debate about whether cancer is inherited or whether external agents cause it, this parallels the nature/nurture controversy in psychology — are schizophrenics born or made?

The former distinction between genetic and environmental causes of cancer has been rendered artificial by the recent advances in **molecular biology** (**p. 203**). It is clear that malignant transformation usually arises as the result of environmental influences acting upon genetic susceptibility. The relative roles of environment and heredity can be regarded as lying along a spectrum, from those tumours, such as **retinoblastoma** (**p. 271**), for which genetic factors usually predominate, to those, such as mesothelioma, in which occupational or environmental factors are dominant. The answer to the questions: why don't all smokers get lung cancer? why do non-smokers get lung cancer? lies in a consideration of the interaction between heredity and environmental insult. Some individuals are predisposed to develop lung cancer, whether or not they smoke. Other individuals, in spite of smoking heavily, have inherited the ability to deal effectively with potential carcinogens. Estimates of the overall environmental contribution to human cancer vary widely — figures of up to 90% have been quoted.

Inherited conditions associated with increased risk of malignant disease

See Table 1 (next page).

Environmental factors implicated in human cancers

See Table 2 (next page).

Caveolin

A protein that is the main constituent of caveolae (little caves) — invaginations of specialised plasma membrane found in smooth muscle, adipocytes, fibroblasts and endothelial cells. These can be used to move large molecules, such as cholesterol, albumin and ceramide, into cells by transcytosis. They are also involved in signal transduction, particularly that involving G-protein-linked receptors and the *src*-family kinases. Caveolin-1 expression is increased in metastatic prostate cancer and may be implicated in the development of androgen resistance in prostate cancer.

CD Antigens

These are differentiation antigens expressed on the surface of white cells: CD stands for cluster of differentiation or cluster designation. They are recognised by specific monoclonal antibodies and this forms the basis for an internationally recognised system for classifying white cells. The classification has been developed in a series of international workshops, starting in 1980, in Paris. The historical development of white cell classifications means that, although the CD designation is the politically correct one, many have alternative names — as if they

Table 1

Syndrome	Molecular mechanism	Associated malignancy
Li–Fraumeni syndrome (**p. 179**)	germline defect in *p53*, loss of tumour suppression	soft tissue sarcomas, breast cancer
Ataxia-telangiectasia (**p. 23**)	defective repair of **DNA damage** (**p. 98**) 11q22–23	lymphoma leukaemia breast cancer
Xeroderma pigmentosa (**p. 329**)	defective repair of DNA damage A-q22	BCC, SCC, malignant melanomas
Retinoblastoma (**p. 271**)	deletion or mutation of RB gene (chromosome 13q14), loss of tumour suppression	retinoblastoma, osteosarcomas, soft tissue sarcomas
MEN syndrome Type 1		insulinoma, gastrinoma, pituitary tumours, soft tissue sarcomas
MEN syndrome Type 2	mutations affecting the ***ret* gene** (**p. 271**), a member of the tyrosine kinase receptor family	medullary carcinoma of the thyroid, phaeochromocytoma
Wilms' (**p. 328**) Aniridia Genital anomalies, mental or motor Retardation (**WAGR syndrome**) (**p. 327**)	abnormalities at 11p13 locus causing loss of tumour suppression	Wilms' tumour
Neurofibromatosis (**p. 213**)	loss of function of NF1 (chromosome 17q11.2) a tumour suppressor gene	Benign and malignant tumours of nerve sheaths and soft tissues
Peutz–Jegher's syndrome (**p. 232**)		
Familial melanoma	defective **p16** (**p. 224**) (mutations involving CDKN2A locus on chromosome 9p21)	multiple primary melanomas
Familial polyposis coli	loss of tumour suppressor gene on chromosome 5 (5q21)	large bowel cancer
HNPCC		colon, endometrium

Table 2

Environmental factor	Associated malignancy
Tobacco use (**p. 309**)	lung cancer, head and neck cancer, bladder cancer
Ionising radiation	leukaemia, thyroid cancer, osteosarcomas, lung cancer
Alcohol	head and neck cancer, liver cancer
Aromatic amines (rubber industry)	bladder cancer
Asbestos	mesothelioma
Aflatoxin	liver cancer
Hepatitis B infection	liver cancer
Human papilloma virus (HPV) infection; sex workers	cancer of the cervix
UV light	lip, skin, melanoma
Schistosoma haematobium	bladder cancer
Excess dietary fat	breast, colon, rectum
Leather goods manufacture	paranasal sinus tumours (especially maxillary antrum)
Helicobacter pylori infection	stomach cancer, MALT lymphomas
HIV (**p. 151**) infection	lymphomas, Kaposi's sarcoma, germ cell tumours, cancer of the cervix
Cytotoxic chemotherapy	leukaemia, bladder cancer (**cyclophosphamide**, **p. 83**)

were palimpsests. CD45 was formerly known as leukocyte common antigen; CD2 as the E rosette receptor; CDw52 as CAMPATH-1, etc. The interposition of the 'w' indicates a preliminary (workshop) designation that will require confirmation. The number of clusters is continually increasing: there are more than 120 at the last count. Some of the more important cells with CD designations are summarised:

- All leucocytes: CD45: expressed by white cells but not erythrocytes, also known as LCA — leucocyte common antigen.
- T-cells:
  - — CD1: thymic cells involved in antigen presentation, CD1 is a β2 microglobulin.
  - — CD2: activated T-cells, also known as the E-rosette receptor.
  - — CD4: T-cells involved in regulating immune responses, specifically killed by HIV. Also known as helper or inducer cells.
  - — CD8: T-cells that prevent over-expression of immune responses — suppressor cells.
  - — CD28: T- and plasma cells that convey an activation signal when stimulated by the B7 antigen.
- B-cells:
  - — CD10: early B-cells and blasts, the antigen is also expressed by some epithelial cells, also known as CALLA.
  - — C22: an adhesion molecule associated with hairy cell leukaemia.
  - — CD77: germinal centre B-cells, Burkitt's lymphoma.
- Natural killer (NK) cells:
  - — CD16: NK cells, but they may also express CD2.
  - — CD56: involved in the binding of NK cells to their target.
- Granulocytes:
  - — CD15: granulocytes, monocytes and Reed — Sternberg cells.
  - — CD35: an antigen involved in phagocytosis.
- Monocytes and macrophages:
  - — CD14: a lipopolysaccharide receptor antigen involved in intercellular signalling expressed by monocytes, macrophages and some B-cells.
  - — CD64: monocytes, macrophages and activated neutrophils.
- Stem cells and progenitors: CD34: immature haemopoietic cells, presence of the antigen is used to select stem cells for transplantation, e.g. in peripheral blood **stem cell** (**p. 298**) transplantation.
- Platelets:
  - — CD31: platelet-related antigen but also associated with endothelial junctions as well as granulocytes and macrophages. Also known as PECAM-1.
  - — CD42a, b: antigens associated with platelet adhesion.
- CD antigens not necessarily restricted to specific lineages:
  - — CDw52: found on lymphocytes and monocytes, also known as CAMPATH-1. Antibodies to CAMPATH-1 have been used in attempts to modify the severity of graft versus host disease in patients receiving allogeneic bone marrow transplants.
  - — CD25: activated T-cells, B-cells and monocytes, CD25 is the **IL-2** (**p. 159**) receptor also known as the Tac antigen, the gene is on chromosome 10.

— CD71: found on dividing cells, is the transferrin receptor.

— CD11a, b, c: integrin α-chains, found on a variety of white cells.

— CD18: found on most white cells, the integrin β-chain.

CD34

The cell surface antigen expressed by haemopoietic stem cells. It is an adhesion molecule, resembling sialomucin, that can be identified in between 1 and 3% of marrow cells.

cdc25

A phosphatase, originally identified in a yeast (*Schizosaccharomyces pombe*), that regulates the **phosphorylation (p. 234)** of p34cdc2. It is, therefore, important in the regulation of the transition to mitosis. The presence of dephosphorylated p34cdc2 and mitotic cyclin (cyclin B) will, through activation of maturation promoting factor (MPF), drive a cell towards mitosis. Other important factors controlling this process are wee1, cdc2 and cdc13.

Ced Family of Genes

These genes have been described in the nematode, *Caenorhabditis elegans*, and are involved in the modulation of programmed **cell death (p. 55)**, or apoptosis. Two genes, *ced*-3 and *ced*-4, are required for cell death and *ced*-9 prevents it. *Ced*-3 encodes a **caspase (p. 47)**, a protease specific for aspartic acid; *ced*-9 prevents activation of this caspase. Mutations causing loss of function of *ced*-3 are associated with loss of the normal apoptosis that occurs during development. The *BCL*-2 family in mammals is homologous to *CED*-9. A mammalian homologue for *CED*-4 has now been described, it has been termed Apaf-1 and, in combination with Apaf-3, dATP and **cytochrome c (p. 84)**, it is sufficient to activate a caspase (procaspase-3).

Cell Cycle

The morphological changes that accompany mitosis are the most dramatic manifestation of the changes occurring during cell division they must, however, be underpinned by a series of kinetic and biochemical events: the concept of the cell cycle usefully summarises these events.

M-phase is **mitosis (p. 201)**; S-phase is the phase during which DNA synthesis occurs and G_1 and G_2 are gaps in the cycle. Interphase is the collective term used to describe phases G_1, S and G_2. The time taken to progress around the cell cycle varies widely between cell types: it takes ~12 h for a mammalian intestinal crypt cell and >1 year for a hepatocyte. The variation arises mainly as a result of varying the length of G_1, the other phases of the cell cycle are relatively constant in duration:

- M: 1 h.
- G_1: 1 h to several years.
- S: 6–8 h.
- G_2: 3–4 h.

An alternative way to look at the range in duration of G_1 is to introduce the concept of the resting cell (G_0): the cells opt out of the cell cycle for varying lengths of time and it is this that produces the apparent variation in the length of G_1.

A variety of techniques have been devised to estimate cell cycle parameters: these include the **stathmokinetic method (p. 297)**; **percent labelled mitoses (p. 201)**; continuous thymidine labelling; halogenated **pyrimidine (p. 252)** incorporation (e.g. BUDR); immunochemical staining;

FACS (**p. 129**) (fluorescence-activated cell sorting); relative movement method; and identification of antigens associated with proliferation. Many of these methods can be used only in culture or in laboratory animals and cannot be applied to patients. There is a series of common-sense criteria to be applied in assessing the clinical usefulness of a method for estimating cell cycle parameters:

- They should not involve sacrificing the patient.
- They should not involve poisoning the patient; they should not involve multiple biopsies.
- Ideally, they could be performed on archival material.
- They should involve only a single procedure.
- They should give results rapidly.
- They should have prognostic relevance.
- They should be affordable.
- They should aid choosing therapy for individual patients.
- They should be precise and representative.

Cell Cycle Checkpoints

These are positions within the cell cycle at which control processes exert their effects — in anthropomorphic terms, they are points at which the cell makes decisions about its own **replication** (**p. 267**). Understanding of **cell cycle checkpoints** (**p. 53**) and their control is, therefore, crucial to any consideration of malignant transformation.

Two main checkpoints in the cell cycle

A cell is constantly receiving and processing information about itself and its environment. It uses this information to govern events at two main checkpoints in the cell cycle: the G_1 checkpoint (syn. START), at the end of G_1, and the G_2 checkpoint, at the end of G_2, immediately before mitosis. Factors necessary for transition through the G_1 checkpoint include:

- Cell should be large enough.
- There should be no evidence of **DNA damage** (**p. 98**).
- The environment should be favourable.

The retinoblastoma gene product is an important component of the G_1 checkpoint. There are only two main criteria to be satisfied at the G_2 checkpoint: the DNA should have replicated and the cell should be large enough. The G_1 checkpoint is the more important of the two. The cell has three main options at this point: proceed to S-phase; opt out of cycle and move into G_0; pause and await more propitious conditions.

A complex system of enzymes (kinases) and proteins (cyclins) controls the cell cycle

The cell cycle is controlled by a system of enzymes, kinases, that catalyse the transfer of a phosphate group from ATP to specific amino acids on the target proteins. These kinases must be activated in the correct sequence for the cell cycle to proceed correctly. Activation of the kinases is the responsibility of a set of proteins, the **cyclins** (**p. 82**), hence the kinases are classified as cyclin-dependent kinases (cdk). The first cyclin — cdk complex to be identified was called M-phase-promoting factor (MPF). This consists of cyclin A or B complexed with cdc-2 (a kinase) and this molecule, through a series of specific **phosphorylations** (**p. 234**) and dephosphorylations, induces **mitosis** (**p. 201**). A similar system with cdk2 (p33), cdk4 and cyclins A, D and E is a component of the G_1 checkpoint and is responsible for the move from G_1 to S-phases.

Cell cycle checkpoints as decision points in the cell cycle

Should the cell commit to chromosome replication? Should the cell commit to mitosis? Should the cell exit from mitosis? With each cycle DNA must be exactly doubled: no more, no less. The important questions at the G_2/M checkpoint are: is replication complete? Are the chromosomes on the spindle? Damaged DNA prevents entry to mitosis: the genes involved include *rad9*; *p53* and *GADD*.

The important factors that influence the transition to G_0, or resting phase, include: low rate of protein synthesis, absence of **growth factors** (**p. 144**), lack of essential nutrients, density-dependent inhibition, anchorage dependence and **senescence** (**p. 282**).

Cell Cycle Kinetics: Definition of Terms

There are several quite specific terms used to describe the various aspects of the proliferative activity of cells.

- Mitosis duration = T_m. The duration of mitosis.
- **Mitotic index** (**MI**) (**p. 201**), the percentage of cells in mitosis: MI = $(T_m/T_c) \cdot \lambda$, where λ is a constant in the range 0.5–1.0 used to define the age distribution of the cell population.
- **Labelling index** (**LI**) (**p. 173**), the percentage of cells in S-phase: LI = $(T_s/T_c) \cdot \lambda$, where λ is a constant in the range 0.5–1.0 used to define the age distribution of the cell population.
- S-phase duration (T_s), the duration of S-phase.
- Total cycle time (T_c), the time a cell takes to complete the cell cycle.
- **Growth fraction** (**GF**) (**p. 144**), the ratio of the number of proliferating (cycling) cells to the number of non-proliferating (resting) cells: GF = fraction of labelled cells ÷ fraction of labelled mitoses.
- **Tumour doubling time** (**T_d**) (**p. 316**), measurement of the time a tumour takes to double in volume.
- **Potential doubling time** (**T_{pot}**) (**p. 241**), a calculation, based on knowledge of growth fraction and cell cycle time, of the time it should take the cell population tumour to double — provided there were no cell loss: $T_{pot} = \lambda(T_s/\mathrm{LI})$, where λ is a constant in the range 0.5–1.0 used to define the age distribution of the cell population.
- **Effective doubling time** (**T_{eff}**) (**p. 111**), defines the ability of a tumour to repopulate in the face of cell depletion — as a result, for example, of treatment with drugs or radiation. T_{eff} is distinct from T_{pot} since it will be influenced by several other factors including the intrinsic sensitivity of the cells to the cytotoxic treatment. In general, T_{eff} is longer than T_{pot}: the data from experimental tumours typically show T_{eff} nearly double T_{pot}. There is no clear correlation between the two parameters and, thus, it would be unwise to use T_{pot} as a proxy for T_{eff}. The position is further complicated by the differences in intrinsic **radiosensitivity** (**p. 260**) between different tumours of the same morphological appearance. In clinical practice values may range from 5 to 50–60 days, even though the corresponding T_{pot} may be similar. Data obtained from a population of patients will, therefore, give a poor indication of T_{eff} for an individual patient.
- Cell loss factor (φ), fraction of new cells that are lost from the tumour: $\varphi = 1 - (T_{pot}/T_d)$.

Cell kinetic parameters for skin

There are important species differences in kinetic parameters for normal tissues, which is not surprising considering, for example, the relative lifespans of human and mouse:

	Mouse	**Human**
Cell layers	10	>15
Transit time	9d (5–9)	14d (12–48)
EPU/mm^2	1397	1124
basal cells/mm^2	14 375	25 000 to 30 000
Labelling Index	9.1%	3%
S phase (hr)	12.6	7 to 14
Cell cycle time (Tc (hrs))	138 (65 to 150)	230 (120 to 325)
Cells per EPU per day	1 to 2	2.5
Cells/hr/ 100 basal cells	0.5 to 0.8	0.44

EPU = epithelial proliferative unit

Cell Death

For a cell, death comes in many disguises, a taxonomy:

- Totally dead: no function at all.
- Reproductively dead: unable to divide but otherwise able to function.
- Doomed: only limited capacity for further division.
- Loss of special function, e.g. hormone biosynthesis.

Cell death, cell survival and reproductive integrity

Cell death can be defined in a variety of ways. One definition would imply that the cell has lost all its functions, it is inert metabolically and in every other sense. Another definition would define any cell that has lost a specific function, e.g. the loss of a follicular thyroid cell's ability to produce colloid, as dead. A third definition would define a dead cell as one incapable of successful division. Different definitions will suit different circumstances and, in any account of cell killing, it is important that 'death' be defined before any description or discussion.

Cell survival and cell killing are terms often used in a specialised sense in oncology. Often, it is the reproductive integrity of a cell that is important: that a cell be capable of division and that both daughters produced by that division be themselves capable of division. This quality of the parent cell, the ability to give rise to a potentially infinite series of generations, is termed clonogenicity. When we refer to cells being killed, we are really talking about loss of clonogenicity. Conversely, when we talks about cell survival, we are really talking about the ability of the surviving cells to reproduce indefinitely, to form colonies.

There is, assuming no cells are lost, a simple mathematical relationship between the number of clonogenic cells (N) in a tumour and the number of generations of cell division (n):

$$N = 2^n.$$

Cells may lose their reproductive integrity and yet remain functionally and metabolically viable. Just as a mule can eat, breathe, excrete and carry tourists into the Grand Canyon — but cannot breed, so a thyroid follicular cell after a lethal (in the sense outlined above) dose of radiation can continue to metabolise and synthesise thyroid hormone. The lethal effects of the radiation will not be expressed until that cell is called on to divide; this may be months or years after the original dose of radiation. This is why hypothyroidism is an important, but late, complication of radiotherapy to the neck.

The loss of reproductive integrity after cytotoxic insult is not always absolute — a reproductively doomed cell may undergo several divisions before its lineage peters out. In a colony assay, the progeny of these cells appear as small colonies, usually smaller than the 50 cell criterion applied to define a colony worthy of being counted.

Cell Loss Factor (φ)

Cell loss factor (φ) is a term used to define the proportion of cells produced by a tumour but which do not persist as clonogenic cells within the tumour itself:

$$\varphi = 1 - (T_{pot}/T_d),$$

where T_{pot} is the potential doubling time and T_d is the observed **tumour doubling time** (**p. 316**).

Tumour type	T_d (days)	T_{pot} (days)	Cell loss (φ) (%)
Head and neck cancers	33–150	2–46	0–94
Colorectal cancer primary tumour	60–170	3	95–98
Lung metastases from colorectal cancer	95	3	97
Lung cancer (undifferentiated)	40–160	2.5	up to 97
Experimental tumours in animals			0–93

Cell loss, in the clonogenic sense, can occur by a variety of mechanisms including desquamation and exfoliation, migration and **metastasis** (**p. 198**) formation, necrosis, apoptosis, and differentiation.

Cell loss, and its implications for the genetic age of tumours, may explain why some tumours are resistant to treatments to which they have not previously been exposed. Simple calculation shows that, starting from a single cell, if all cells produced by a tumour were retained within that tumour, it would only take ~40 generations to produce a tumour weighing 1 kg: $2^{40} = 10^{12}$; 10^{12} cells weigh 1000 g. On this basis it would take 46 generations to produce a person-sized tumour. For a tumour with a high cell loss factor, it could take >1000 generations for a tumour to reach a size at which it could be detected clinically. This provides ample opportunity for mutations to accumulate within a population of tumour cells.

Cell Renewal Systems

Normal cells of the body are being lost and replaced all the time. In accordance with homeostatic principles, loss is balanced by production and a variety of different organisational patterns has arisen to accomplish this.

Simple renewal, also referred to as flexible renewal (F-type) system

An adjacent cell simply divides to replace a missing neighbour, examples include:

- Liver: cells divide every 1–2 years and can divide in response to insult (unless previously irradiated).
- Endothelium: there is usually slow turnover of cells, except at sites of turbulence. Important changes occur during tumour angiogenesis, neovascularisation or wounding.
- Thyroid: turnover is slow (9 weeks) and this is reflected in the estimates of cell cycle parameters LI 0.78%, MI 0.035%.

Hierarchical (H-type) systems

Hierarchical systems in which there is an architecture, in both space and time, that governs how loss is balanced by gain. Classically there is a small pool of uncommitted pluripotential precursors (stem cells) that have the ability to produce progeny which can then differentiate in a hierarchical fashion to produce the differentiated progeny required. This process can involve amplificatory divisions, so that one stem cell committing to a line of differentiation can produce a vast number of differentiated progeny in a relatively short period of time. Examples include:

- Bone marrow.
- Gut.
- Skin/mucosa.
- **Testis** (**p. 306**).

Cell Survival Curve

A cell survival curve is simply a graphical method of presenting data to describe the relationship between different doses of a cytotoxic agent and the number, or proportion, of cells that survive after each dose. Relationships are complex, both for **cytotoxic drugs** (**p. 84**) and for radiation: straightforward linear relationships are uncommon. The shape and calibration of a cell survival curve can often be characterised mathematically, but the curve itself is not a mathematical construct, for it represents a series of biological measurements. The mathematics supply a useful shorthand for comparing different curves, and might even suggest possible biological mechanisms for cell killing, but will not fully describe the curve. A cell survival curve may be viewed, therefore, in various ways:

- As a simple act of observation, with no interpretation.
- As observations to which a mathematical form can be fitted; a form that can be used to predict the consequences of changing the parameters that, in the model at least, define the curve.
- As an observation that requires biological explanation: why is the curve that particular shape, and no other? If altering the experimental conditions alters the shape of the curve, then this requires explanation: why does irradiation under hypoxic conditions change the shape of the radiation cell survival curve?

Radiation cell survival curves provide a useful illustration of the uses and interpretation of cell survival curves in general. The typical radiation cell survival curve looks like that shown in figure 1.

The **surviving fraction** (**p. 302**) is usually plotted on a log scale, the dose as an arithmetic scale. A straight line on such a plot describes exponential cell killing. Under normal conditions, cell killing is exponential at doses >3–4 Gy, but at the doses used in conventionally fractionated radiotherapy (2–3 Gy) the relationship between dose and survival is non-linear. This portion of the survival curve is termed the shoulder.

There are several possible biological explanations for the presence of the shoulder on the mammalian survival curve, explanations that are not necessarily mutually exclusive (figure 2).

The simplest explanation for the presence of the shoulder has its origins in **target theory** (**p. 304**). If it takes a certain type of damage to sterilise a cell reproductively then it is possible to imagine two ways in which such damage might arise. A single ionisation along the electron track could hit a single vulnerable target (**single-target, single-hit inactivation**). The likelihood of this event will be proportional to dose. This type of damage would be linear on a conventional cell survival plot. Alternatively there could be several targets within the cell, each of which needs to be hit, before the cell is killed. The likelihood of the cell being killed will be proportional to $(\text{dose})^n$, where n is

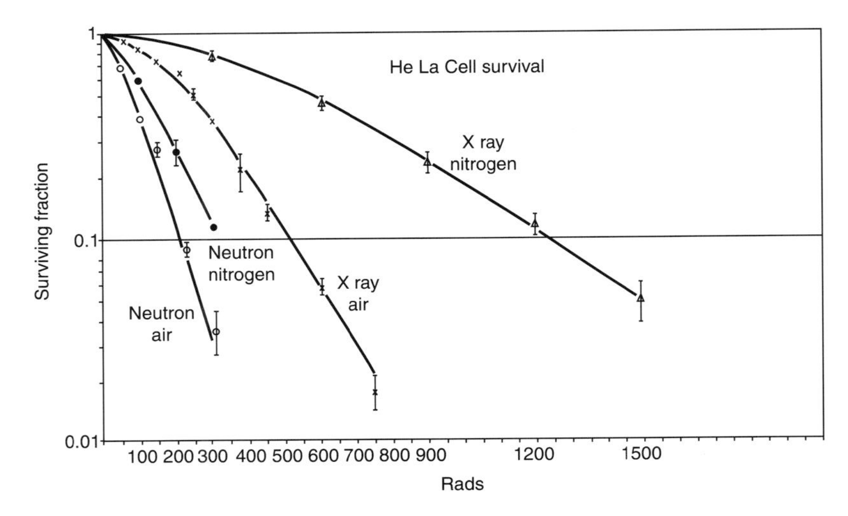

Figure 1 — Experimental data for survival after irradiation of mammalian cells. Cells irradiated in nitrogen are relatively hypoxic, while those irradiated in air are well oxygenated. X-rays are of low LET, while neutrons are an example of high LET irradiation. Radiation-induced cell killing by X-rays is less efficient when cells are hypoxic, but this effect is much less apparent for high LET radiation.

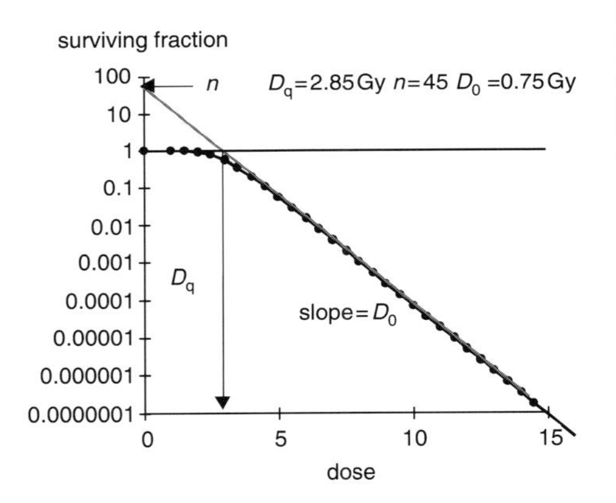

Figure 2 — Cell survival curve based on the multiple-target, single-hit model.

the number of targets (**multiple-target, single-hit inactivation**). This type of damage would produce a survival curve with an initial shoulder and the width of the shoulder will be related to the size of *n*. The smaller *n*, then the narrower will the shoulder be on a log-linear plot.

Neither of these target theories provides a satisfactory model for the radiation — cell survival curve that is actually observed. The single-target, single-hit model predicts a straight line, a cell survival curve that is only observed for **high LET radiation (p. 151)**. The multiple-target, single-hit model predicts a curve with an absolutely flat initial portion before the bend of the shoulder and, ultimately, the straight line.

The single-hit, single target model has a simple mathematical representation:

$$\text{Surviving fraction} = e^{-(\text{dose}/D_0)}.$$

For the multiple-target, single-hit model the formula is:

$$\text{Surviving fraction} = 1 - (1 - e^{-(\text{dose}/D_0)})^n,$$

where D_0 is the slope of the exponential part of the survival curve and n is the **extrapolation number (p. 124)**. The extrapolation number is an indication of the number of targets per cell that need to be inactivated before the cell is, in the reproductive sense, killed. It can be estimated by (unsurprisingly) extrapolating the linear part of the curve to the y-axis. Another term has been used to describe the width of the shoulder on the cell survival curve: D_q, the quasi-threshold dose. This is the dose at which the backward extrapolation of the linear portion of the curve intersects with a surviving fraction = 1.0.

There is a mathematical relationship between D_0, D_q and n:

$$D_q = D_0 \cdot \log_e n.$$

Observation shows that there is no such initial flat portion to the cell survival curve for radiation. If both mechanisms are operating simultaneously, then a hybrid curve, with both linear and bending components, will be produced. The problem is, however, that there would still be an initial linear segment to the curve and this does not correspond to experimental observation.

Recent analysis of the shoulder region of the curve using fractional doses of less than 2 Gy shows the position to be even more complex. There is a dip within the shoulder, indicating increased radiosensitivity at very low fractional doses (0.5 Gy).

The linear–quadratic (LQ) model provides the most accurate representation of the cell survival curve for radiation:

$$f(d) = e^{-\alpha d - \beta d2},$$

where $f(d)$ is the fraction of cells that survives after dose d, α is a constant defining the component of cell killing, which is directly proportional to dose, and β is a constant describing that component of cell killing which is proportional to $(\text{dose})^2$ (figure 3).

The equation is simple, can be manipulated usefully (see **Linear Quadratic Model**) (**p. 180**) but it is not necessarily correct. For example, if, as the model would predict, cell killing at high doses is dominated by the β component, the cell survival curve at high doses would continually bend. However, at doses >5 Gy the experimental data suggest

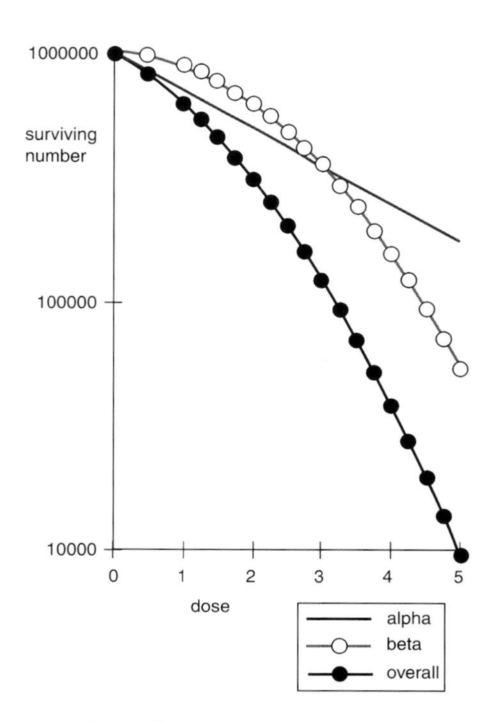

Figure 3 — Cell survival curve based on the LQ model showing the composite curve produced when there simultaneously exists a component of cell killing directly proportional to dose and a component proportional to (dose)^2.

that the relationship is linear. The LQ model may, therefore, be useful at the doses per fraction used clinically but may be unreliable at very high fractional doses. This is the converse of the hybrid model described above, which is inaccurate at doses <2 Gy per fraction but is reasonably reliable at the higher fractional doses (5 Gy) often used in experimental radiobiology.

If the LQ model fits the data, at least at lower fractional doses, then what is the biological explanation? Why is there a component of damage directly proportional to dose and another component that is proportional to $(\text{dose})^2$? One explanation, with its origins in target theory, would be that there are two important components to radiation-induced cell killing. One is proportional to dose and simply represents single-target, single-hit inactivation. The other, the $(\text{dose})^2$ component, is due to inactivation that requires two separate tracks to produce ionisations close together both in time and space. Alternative, or supplementary, explanations of the biology behind the LQ model deal with repair mechanisms. If there are two types of damage, that which is inevitably lethal and that which can be lethal, but may not be if repair is allowed to occur, then this could produce a response to irradiation that contained elements that were directly proportional to dose (the inevitably lethal damage) and to $(\text{dose})^2$, the potentially lethal damage.

The situation, for now at least is best summarised as: the LQ model is a model whose descriptive powers are adequate but for which the underlying biology is not yet completely explained.

Cell survival curves for high LET radiation

High LET (linear energy transfer) radiation is densely ionising and cell survival curves for high LET radiations (e.g. neutrons, pions, neon nuclei, α-particles) are very different from the typical survival curves obtained with X-rays. The survival curves for high LET radiations have little or no shoulder and steep slopes. Cell killing is virtually independent of oxygen levels at the time of irradiation (figure 1).

Cellular Optical Coherence Tomography (OCT)

This offers the elusive prospect of a diagnosis without histology. The principle is similar to that of ultrasound but, rather than measuring reflected sound, it is the intensity of reflected infrared light that is measured. It allows cells to be imaged within living tissues. In theory, and perhaps in practice, the technique offers the prospect of producing images very similar to those of a conventional histological section. The technique has a range in tissue of ~3 mm and can be adapted to endoscopic use. The procedure has obvious potential in the diagnosis of malignant and premalignant conditions of the skin, genitourinary and aerodigestive tracts.

Censoring

This term is confusing. In its everyday usage it implies the conscious desire to remove something considered unacceptable; used in its statistical sense it describes a passive, and potentially undesirable, process whereby subjects are lost from studies for reasons beyond the control of those conducting the study. Good **experimental designs (p. 122)** will include procedures designed to keep censoring to a minimum. Three types of censoring have been defined. In Type I, an experiment runs for a fixed duration and any subjects who have not experienced the event of interest are all censored at the same time, that chosen for the experiment's duration.
In Type II, which is occasionally encountered in studies of adverse effects, the study is stopped when a certain proportion of subjects have suffered the adverse event. Those subjects who have not experienced the event are censored, all at the same time. In Type III, which is that which causes the most difficulty in clinical trials, patients in a prospective study are lost to follow-up for a variety of reasons that may have little to do with the event of interest. In a trial of cancer treatment the main event of interest will be death from cancer, but patients will drop out of the study for other reasons: death from strokes, heart attacks, etc.; they may refuse to attend for hospital visits; they may move abroad. All of these other events must be handled as censored data.

Counting and attributing events in clinical studies is often difficult and can lead to problems with censoring. Is a death from pneumonia in a study of synchronous chemotherapy and radiotherapy for head and neck cancer to be censored as an incidental, intercurrent, event? Is it related to treatment (post-chemotherapy neutropenia), or is it related to the cancer itself (aspiration because of disturbed laryngeal anatomy)? **Bias (p. 28)** will arise if the problem of censored data is approached in a haphazard fashion. The most obvious example would be a different approach to the attribution of potentially censored data between the control and experimental groups. If deaths in the control group were not properly investigated, and simply assumed to be due to cancer, and if deaths in the experimental group were more rigorously investigated, and many that appeared to be due to cancer in fact turned out to be for other reasons, then the effectiveness of the intervention may be grossly overestimated.

Information on censoring can be easily incorporated into a Kaplan–Meier survival curve. Censored patients (i.e. those who are no longer available for follow-up but who have not experienced the event of interest) are indicated by ticks on the curve.

Centimorgan

A unit used in genetic mapping. A gene map can be obtained from a **linkage (p. 181)** map — in which the frequency of recombination events for individual genes can be assessed and used to calculate distances between them, in units called centimorgans (cM). If genes are not physically close to each other, are unlinked, then they have a 50% chance of being assorted separately at **meiosis (p. 194)**. If they are physically close to each other, then this probability falls: 1 cM is equivalent to a recombination frequency of 1%. The LOD score is a statistical measure of the degree of linkage between two loci.

Mapping functions can be used to convert distances in cM (genetic distance) and distance in kilobases (base pairs, physical distance). Three main eponymous functions are used: Haldane's, Ott's and Kosambi's. As a rough guide, 1 cM = 1000 kb = 10^6 bp. The unit is named after Thomas H. Morgan, a University of Columbia geneticist whose studies on mutations in fruit flies (*Drosophila*) formed a bridge between the classical observations of Gustav Mendel and the structure and function of chromosomes. His student, A. H. Sturtevant, drew the first gene map in 1913. It showed the linear disposition of genes along a short stretch of *Drosophila* chromosome

Central Limit Theorem

It is a remarkably, and exceptionally useful, fact that, if one takes repeated samples of the mean of a population (mean of x, i.e. x/n, where n is number of observations), then the distribution of the means will follow a Normal distribution. This statement is true even if the variable itself (x) is not Normally distributed, provided the size of the sample is large enough. Put more formally, the central limit theorem states that the distribution of the sample means will be nearly Normal, whatever the distribution of the variable in the population, provided the samples are sufficiently large. The theorem provides the basic foundation for the calculation of **confidence intervals** (CI) **(p. 70)** from a sample mean.

Centromere

The specialised region of a chromosome that interacts with the **kinetochores (p. 171)** during mitosis and ensures that the chromosomes segregate accurately.

Centrosome

The region within the cytoplasm within which microtubules are produced. It usually lies close to the nucleus and controls not only just the production of microtubules, but also their orientation within the cell. The centrosome is prominent during mitosis and, morphologically appears as the aster of the mitotic cell.

Chaperones, Molecular Chaperones

These are proteins that marshal the correct assembly of other proteins and their polypeptide chains but do not themselves appear in the final product. They function as both a transport system, making sure that components go where they should go, and as an introduction agency, making sure that entities that need to go together end up in close proximity to each other. They also produce changes in the higher structure of polypeptides to facilitate the assembly of macromolecules. To describe proteins with such diverse, and permissive, functions as chaperones is a misnomer. A chaperone exists primarily to prevent progress — molecular chaperones do the opposite. A wide variety of chaperones has been described: nucleoplasmins, chaperonins, **heat shock proteins (p. 149)** and **ubiquitinated proteins (p. 319)**. **Prions (p. 246)** may represent chaperones that have gone haywire.

CHART (Continuous Hyperfractionated Accelerated Radiotherapy)

CHART is an attempt to combine the advantages of acceleration (less opportunity for **repopulation (p. 267)** of tumour) with those of **hyperfractionation (p. 156)** (relative sparing of late damage to normal tissues). The schedule was developed at Mount Vernon Hospital by Dische and Saunders and the results of a multicentre randomised trials have now been published. There was no clear advantage, in terms of survival, when CHART was compared with conventional (6.5 weeks) treatment for head and neck cancer. There was a statistically significant survival advantage from CHART therapy for patients with lung cancer. The regimen is difficult to accommodate within a conventionally operated radiotherapy department since it uses thrice-daily fractions and patients are treated at weekends. The regimen is 54 Gy in 36 fractions over 12 consecutive days (at least 6 h must be allowed between each of the three daily fractions).

The CHARTWEL schedule represents an attempt to retain the radiobiological advantages of CHART while using a schedule that can be more easily accommodated within departmental routine. Patients are not treated at weekends, but thrice-daily fractions are used. Schedules of 60 Gy in 36–40 fractions over 18–20 days are feasible for the treatment of lung cancer.

Chemical Equivalence

Two formulations that contain the same amounts of drug when assayed according to their chemical activity are said to be chemically equivalent. Chemical equivalence, because of the possible interactions between the active chemical and other constituents of the formulation, does not inevitably imply **bioequivalence (p. 29)**.

Chi-Squared (χ^2-) Test

The χ^2-test is a useful test for statistical significance is. It is based on a contingency (syn. frequency) table and is widely used in epidemiology as well as in comparisons involving treatment or potential prognostic factors. Its foundation is the comparison of the observed, as opposed to the expected, values within each group; the expected values are calculated using the data from both groups combined. The test statistic is calculated and the value compared with a table of χ^2 values and their corresponding probability (p).

Chimera

A chimera (the term derives from the Greek for a monster) is an animal that contains cells from two distinct genetic lineages. The early chimeric animals (radiation chimeras) were devised as a necessary preliminary to the development of **bone marrow transplantation** (**p. 34**) in humans. Lethally irradiated recipient mice received bone marrow from a donor mouse and, as a result, the bone marrow was of donor lineage, the cells in the rest of the animal maintained the host genome. Chimeric animals are now ubiquitous and are part of the development of the knockout mouse, chimeric humans are also ubiquitous and would include almost all transplant recipients — the exception would be those rare bone marrow transplant patients whose marrow regenerates from host, as opposed to donor, stem cells. Fifty years ago chimeras were a science fiction fantasy, now they are an everyday reality. The term can also be applied to biological products: a chimeric antibody, such as **ritumixab** (**p. 273**), contains a mixture of mouse antibody (the variable region) and human antibody (the constant region).

2-Chlorodeoxyadenosine (2-CdA)

This drug is used for treating hairy cell leukaemia. It is phosphorylated by deoxycytidine kinase, which forms triphosphates that then cause DNA strand breaks.

Chromatids

The term given to the copies of the chromosomes produced during the S-phase of the **cell cycle** (**p. 52**) and which persist during G_2 but are then lost, to the daughter cell, at mitosis.

Chromatin

During G_1-, S- and G_2- (interphase) phases the chromosomes cannot be identified morphologically. During this period the genetic material exists as a complex of nucleic acids and protein called chromatin. The structure of chromatin can be deduced using electron microscopy. A string of nucleosomes forms a fibre of 10 nm diameter. Under the influence of histones this fibre coils so that there are about six nucleosomes per turn of the helix. This produces a fibre of 25–30 nm diameter. At and around the time of mitosis there are further changes in the higher structure of the chromatin that produce the structures that, using light microscopy, are visible as chromosomes. Coils are, themselves, coiled.

Chromosomal Heterogeneity

This is a specialised term used to describe a situation in which a gene that causes disease may be on more than one chromosome.

Chromosomal Marker

An identifiable chromosomal abnormality associated with the presence of, or increased susceptibility to, a particular disease. Examples of relevance to oncology:

- Chromosome 11,22 translocation: Ewing's sarcoma, desmoplastic small round-cell tumours.
- Chromosome 1,19 translocation: melanoma.
- Chromosome 9,22 translocation (**Philadelphia chromosome**) (**p. 233**): chronic myeloid leukaemia.
- Chromosome 14,18 translocation: follicular lymphoma (involves the increased expression of *bcl-2*, and, therefore, inhibition of apoptosis).

Chromosome

A chromosome is an organisational unit of the genome that can only be recognised morphologically at mitosis. Each chromosome contains many genes. In each human cell there are 46 chromosomes: one pair of sex chromosomes and 22 other chromosome pairs. Each chromosome includes along its length a **centromere (p. 61)** that divides it into two unequal arms. The shorter arm is designated 'p' and the longer arm 'q'. Chromosomes can be stained with Giemsa to produce a banding pattern characteristic for each pair of chromosomes. Each band contains ~10^7 bp. The bands are defined centrifugally from the centromere. Figure 1 shows the human X chromosome with the banding pattern of the short arm defined.

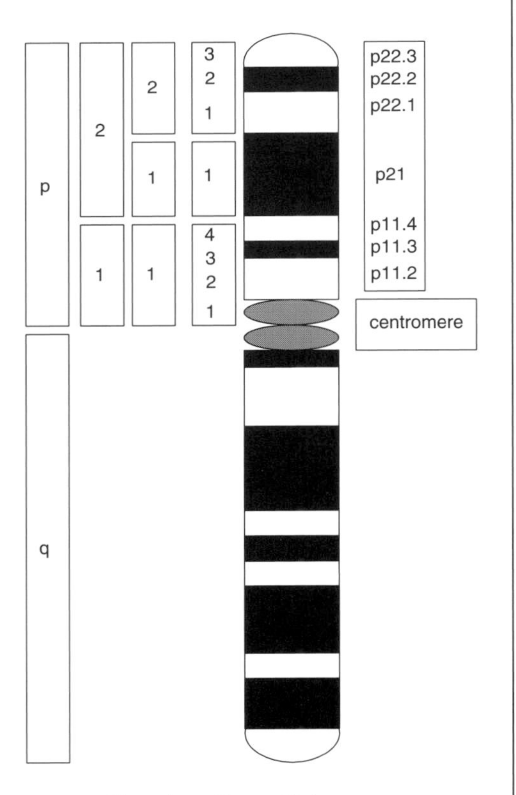

Figure 1 — Human X chromosome.

Chromosome Walking

An approach to finding the position of a gene of interest with a view to cloning the gene. This is a fundamental step in reverse genetics. An initial **linkage (p. 181)** analysis can define a probe located near to the gene of interest. The chromosome, or part thereof, is cut up into fragments of DNA to form a genomic library. The original probe hybridises with a fragment of DNA that contains the sequence of DNA complementary to that in the probe but which also contains a contiguous, but as yet unidentified, sequence.

A new probe is formed using the contiguous unidentified sequence and this is then used to hybridise with a further fragment from the genomic library. By repeating this cycle, a series of overlapping DNA fragments is identified that extends from the site recognised by the original probe up to, and perhaps through, the DNA sequence that codes for the gene of interest. The overlapping fragments are used to 'walk' along the chromosome. In 1996, the rate at which it was possible to walk a chromosome was ~100 kb month^{-1}. Using larger, not necessarily contiguous, fragments can speed up the rate of progress, e.g. those generated by **pulsed-field gel electrophoresis (p. 251)**: skipping or jumping rather than walking.

The contiguous clones (contigs), in their recombinant hosts (plasmids or phage), can then be used to deduce the position and nucleotide sequence of the gene of interest. From the nucleotide sequence, the protein structure, and possibly function, can be deduced. This approach, which starts from knowing roughly where the gene is located but not knowing its product, and which ends with a knowledge of both the precise position of the gene and the amino acid sequence of the gene product, is known as **reverse genetics (p. 272)**.

Chronobiology

Chronobiology is the study of the effects of time, of day or of season on biological processes. The effect that has been most widely investigated is that of diurnal variation, circadian rhythm. Seasonal variations may also be important, particularly in terms of mood and emotion — seasonal affective disorder, for example.

Chronotherapy, Chronochemotherapy

Tumour cells exhibit circadian activity, often proliferating most rapidly in the early hours of the morning. This has led to the concept of chronotherapy, in which the intensity of treatment is tailored to the activity of the tumour. Infused chemotherapy can be delivered using pumps programmed to increase the dose per unit time to correspond to the time at which the tumour cells most actively proliferate. This approach only improves the therapeutic ratio if the tumour cells have a different diurnal pattern to that of the normal tissues.

Circadian Rhythm

The term is derived form *circa die* (around the day) and describes the fact that biological activities do not continue at a constant rate throughout a 24-h period but are subject to diurnal variation. Anyone who has ever experienced jetlag can appreciate the importance of circadian rhythms to everyday life.

These variations are important in oncology for two main reasons. There may be diurnal variations in how exogenous agents such as cytotoxic drugs, biological response modifiers and carcinogens may be handled by the body, the rate of 5FU metabolism, for example, exhibits a diurnal pattern. Similar patterns may also affect tumours and normal tissues. If the patterns are completely synchronous in both tumour and normal tissues then there will be little opportunity for improving the **therapeutic ratio** (**p. 307**). If, however, there is either a loss of circadian rhythm in tumour, but not in normal tissue, or if the rhythms are not in step, then this might be exploited therapeutically. The rapid re-establishment of underlying diurnal patterns may explain why, by and large, the attempts made to manipulate tumour kinetics using **cytotoxic drugs** (**p. 84**) have been largely unsuccessful.

The control mechanisms involved in circadian rhythms are complex and predominantly driven by alternating light — dark cycles. The retina provides information on light intensity to the suprachiasmatic nucleus (SCN) in the hypothalamus. The SCN contains a circadian oscillator that governs melatonin secretion by the pineal gland. The circadian oscillator involves clock genes, rhythmic expression of which is governed by feedback loops. The three mammalian clock genes are *CLOCK*, *BMAL*1 and *PER*.

The diurnal variation in the proliferative status of normal human mucosa correlates with levels of cyclins and cyclin-dependent kinases. Mitotic activity is greatest between midnight and 02:00 hours, the level of cyclin B1 is maximal at ~21:00 hours. Cyclin A, expressed during the G_2-phase of the cell cycle, is maximal in the late afternoon. These observations have potentially important clinical consequences: they imply that there may be a circadian sensitivity to the adverse effects of radiotherapy since cells are most sensitive to radiation at the G_2/M boundary, and least sensitive during late S-phase. It is much more difficult, mainly because of heterogeneity, to assess the existence and pattern of circadian rhythms in tumours. Even if there were no such variation, the existence of diurnal fluctuations in the sensitivity of normal tissues would still offer a simple means of improving the therapeutic ratio. It is possible that some of the benefits, and some of the problems, encountered in radiation therapy using multiple fractions per day, may be mediated by circadian effects.

Cisplatin

Cisplatin is a cytotoxic drug with a broad spectrum of activity. It is a bifunctional alkylating agent and is **cell cycle (p. 52)** non-specific. Chemically it is *cis*-diammine-dichloroplatinum: effectively this means that it presents the platinum ion in an organic form. It was discovered serendipitously during studies on the effects of electric current on bacterial growth. There was inhibition of growth around the platinum electrode depending on whether the current was switched on. The discovery of cisplatin and its introduction into cancer treatment revolutionised the management of patients with metastatic germ-cell tumours of the testis. Within 5 years in the mid-1970s the prognosis changed from cure rates of <5 to 80% or more. Cisplatin is also used in the treatment of many other types of tumour including lung cancer (both small cell and non-small cell), ovarian cancer, stomach cancer and lymphomas.

Cisplatin acts by binding to DNA and RNA and forming inter- and intrastrand cross-links. The intrastrand links are dominant, particularly involving the guanine and adenine bases. The **adducts (p. 7)** formed as a result of the action of platinum can be measured electrophoretically and may have some value in predicting response to treatment.

The main toxic effect of cisplatin is renal damage. It causes damage to the tubules so that sodium reabsorption in the descending limb of the loop of Henle is decreased; this in turn causes a fall in renal blood flow and an increase in vascular resistance within the kidney. It is essential to ensure that patients treated with cisplatin are adequately hydrated before, during and after treatment: this usually means at least 48 h receiving intravenous fluids. Hypertonic solutions and, possibly, **amifostine (p. 12)** may protect against some of this damage but cisplatin usually causes a fall in glomerular filtration rate (GFR). The magnitude of the fall is related to the total cumulative dose of platinum. Cisplatin also damages Schwann cells and the main clinical manifestation of this is damage to the auditory division of the VIIIth nerve. The toxicity of cisplatin has stimulated the development of other, potentially less toxic, derivatives such as **carboplatin (p. 46)** and **oxaliplatin (p. 221)**. Alternative formulations are also being investigated: liposomal cisplatin and a cisplatin/adrenaline gel for direct injection into tumours.

C-Kinase

This is activated by the intracellular messenger diacylglycerol, produced in response to **ligand (p. 180)** binding by the G protein-linked receptor. C-kinase has two main pathways of action. It can initiate a cascade of phosphorylation that activates MAP kinase and ultimately affects *Elk*, a gene regulatory protein. It can phosphorylate I-κ B in the cytoplasm, which releases a protein, NF-κ B, which migrates to the nucleus and acts as a specific **transcription factor (p. 313)**.

Classification of Adjuvant Cancer Treatment According to Timing

Neoadjuvant

Treatment given before any local or locoregional treatment, e.g. treating large breast cancers with chemotherapy before surgery; treating germ cell tumours with chemotherapy before any surgery to the retroperitoneal nodes; preoperative radiotherapy for rectal cancer. There are several mechanisms by which this approach might improve results:

- Sterilisation of malignant cells at the periphery of the tumour might make surgical margins less critical.
- Shrinking the tumour might:
  - improve resectability
  - permit a less radical operation, e.g. wide local excision rather than mastectomy.

Synchronous

Treatment given during definitive local treatment, e.g. giving synchronous chemotherapy and radiotherapy for a patient with head and neck cancer. The aim is to produce true synergy in which the effect of the two treatments may be more than additive (i.e. $1+1 \geq 2$). The problem here is that what applies to the tumour may also apply to the normal tissues and the net effect may be no improvement in therapeutic ratio.

Adjuvant (subsequent)

Treatment given after definitive local treatment, e.g. chemotherapy for node-positive breast cancer; postoperative radiotherapy for head and neck cancer. The aim is to treat any residual disease at a time when it will be most vulnerable to treatment. This is adjuvant therapy in its original, classical, sense.

Classification of Cancer Treatment According to Purpose

Radical versus palliative — the crucial decision

It is ironic that one of the most important decisions in oncology should have received so little direct attention. **Radical treatment** (**p. 259**) implies that treatment is given with the hope, if not the expectation, of cure. **Palliative treatment** (**p. 226**) implies that cure is not considered possible and that relief of symptoms is the main therapeutic aim.

Palliative treatment

There is no justification for a policy of routinely treating all patients simply because they have cancer. If a patient has no symptoms to relieve, then treatment may simply be meddlesome and may make the patient feel worse rather than better. In some circumstances, however, it may be sensible to give treatment to an asymptomatic patient to forestall symptoms that will almost inevitably develop. An example would be to give local radiotherapy to a patient with a lung cancer that is severely narrowing a bronchus and which, were it allowed to progress, would obstruct a major airway.

Ideally, palliative treatment should relieve symptoms without itself producing any side-effects. This ideal is often difficult to achieve. If there is an art to caring for patients with cancer then this is where it lies — there can be no off-the-shelf approaches, each patient's problems and priorities are unique and require management that is tailored to that individual.

Radical treatment

Radical treatment is given in the hope that it might be possible to cure the patient. It is accepted that some normal tissue damage will inevitably occur — it is a matter of experience and judgement to choose a treatment that will have a good chance of curing the patient without producing toxicity that, either in the short- or long-term, is unacceptable. It is interesting that studies of patients' attitudes suggest that

they might be prepared to accept greater toxicity for lower chances of cure than their doctors might assume. This observation emphasises that it is vital to involve the patient in any decision-making about treatment. The doctor has the duty to explain and perhaps to recommend, but the right to decide must always rest with the patient.

Radical treatments can be divided into definitive and adjuvant. Definitive treatment is intended to extirpate the known tumour: mastectomy for breast cancer; chemotherapy for leukaemia. Adjuvant treatment is treatment given in addition to definitive treatment in an attempt to lower the risk of relapse. The concept is based on the following assumptions and observations:

- There has been abundant time during the natural history of a cancer for malignant cells to spread beyond the clinically demonstrable tumour (micrometastatic disease).
- There may be microscopic local or disseminated disease remaining after apparently successful treatment for cancer.
- No imaging technique currently available can detect single cancer cells.
- Accurate tumour markers are available only for a few rare tumours (e.g. choriocarcinoma).
- Microscopic residual disease is undetectable using currently available technology.
- A single cancer cell persisting after initial treatment is sufficient to cause recurrence.
- Small volume tumours are more sensitive to treatment than bulky tumours.
- Treatment given earlier is more effective than treatment given later: cells are more actively proliferating; drug resistance is less likely.

It is implicit within the concept of adjuvant treatment that there will be three categories of patient:

- Patients with no residual disease after their initial treatment and for whom adjuvant treatment can confer no benefit.
- Patients with residual disease but whose disease is refractory to therapy and who are destined to recur in spite of adjuvant treatment.
- Patients with residual disease and whose disease will be eradicated by adjuvant treatment.

It is only the latter category who receive personal benefit from adjuvant treatment.

Classification of Cancer Treatment According to Scope

Regional in this context means the lymph nodes that drain the site of the primary tumour:

- Local:
  - — radiotherapy
  - — surgery.
- Local + regional:
  - — radiotherapy
  - — surgery.
- Systemic:
  - — chemotherapy
  - — hormone treatment
  - — 'biological' treatments.

Clearance

The clearance of a drug or compound is calculated as:

$$\frac{\text{Rate of elimination}}{\text{plasma concentration}}.$$

Clearance of endogenous creatinine is widely used as a measure of renal function. It can be calculated, using the Cockroft–Gault formula, from serum creatinine or estimated using simultaneous measurements of 24-h urine creatinine and serum creatinine.

Clinical Effectiveness

A term used in healthcare management: it combines several concepts, each of which is necessary for clinical practice to be considered 'effective':

- Clinical **guidelines** (**p. 145**) based on appropriate research of high quality.
- Provision of high clinical standards of care.
- Systematic clinical **audit** (**p. 24**).
- Defined indicators of clinical outcome.
- **Peer review** (**p. 229**) critical review of performance.
- Sharing and dissemination of information.

It is interesting, but alarming, to note that the concept of clinical effectiveness pays no explicit or direct attention to the views of the most important collaborators in the clinical process — the patients themselves.

Clinical Governance

A concept recently introduced into the language of healthcare management. It has three main components:

- Corporate accountability for clinical performance.
- Mechanisms for improving clinical performance:
  — external
  — internal.

The aim of this concept is to make managers responsible for clinical care and to give them the power to sort out any problems identified. It is implicit within this concept that no doctor can be better than the environment within which they operate; it is also recognised that some departments fail to deliver adequate clinical care and that this cannot be allowed to continue. The resources to address those inadequacies that have arisen as a result of insufficient funding are not, in the UK at least, readily available.

The NHS definition of clinical governance is 'A framework through which NHS organisations are accountable for continually improving the quality of their services and safeguarding high standards of care by creating an environment in which excellence in clinical care will flourish.' This poorly written, unpunctuated, definition demands improvements in care and allows no opportunity cost for such improvements — 'high standards of care' must be safeguarded. There is no recognition that improvements might actually cost money. 'Excellence', the management buzzword of the 1980s, is assumed to be cost neutral. A more succinct definition has been provided recently by a Scottish Secretary of State for Health, 'It means corporate accountability for clinical performance.'

The origins of the concept of clinical governance may be traced back to Donabedian who pointed out that quality rested on a tripod of structures, processes and outcomes.

Complementary Log Transformation

This technique is used in the analysis of survival data to check whether the hazard rate is constant over time. The complementary log transformation:

$$\log\{-\log[S(t)]\}$$

is plotted against log t.

If the **hazard rate** (**p. 148**) changes only slightly with time, then the plot will be linear. If, for two groups of subjects, the plots are parallel but non-linear, then the hazard rates are proportional even though the individual rates are not constant over time. This concept, of proportional hazards, underpins the Cox model used for multivariate analysis of survival and the construction of prognostic indices.

Complete Response

This term is used in the evaluation of cancer treatment. It is defined as: 'The complete disappearance of all measurable disease, both clinically and radiologically, for at least four weeks following treatment. No new lesions develop during this period.' A pathological complete response fulfils the above criteria, with the additional requirement that a biopsy from the site of the tumour, after treatment, should show no microscopic evidence of active cancer.

Confidence Interval (CI)

A CI indicates a range of values within which we can, with a specified degree of certainty, be sure that the true value will lie. The specified degree of certainty is usually 95% (95% CI) or 99% (99% CI). The CI can be calculated for a single parameter (age or height), for a rate or proportion (survival rate, remission rate) or for the difference between two sets of values (difference in survival rate). In the latter case, if the CI for the difference includes zero, then we cannot exclude the possibility that any difference observed has arisen simply through the play of chance. A variation on Bernoulli's pebbles can illustrate the principles underlying the calculation of a CI.

Imagine an urn filled with 1000 pebbles. The pebbles come in four sizes: there are 500 pebbles 1 cm in diameter; 200 pebbles 2 cm in diameter; 150 pebbles 3 cm in diameter; and 150 pebbles 5 cm in diameter. We can calculate, with absolute precision and certainty, the average (mean) diameter of a pebble. We are omniscient and know all we need to know about this particular universe of observations: the precise value for the mean diameter is 1.35 cm, and is calculated as:

$$\{(500 \times 1) + (200 \times 2) + (150 \times 5)\}/1000.$$

Our servant is not so lucky. We, as gentleman-amateur scientists of the Enlightenment, set him the task of estimating the average diameter of a pebble. Of course, he could empty out the urn and measure each pebble individually, but we want supper in an hour and he has to cook it. He only has time to sample the urn and thereby estimate the true state of affairs. Each evening, before supper, he measures 100 pebbles drawn at random and calculates the average diameter. He does this for 1000 consecutive days and the means he calculates are plotted. A frequency distribution is obtained. The plot is Gaussian, for it follows a Normal distribution. This occurs even though the distribution of pebble size is neither uniform nor Gaussian. The fact that the distribution of the means is Gaussian, even though the distribution of variable within the population being sampled is not Gaussian, is known as the **central limit theorem** (**p. 61**). It is this that enables the calculation of the CI.

The servant is sacked because he spends more time measuring pebbles than looking after us. A new servant is hired and she is set the same problem as that which tortured her predecessor. She is shown the normal distribution of the means that the previous servant had estimated. She measures 25 pebbles, calculates the mean and then tells us that she can be 95% certain that the true mean lies within the range between x and y. She is asked how she can be sure of this and she points out that one of the features of a Normal distribution is that 95% of observations will lie within ±1.96 SD of the mean. She calculated the SD for her mean, from the sample of 25, and used the properties of the Normal distribution to calculate a 95% CI for her estimate of the mean.

The pebble problem also illustrates a limitation of statistical methods for summarising data: does the mean diameter convey useful information when the pebbles come in three discrete sizes? Does quoting a mean survival help an individual patient make an informed decision about treatment when there are in fact three main possible outcomes each with a discrete probability of occurrence: early death from treatment-related toxicity; cure, with prolonged survival; initial response to treatment followed by relapse and death? Statistics apply to populations, but clinicians deal with individuals and an average effect may not be a useful or interpretable measure of outcome.

Confidence intervals versus significance tests

The heyday of the significance test is past: some journals will not publish comparisons based solely on significance tests. CI are now much more popular, not without reason:

- A significance test, by itself, says nothing about the magnitude of any observed difference.
- Statistical significance says only that what has been observed is unlikely to have arisen by chance.
- Concentrating on p as an outcome measure may lead to publication bias.
- Worship of p leads down the road to multiple testing: take multiple peeks at data on work in progress. Stop the study and publish when a 'significant' p is achieved. If we take 30 peeks at a negative study, it is highly likely that one of the peek will yield a $p < 0.05$.
- CI provide an easy means for assessing visually the power of a study: wide intervals indicate low power.
- The magnitude of any difference can be easily assessed using CI.
- The **central limit theorem** (**p. 61**) implies that it is entirely proper to use CI to compare non-normally distributed data.

Conformal Therapy

This term is used to describe a repertoire of techniques used in planning radiotherapy treatment: it implies that the **irradiated volume** (**p. 169**) conforms to the target, the tumour, in three dimensions. The volume is not just a cube or rectangular box, as in traditional treatment planning, but can be irregularly, indeed sinuously, shaped. The goal is to increase the volume of tumour that is homogeneously irradiated to the desired therapeutic dose while limiting the volume of normal tissue exposed to radiation. A number of technological developments, including multileaf collimation and the development of appropriate algorithms for planning computer software, have converted what was once a somewhat impractical dream into something that can be incorporated into the daily practice of radiation therapy. Whether the theoretical advantages of conformal therapy will translate into practical benefit remains to be proven. There are two main outcome measures that should be scrutinised when assessing conformal therapy in relation to more traditional

methods of treatment planning: was it possible to increase the dose to the tumour, and were local control and survival improved thereby? Was there, for any given local control rate, a decrease in damage to normal tissues as a result of using conformal treatment?

Confounding

The process whereby the true effect of a major variable can be changed, or obscured, by other factors present in the population under study. These factors are termed confounding variables and will have their effect because they are associated with both the disease and the exposure. In studies of racial differences in outcome following treatment for cancer, socio-economic status is a confounding variable: blacks tend to be poorer than whites; poorer patients have lower survival rates than the well off. Is any decreased survival observed in blacks due to race, *per se*, or does it simply reflect their lower socio-economic status?

Confounding variables, which act simultaneously on both condition and exposure, should be distinguished from interactions (syn. effect modification): in which variables act sequentially along a causal pathway. For example, lung cancer may be less common in subjects who have enzyme systems readily inducible by debrisoquine. The effect here is sequential:

Good enzyme induction → fewer cases of lung cancer.

Poor enzyme induction → more cases of lung cancer.

Debrisoquine inducibility affects the development of disease in patients who smoke, but it does not affect the exposure, cigarette smoking, itself. There are several methods for coping with confounding variables:

- **Randomisation** (**p. 260**): in prospective studies, random allocation should ensure uniform distribution of potentially confounding variables, both known and unknown, between the groups in the study.
- Elimination: if there are several possible confounding variables, choose one and eliminate from the study all subjects who have any of the others. In a study of the relationship between oral cancer and **tobacco use** (**p. 309**), this might involve restricting the study to patients who smoke, but who do not drink alcohol. The problems with this approach are: (1) decreased population available for study and (2) decreased applicability of study results — since the study population is no longer typical of the general population, most smokers drink alcohol.
- Matched pair: this can be done both retrospectively and prospectively. The control and experimental subjects are matched as pairs for known confounding variables. This is tedious and time-consuming. Subjects for whom matches cannot be found must be excluded and this could introduce **bias** (**p. 28**).
- **Stratification** (**p. 298**): if the necessary data are available, then it is possible, after the study has been completed, to perform an analysis stratified for confounding variables. For example, in a study of dietary fat intake and the risk of breast cancer, it might be appropriate to look at the effect of parity on risk within both the high and low fat intake groups. This would look for any confounding due to socio-economic status, since high parity and high fat intake might be related to lower income.
- Multivariate analysis: this permits the extrication of the individual effects of each variable from the effects of variables acting together so that those variables which, in their own right, affect outcome can be identified.

Residual confounding

- Direct: occurs when the effect of an identifiable confounding variable is dealt with only partially. There is, for example, a dose — response relationship between cigarette smoking and carcinogenesis. In a study of alcohol intake and oral cancer, smoking will be a confounding variable. To allow for this, adjustment according to smoking habit could be used. If the subjects were simply categorised as smokers or non-smokers this would not adequately adjust for confounding. Within the group of smokers, the heavy smokers will have a higher risk of cancer than those who smoke less. The solution to the problem is to use a more fine-grained approach, e.g. dividing into groups according to number of cigarettes smoked per day: 0/1–9/10–19/20–29/>30.
- Latent: arises when an unidentified variable is causing the confounding. Since the variable is unidentified, perhaps unidentifiable, it is difficult to deal with the problem. How can we cope with that which we cannot recognise? In prospective studies, randomisation should deal with the problem since the process of randomisation should distribute any unidentified variables evenly between groups.

Part of the art, if not the science, of assessing evidence is to develop the ability to identify potentially confounding variables. Interpretation of the value of a study should be influenced by whether the authors identified, and admitted the potential influence of, possible confounding variables.

Conjoint Analysis

A technique used in decision-making and **health economics (p. 149)** to elicit subjects' preferences concerning alternative policies and choices. It is particularly useful when, as in the NHS, choices have to be made under conditions of constraint and uncertainty.

The possible outcomes or choices are divided into attributes, each of which is further divided into levels. All possible combinations of attributes and levels are compared with a baseline choice, often the status quo. **Regression (p. 264)** techniques can then be used to ascertain which attributes are considered the most important.

The method could be used to analyse how patients balance the competing considerations of the inconvenience of travel to a distant centre for treatment, when there is a waiting list at their local centre, and the anxiety and distress associated with delay in starting treatment. The attributes and levels could be:

Attribute	Levels
Place of treatment (miles away)	<10
	11–50
	51–100
	>100
Waiting time for Rx (weeks)	<4
	5–10
	11–25
	>25

This would produce 16 potential scenarios for the subjects to assess and compare with the current arrangements (e.g. a 12-week wait for treatment within 10 miles of home).

Consequential Late Effects

A term used in radiotherapy to describe acute reactions that do not heal but which progress to necrosis. These effects were first described as radionecrotic ulcers in patients treated with unconventional fractionation, particularly **accelerated fractionation (p. 1)**, for head and neck cancers. The pathogenesis presumably involves the accumulation of too much damage to epithelial cells in too short a time. The pathogenesis is distinct from that of classical late effects.

Consistency

Consistency and reliability are synonymous: they indicate the extent to which the results of repeated measurements of the same phenomenon agree with each other. If the height of a patient is measured, on various occasions, as 160, 134 and 178 cm, then this shows a suspicious lack of consistency. Variation represents a lack of consistency. Lack of consistency can arise in several ways: because different observers measure the same thing differently (interobserver variation); because the measurements made by an individual observer vary over time (intra-observer variation). In both these examples there is no true change in the value being measured. Lack of consistency will also arise if that which is being measured is itself changing: if the weight of a patient were measured on three separate occasions as 65, 55 and 75 kg, then this does not necessarily imply that the measurements are inaccurate.

CONSORT Statement

The CONSORT statement concerns the design and reporting of a randomised controlled clinical trial. It describes an internationally agreed criteria which establish what should be regarded as the basic standards for this type of research. It covers issues such as: adequate randomisation procedures; intention-to-treat analysis and reporting; how to deal with subjects who fail to complete or comply with treatment; and appropriate statistical methods for analysing and reporting trial data.

A well-conducted, well-reported trial should, by these standards, contain:

- Title: identifying study as a randomised trial.
- Abstract: using structured format.
- Methods:
  — protocol describing study population with criteria for exclusion and inclusion
  — planned interventions
  — outcome measures, both primary and secondary
  — methods of statistical analysis used, and why they were used
  — stopping rules
  — assignment
  — randomisation method and units of assignment (individual, group, cluster)
  — degree of blinding and separation between the person assigning the intervention and the person carrying it out
  — degree of independence of the individuals performing the outcome assessments.
- Results
  — trial flow chart showing the eligible, exclusions at each stage, the assignments and the follow-up. The final row shows those completing the study, as allocated, and evaluable for all outcomes
  — analysis of results according to the outcomes previously specified, with calculation of confidence intervals. Numbers should be expressed as absolute values, rather than percentages. Detail sufficient to allow independent recalculation, particularly when significance tests have been used
  — prognostic factors should be identified and their possible confounding effect upon outcome according to allocated treatment should be assessed

— deviations and exclusions should be accounted for, their possible effects upon conclusions should be explored.

- Comment: should include, in any interpretation of the results, an account of possible sources of **bias** (**p. 28**) and how these may have affected the conclusions. Relationship to previous work, consistencies and inconsistencies should be discussed. The overall conclusion should take account of all the available evidence, not just that from the study itself.

The above summary is based on a checklist and a flow diagram for assessing a clinical trial, which can be found via the *JAMA* web site: www.ama-assn.org

Constitutive Expression

Constitutive expression describes a gene expressed without the need for activation or induction. This is in contradistinction to genes only expressed when an inducing factor is present: cyclooxygenase 1 is constitutively expressed, **cyclooxygenase 2** (**p. 83**) is not.

Contortostatin

An anti-angiogenic factor isolated from the venom of the copperhead snake that has shown some activity against human breast cancer xenografts. It is a member of the disintegrin family of proteins.

Correlation

Correlation is a measure of the extent to which two separate variables have a linear relationship. It is measured as the correlation coefficient (Pearson's r, the product moment coefficient): 0 indicates no linear relationship; -1 a perfect negative correlation (as one variable falls, then so the other increases); and $+1$ a perfect positive correlation (as one variable increases, then so does the other). Put another way, an r approaching 1.0 or -1.0 indicates a strong association between the two variables; $r=0$ indicates no association. Correlation can only be used when all the observations are independent — in practice this usually means one pair of observations per subject. If multiple paired observations were made, e.g. in a time-series analysis, then the criterion of independence would be violated and correlation should not be used.

Correlation does not indicate cause: storks and birth rate are correlated; frogs and rain are correlated. Storks do not cause babies, frogs do not cause rain, nor vice versa. The area of the polar icecaps has decreased between 1950 and 1997, this decrease correlates reasonably well with the distance from Earth of the Hale — Bopp comet. It would be foolish to claim that the heat from the comet is melting the polar ice.

$r=0$ does not imply that two variables are unrelated, simply that any relationship is non-linear. Other relationships, J- or U-shaped, are not excluded.

By squaring r, we obtain the **coefficient of determination**, which indicates the proportion of the original variance that has been explained by adopting a linear model. Statistical **significance testing** (**p. 286**) is often applied to correlation coefficients. This can be very misleading. There are only 2 **degrees of freedom** (**p. 91**) (d.f.) and so, with large sample sizes, even weak correlations will achieve statistical significance: $p<0.05$ for a sample size of 100 and $r=0.2$. In spite of this 'statistically significant' result, only 4% of the original variance is explained by assuming a linear relationship between the two variables, 96% of the variance is left unexplained. As a general rule: for assessing correlation, the r is important, p is not.

The crucial difference between correlation and regression is that correlation simply describes the extent to which the relationship between two independent variables is linear; regression seeks to explain how one variable (explanatory variable) influences another (dependent variable), regression models need not necessarily be linear.

Cosmids

A cosmid is a cloning vector used in DNA libraries. It is a hybrid between DNA from a plasmid and the *cos* region of the λ (phage) genome. The *cos* region confers the ability to insert long segments (up to 45 kb) of DNA. The sequences, once incorporated within the cell, propagate as plasmids.

Cost–Benefit Analysis

Cost–benefit analysis is a form of economic analysis in which both the outcome and the costs of an intervention are measured in monetary terms. For example, an analysis that looked at screening for lung cancer and measured both the costs of the programme and the potential savings (£) would be a cost–benefit analysis. By investing £3 million in a screening programme, it might be possible to save £500 000. The cost–benefit analysis suggests that the programme may not be worthwhile.

Cost Containment

The purpose of cost containment is to constrain the amount of money spent so that it is within limits regarded as reasonable or affordable. There are many mechanisms used to limit the costs of medical care, Some are explicit, others less so. The more overt mechanisms include:

- Rationing.
- Decreased choice.
- Shifting costs onto other budgets (using lottery money to buy linear accelerators; relying on charities to provide services that should be available under the NHS).

There is concern that other influences on clinical practice such as practice guidelines, consensus statements and the cruder forms of 'evidence-based' oncology represent more covert attempts to save money — without, necessarily, ensuring that standards of care are maintained. The premise underlying cost containment is that much of the money spent on healthcare is wasted and that it is possible to make savings without jeopardising outcomes. This premise is not always valid and application of the principle of cost containment may mean that by spending less, less is obtained. Economies are often false: a particular catheter may be cheaper but it may be less robust. More of the cheaper catheters will be required and so the overall cost of the 'cheaper' option may be greater. Of course, if the original catheter is inserted in hospital, and the replacement of such catheters is a responsibility for primary care, then the hospital will save money by using the cheaper catheter. The true cost is shifted onto another budget.

Cost-Effectiveness Analysis

An economic analysis in which alternative strategies are compared in terms of a common effect. Costs are usually measured in monetary units but outcomes can be measured in any common unit appropriate to the strategies under comparison. For example, two adjuvant chemotherapy regimens could be compared according to monetary cost per recurrence-free year gained. The most commonly used measure is cost per life-year, **marginal cost-effectiveness** (**p. 188**) is used when two policies, e.g. policies A and B, are compared.

Table 1 — Cost-effectiveness of some typical medical and social interventions intended to save lives

Intervention	Cost per life year 1993 (US$)
Nicotine gum plus smoking cessation advice men aged 35 to 69	7 500
Nicotine gum plus smoking cessation advice women aged 35 to 69	11 000
Smoking advice for pregant women who smoke	0
Chemotherapy for adult non-ALL	27 000
Captopril for 35 to 64 yr olds with no cardiac disease diastoloic >95 mmHg	93 000
Beta blocker for 35 to 64 yr olds with no cardiac disease diastoloic >95 mmHg	14 000
Hemoccult screening for colorectal cancer asymptomatic 55 yr olds	1 300
Annual mammography plus breast exam women 40 to 49	62 000
Postoperative adjuvant chemotherapy for premenopausal breast cancer	18 000
Compulsory annual motor vehicle inspection	20 000

The marginal cost-effectiveness is simply the cost per extra life-year gained with policy A compared with B. A negative marginal CE would imply a cheaper cost per life-year for that particular policy. The cost-effectiveness of some typical medical and social interventions intended to save lives is compared in table 1.

See also Cost–Benefit and Cost–Utility Analysis.

Cost–Utility Analysis

The usual outcome measure in a cost–utility analysis is quality-adjusted survival. This type of analysis differs from a cost-effectiveness

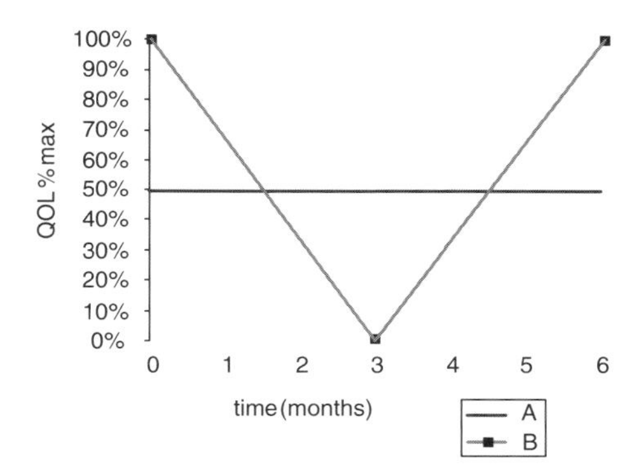

Figure 1 — Problem of duration and intensity in the assessment of utility: patient A has exactly the same total utility (three quality-adjusted months) as patient B, but their experiences over the 6 months are very different. Patient A has 'enjoyed' chronic ill-health; patient B has suffered a catastrophic, but transitory, disruption to quality of life and, by the end of the study period, has recovered completely.

analysis in that a **quality factor (p. 253)** is applied to the outcome. Implicit within this strategy is the recognition that some interventions may well lengthen survival but the quality of that survival may be compromised as a result of treatment-related toxicity. The cost–utility analysis explicitly allows for this type of trade-off. Options are compared in terms of monetary cost per quality-adjusted life-year gained. This approach is enticing in theory but tricky in practice — how do we elicit utilities for differing health states, and whose utilities do we use? A young professional footballer may have a very different attitude to paraplegia than a 78-year-old retired pianist. There is also a problem with intensity and duration when different health states are compared. When quality is plotted against time and the area under the curve (AUC) is calculated, the utility over time may average out as equivalent, but two very different states are being compared. Which would you rather have: a hard bang on the thumb with a hammer or a mild backache for a month? (see figure 1).

The unit usually used in a cost–utility analysis is **QALY (quality-adjusted life-year) (p. 253)**.

Costs of Cancer to a Developed Country

These are the estimates of the costs (US$) of cancer to the USA in the late 1990s:

- Direct medical costs: $37 000 000 000. Direct costs include the cost of investigations and of drugs, surgery and radiotherapy, the 'hotel costs' of in-patient treatment and all the other immediate expenses incurred in treating cancer.
- Indirect costs: those borne by society as a whole rather than by the providers and purchasers of cancer care. They include the costs of time off work, both to patients and their relatives, and the costs of lost productivity due to premature death. Morbidity: $11 000 000 000; mortality: $59 000 000 000.

Note: data do not include costs of screening for cancer that probably add at least another $3 000 000 000 (to direct medical costs).

Cancer accounts for ~11% of all medical care expenditure in the USA and the total cost of cancer to the nation is ~$107 000 000 000.

Coterie Effect

The coterie effect has to do with closed shops, mutual admiration and the propagation of constraining **paradigms (p. 226)**. As such, it is a perverted version of peer review. Journals may have their usual suspects, whom they round up to write editorials and commentaries as well as to referee papers. The result can be the domination of an area of investigation by a small group of active individuals who happen to agree with each other. 'Your editorial accompanies my paper and my editorial complements your paper.' Oncology may start to resemble the mutual admiration societies that grace the literary pages of the Sunday newspapers.

Cotswold Staging

The Cotswold Report suggested modifications to the **Ann Arbor (p. 16)** system used for the clinical staging of Hodgkin's disease. The basic four-stage system with A, B and E suffixes is retained, but additional information is included and some of the definitions tightened.

'E' is defined as: limited direct extension from an adjacent nodal site with the implicit expectation of a prognosis equivalent to that of nodal disease of the same anatomic extent.

The criterion of 'liver involvement' demands that multiple focal defects that are neither cystic nor vascular must be demonstrated using at least two imaging techniques. The older criterion of clinical hepatomegaly ± abnormal liver function tests is no longer valid.

There is, in general, an increased reliance on imaging with MRI and/or CT to define other sites of involvement: bone, spleen, CNS.

A PS (pathological stage) has been introduced to annotate sites of histologically proven involvement (M = marrow, H = liver, L = lung, O = bone, P = pleura, D = skin).

The bulk of disease is more clearly defined: any nodal mass >10 cm in its largest dimension is defined as bulky. Bulky mediastinal disease is strictly defined: on CXR, taken in full inspiration, in upright position at a source skin distance of 2 m, the maximum width of mediastinal nodal disease is >1/3 the transverse thoracic diameter measured at the level of T5/T6.

Bulky disease is indicated by the subscript 'X'. The number of sites of disease is indicated by a numerical subscript. $CSII_{7XE}A$ indicates clinical stage II (i.e. disease confined to one side of the diaphragm), which is bulky, involves seven different anatomical sites and there is extranodal extension. The patient has no systemic symptoms. $PSIVB_{LHMX}$ indicates that a patient has systemic symptoms and that there is a pathologically proven involvement of the bone marrow (M), that nodal disease is >10 cm in maximum diameter (X) and that there is radiological evidence of parenchymal lung involvement (L) as well as liver involvement (H).

The other contribution of the Cotswold Report was to specify criteria for evaluating the response of Hodgkin's disease to treatment:

- CR: no clinical radiological or other evidence of HD.
- CRu: normal health no clinical evidence of Hodgkin's disease but persisting radiological abnormality, not explicable by treatment-related change, is present at site of previous disease.
- Partial remission (PR): decrease by a least 50% in the sum of the largest perpendicular diameters of all measurable lesions. B symptoms should have resolved and there should be subjective improvement of those lesions not amenable to formal evaluation
- Progressive disease (PD): ≥25% increase in size of at least one measurable lesion or appearance of new lesion or development or reappearance of B symptoms.

Courtenay–Mills Assay

A method for assessing cell survival after irradiation. A cell suspension is made from a biopsy of an unirradiated tumour. The suspension is irradiated *in vitro* and then plated into semisolid agar with **growth factors** (**p. 144**). Colonies are counted after a 4-week incubation. Results are usually expressed as SF_2 (**surviving fraction after 2 Gy**) (**p. 302**).

Cowden Syndrome

A syndrome associated with an inherited predisposition to cancers of the thyroid and breast. It has been linked to abnormalities of the ***PTEN*** gene (**p. 251**), on chromosome 10q23. The evidence suggests that *PTEN* functions as a tumour suppressor. The clinical syndrome features multiple hamartomas with warty skin lesions, especially on the face. Affected individuals often have large heads: macrocephaly. There are cobblestone-like mucosal abnormalities affecting the oral cavity. Gastrointestinal polyps also may occur. In its classical form the syndrome is inherited as an autosomal-dominant. It is named after the family of the first patient, Rachel Cowden. It is related to a syndrome of cerebellar abnormalities and seizures — Lhermitte–Duclos disease — in which hamartomas and breast cancers also occur.

Cox's Proportional Hazards Model

This technique is used in the analysis of survival data and can be used to assess potential prognostic factors. It is possible to define for a group of subjects an average hazard: that is the average instantaneous event rate for the whole group. It is then possible to calculate, for specific subgroups of patients, the average hazard associated with membership of that subgroup and to compare this with the average for the whole group. Provided that the hazards in the various groups are in constant proportion over time, and this can be checked using **complementary log transformation (p. 69)**, there is no requirement that the **hazard rates (p. 148)** be absolutely constant over time.

In its simplest form the Cox model looks at one variable (e.g. treatment) to see whether it has a significant effect on the average hazard rate. The result is usually expressed as the hazard ratio:

$$HR = \frac{\text{hazard rate with treatment}}{\text{hazard rate without treatment}}.$$

If HR for death is <1.0, then this suggests that treatment is beneficial; HR >1.0 indicates that treatment is harmful. The Cox model can be extended to multiple variables and can be adapted to accommodate both categorical and binary variables as well as continuous variables. The influence of multiple variables on outcome can be considered simultaneously and, used in this way, the Cox model is an example of multivariate analysis.

The temptation is to use the Cox model to look at a large number of potential prognostic factors simultaneously. The problem is that the data can only support so much interpretation. There are two approximations that can be used to define the appropriate number of variables for a Cox model, and these methods do not necessarily agree. The first method is simply to use the fourth root of the number of events as the maximum number of permissible variables: if there were 1000 events then 10 variables could be evaluated; if there were only 75 events then no more than three variables should be investigated. The other approach is to demand that each variable be associated with at least 15 events before it could be considered for inclusion in a Cox model. Thus, if there were only five deaths in patients >60 years of age, it would be inappropriate to use age >60 as an independent variable in a Cox model.

The main use of the Cox model is in assessing factors, including allocated treatment, that can influence outcome. The model generates a term (usually notated 'b') for each variable that indicates the amount of effect that variable has on outcome: variables with a high b are highly influential. The 95% confidence interval around the hazard ratio will give an estimate of whether the influence of a putative **prognostic factor (p. 248)** is statistically significant.

CREB-Binding Protein (CBP)

CREB is a **transcription factor (p. 313)**; CBP is its binding protein. CREB acts a regulatory protein activated by the cAMP/A-kinase pathway and which recognises the cAMP responsive element (CRE) on the gene for somatostatin. Binding of CREB to CRE stimulates production of somatostatin.

CBP is large, with a relative mass of ~250 000. CBP interacts, via a region near its C-terminus, with a basal transcription factor, TFIIB. TFIIB itself interacts with TBP1, a TATA box-related protein. The net result is increased activation of genes that respond to cAMP. The *CBP* gene is at 16p13.3. Up to 70% of the leukaemias associated with treatment using inhibitors of topoisomerase II are associated with translocations involving 11q23. The malignant cells in this particular type of leukaemia have a translocation t(11;16)(q23;p13.3) that produces a fusion gene *MLL-CBP*.

Crude Rate

Data on incidence or **mortality rates** (**p. 207**) are often presented as rates. For a specific tumour and population, a crude rate is calculated simply by dividing the number of new cancers or cancer deaths observed during a given period by the corresponding number of people in the **population at risk** (**p. 239**). For cancer, the result is usually expressed as an annual rate per 100 000 persons at risk. The problem with crude rates is that comparisons may be misleading, e.g. if the age distributions of the populations under comparison are very different then comparison of crude rates may suggest an environmental cause for differences in incidence between the developed and the developing world.
In reality, the difference may be explained by increasing incidence of certain cancers with increasing age and the fact that many citizens of developing countries do not survive for long enough to develop these tumours. The solution is to use incidence rates related to a standardised population of specified age distribution, e.g. the world standardised distribution (WSR), and to compare WSR and world standardised rates. Similar, but more complex, difficulties apply to the use of crude rates for survival analysis.

Cumulative Rate

Cumulative incidence is the probability or risk of individuals getting the disease during a specified period. For cancer, it is expressed as the number of newborn children (out of 100, or 1000) who would be expected to develop a particular cancer before 65 years of age (or 70, or 75) if they had the rates of cancer currently observed. Like the age-standardised rate, it permits comparisons between populations of different age structures.

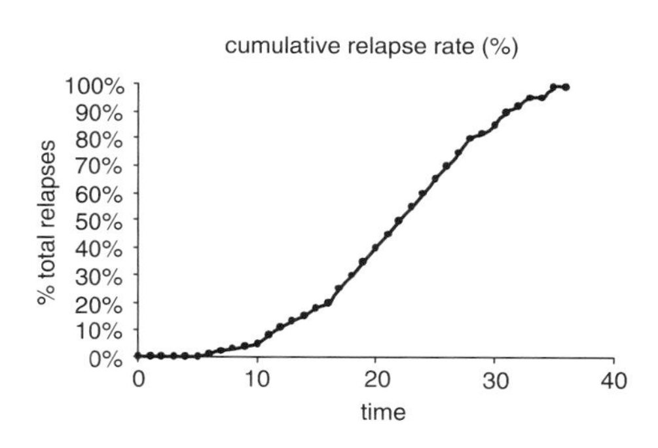

Figure 1 — Cumulative time-to-relapse data. The intensity of follow-up should be greatest between 10 and 36 months. Thereafter, relapse is so unlikely that follow-up, if it is simply to detect recurrence, might be abandoned.

Cumulative rates can also be used in the analysis of survival or relapse data.
The cumulative time to relapse curve can be used to design rational follow-up schedules after treatment for cancer. If the aim of follow-up is to detect recurrences promptly, then the intensity of follow-up should be maximal when the time-to-recurrence curve is steepest and can be relaxed once the majority of those destined to relapse have done so.

Cure

Although survival at 5 years after diagnosis is often equated with the cure of cancer, this may not be entirely accurate, particularly for tumours such as breast cancer, which have a long natural history. A better definition is that proposed many years ago by Eric Easson: we may speak of cure when, some reasonable time after treatment, there exists a population of treated patients whose survival expectancy parallels that of a normal population of similar age and sex distribution. Easson's definition is useful because it recognises several important points: some types of tumour will relapse sooner than others; cancer does not confer immortality; a pragmatic definition is essential — there is no possible way of telling, for most tumours, whether every last cancer cell has been eradicated.

Cyclic AMP (cAMP)

cAMP is an intracellular mediator of hormone action. Its behaviour is partly controlled by the G-protein-linked receptor/trimeric G-protein system.

cAMP is formed when adenylyl cyclase acts on ATP; it is rapidly broken down by cAMP phosphodiesterases. Regulation is mainly through effects on adenylyl cyclase. Different species of trimeric G-proteins will have different effects, some stimulatory, some inhibitory. β-Adrenergic stimulation causes production of a trimeric G-protein that stimulates adenylyl cyclase while stimulation of α-2-adrenergic receptors inhibits adenylyl cyclase and lowers cAMP levels.

cAMP acts predominantly by activating a kinase (A-kinase; cAMP-dependent protein kinase) that transfers phosphate from ATP to individual serine and threonine residues on target proteins. Responses mediated by cAMP include:

- Response of thyroid to TSH.
- Response of adrenal cortex to ACTH.
- Response of bone to PTH.
- Response of ovary to LH.

cAMP can also directly affect genes. The gene that regulates somatostatin production contains a cAMP-responsive sequence controlled by CRE-binding protein, a protein activated by the cAMP/A-kinase pathway. Regulation of cAMP is extremely important. Some bacterial toxins, such as those of *Pertussis* and *Cholera*, produce their toxic effects by interfering with cAMP.

Cyclins

Cyclins are proteins involved in the control of the **cell cycle** (**p. 52**). They accumulate steadily during the cycle and then rapidly disappear during mitosis. They form 1:1 molecular complexes with subunits of the cyclin-dependent kinases. These protein kinases, which include enzymes such as $p34^{cdc2}$, control the phosphorylation of the cyclins and thereby control their level and function. There is, therefore, a system for producing a periodic waxing and waning of cyclin activity and, ultimately, it is this system that controls the cell cycle and, subsequently, mitosis. There are two main cyclins, A and B, that are structurally similar, with ~30% of their amino acid sequence in common. The role of cyclin B is more clearly defined than that of cyclin A.
A complex series of control mechanisms is involved in regulating the activity of the cell cycle: $p34^{cdc2}$ is activated by **cdc25** (**p. 52**) and CAK (cdc2-activating kinase) and inhibited by *wee*1. *Wee*1 is a gene isolated from yeast whose product has tyrosine kinase activity and which inhibits the activation of cyclin B by $p34^{cdc2}$. Cells deficient in *wee*1 are smaller than normal cells, because they can enter mitosis prematurely and, therefore, do not reach their normal size before dividing — they are puir wee cells. Activated $p34^{cdc2}$ and cyclin B together will stimulate **MPF** (**p. 208**) and thereby trigger mitosis. Once this has occurred, the cyclin is rapidly deactivated by proteolysis. The cyclin box is a conserved sequence of 150 amino acids found near the N-terminal ends of both cyclins A and B, which, in the presence of the protein **ubiquitin** (**p. 319**), serve to identify the cyclin as a target for destruction.

There are many different types of cyclin involved in the regulation of the mammalian cell cycle. B1 and B2 are mainly concerned with promoting mitosis; A and E are important during G_1- and S-phases. D is possibly important in the regulation of cell growth. E is maximal during late G_1; A appears during S-phase and is at its highest level in early G_2; B1 is maximal at mitosis. Diurnal patterns of cyclin expression have been observed in normal tissues: B1 is highest at around midnight; E peaks in the early afternoon; A peaks in the later afternoon. These fluctuations in cyclin activity correlate with the circadian fluctuations in proliferative activity.

Cyclooxygenase 2 (COX-2)

An enzyme, particularly high levels of which are found in the colonic epithelium, capable of induction and which can produce metabolites that increase endogenous **oxidative stress (p. 221)**. The *Ptsg*2 gene encodes COX-2. Cyclooxygenases are involved in prostaglandin synthesis: in theory, and perhaps in practice, inhibition of prostaglandin synthesis by non-steroidal anti-inflammatory drugs (NSAID) might protect against large bowel cancer. Similar endogenous oxidative stresses may be important in the aetiology of prostate cancer. There is an important interaction between oxidative stress and abnormalities of *HNPCC* genes, including *APC* (in the colon) and *GSTP*1 genes (in the prostate). Specific inhibitors of COX-2 may be useful in cancer prevention. An important difference between cyclooxygenases 1 and 2 is that COX-1 is constitutively expressed while COX-2 is only fully expressed under appropriate environmental stimulation.

Cyclophosphamide

Cyclophosphamide is a cytotoxic alkylating agent. It is a **prodrug (p. 246)** activated by the hepatic P450 cytochrome system to aldophosphamide and 4-hydroxycyclophosphamide. Further metabolism yields the multifunctional alkylating agent phosphoramide mustard and acrolein. Acrolein is believed to be responsible for the haemorrhagic cystitis, which is the characteristic toxicity of cyclophosphamide. **Mesna (p. 195)** can be used to protect the urothelium from cyclophosphamide-induced damage.

The action of cyclophosphamide is independent of the **cell cycle (p. 52)** phase and it has a broad spectrum of activity: lymphomas, breast cancer, ovarian cancer, myeloma and sarcomas. It is also used as an immuno-suppressive agent in the treatment of non-malignant conditions such as polyarteritis nodosa, systemic lupus erythematosus and the nephrotic syndrome. It is active orally as well as intravenously. Although its main effect is to prevent proliferation of actively dividing malignant cells, there is some evidence to suggest that continuous treatment with low doses of cyclophosphamide might, by inhibiting endothelial cell division, have an anti-angiogenic effect. Unfortunately, prolonged administration of low doses of cyclophosphamide may cause interstitial pneumonitis.

Cylindroma

A irritatingly ambiguous pathological term. Today, it describes a benign skin tumour. Formerly, it described adenoid cystic carcinoma, an unquestionably malignant tumour of salivary origin. Given these ambiguities, the term would be best abandoned.

Cytarabine

Cytarabine (also known as ara-C) is an anti-metabolite used predominantly in the treatment of leukaemias and lymphomas. It is similar to cytidine (and deoxycytidine), with arabinose substituted for ribose (or deoxyribose). It is metabolised to ara-CTP, which acts as a competitive inhibitor of DNA polymerase. Once ara-CTP has been incorporated into DNA, the DNA cannot replicate, nor can the molecule be lengthened. Ara-C is S-phase-specific and is only active against dividing cells. Strand reduplication is a common feature of **DNA damage** (**p. 98**) induced by ara-C. The drug has to be given parenterally as it is inactivated, by deamination, in the gut lumen.

Cytochrome P450 (CYP)

P450 is a system of more than 100 enzymes located in the rough endoplasmic reticulum of the liver and which, on centrifugation of liver homogenates, separates with the microsomal fraction. It is named for the fact that, after treatment with carbon monoxide, it absorbs light of 450 nm.

The system is crucial in the initial metabolism of exogenous chemicals, particularly drugs and other ingested agents. The activity of many potential environmental carcinogens may be influenced by polymorphisms in the P450 system: individual ability to activate carcinogens may depend on the precise configuration of the P450 enzymes. The same argument also applies to drug activation and metabolism. The **clinical effectiveness** (**p. 69**) of drugs may, again, depend on the inherited pattern of genes controlling the P450 system. Biochip technology using arrays to assess the pattern of gene expression is increasingly being used in an attempt to define individuals at risk of developing environmentally induced cancer as well as to predict the clinical utility of specific therapies for individual patients.

Cytostatic

An agent is said to be cytostatic when inhibits cell growth or division but without actively killing cells. Many hormone manipulations used in treating cancer (**tamoxifen** (**p. 304**) in breast cancer; **LHRH** (**p. 178**) analogues in prostate cancer) have a cytostatic, rather than a cytotoxic, effect. The easiest analogy is to envisage cancer as a citadel: cytotoxic agents bomb the citadel and destroy it directly; cytostatic agents lay siege to the citadel and starve the inhabitants to death.

Cytotoxic

An agent is defined as cytotoxic when it is capable of directly causing **cell death** (**p. 55**).

Cytotoxic Drugs

These are drugs that specifically kill cells, particularly cancer cells. A variety of mechanisms may be involved; not all are, for instance, specifically toxic to dividing cells. These drugs can be classified in a number of ways: biochemical mode of action; action interpreted in cell kinetic terms, e.g. the Bruce model; origin, e.g. natural compounds, artificially synthesised, etc.

A vast amount of money and effort has been expended on the development and testing of drugs for the treatment of cancer. Hundreds of thousands of potential compounds have been screened but only ~50 have a role in the treatment of malignant disease. There have been several shifts in strategy and emphasis since the introduction of the first active anti-cancer drugs in the late 1940s. Many of the drugs currently in clinical use would not have made it through the systems currently employed. Conversely, there is the concern that some discarded compounds may have been inappropriately rejected: AZT, now a major component in antiretroviral therapy for AIDS, started out as a cancer drug that was considered too toxic for clinical use.

Only ~5–10% of patients with cancer can be cured by chemotherapy alone; a further 30–40% have some benefit in terms of tumour response. This leaves >60% of patients with cancer for whom cytotoxic treatment can be of little or no benefit. When a vast amount of investment produces so little return, it is always worth questioning the premises on which that investment has been made. One obvious criticism of current practice is the insistence on measurable tumour regression after treatment as a criterion for accepting a compound as clinically useful. At first sight this seems a fairly obvious requirement — without the shrinkage of a tumour, how can there be a therapeutic benefit? This, however, ignores the importance of static or **stable disease** (**p. 294**). If a patient's disease, particularly if their tumour is slow growing, can be stabilised, then this might offer significant benefit in terms of prolongation of life and palliation of symptoms. It also ignores the somewhat arbitrary criteria and methods used to measure 'response'. Why choose one particular definition of 'partial response'? There is also an in-built **bias** (**p. 28**) towards assessing response according to the changes in accessible disease, simply because this may be more easily measured. However, it could be the response of the less accessible disease, e.g. liver metastases rather than skin nodules, that is clinically more relevant. As we move into an era of gene therapy and modulation of biological responses, we are going to be ill served, and undoubtedly misled, by the current criteria for discriminating between useful and useless therapies.

Cytotoxic Drugs According to Mode of Action

One method of classifying cytotoxic drugs is according to their predominant mode of action. The system is useful in designing rational combinations of drugs for therapeutic use. It is sensible to combine drugs with differing modes of action so that different targets within the cancer cell are being simultaneously attacked.

Table 1

Class of drug	Mode of action	Examples
Antimetabolites	prevent the synthesis of DNA, most active against cells in S-phase	**methotrexate** (**p. 199**), 5-fluorouracil (5FU)
Alkylating agents	form cross-links between DNA base pairs so that separation of the DNA strands cannot occur	cyclophosphamide, nitrogen mustard, chlorambucil
Intercalating agents	insert themselves into the coils of the DNA double helix so that separation of the DNA strands cannot occur	adriamycin (**doxorubicin**) (**p. 107**), **daunorubicin** (**p. 88**), **etoposide** (**p. 119**)
Spindle poisons	affect the **mitotic spindle** (**p. 201**) and thereby prevent the segregation of chromosomes, which is essential for mitosis	**vincristine** (**p. 324**), **vinblastine** (**p. 324**), **taxanes** (**p. 305**)

(Continued)

Class of drug	Mode of action	Examples
Topoisomerase inhibitors	prevent the uncoiling of DNA necessary for **replication** (**p. 267**) to occur	**irinotecan** (**p. 168**), **topotecan** (**p. 310**)
Others	a miscellany of drugs, many of which have more than one mode of action	antitumour antibiotics (e.g. actinomycin D, **bleomycin** (**p. 32**)), platinum and its analogues (*cis*-platinum, **carboplatin** (**p. 46**), **oxaliplatin** (**p. 221**))

Data Dredging or Trawling

A reprehensible procedure occasionally employed when a study has failed to achieve its stated aim. The raw data are, in the absence of any prior hypothesis or belief, inspected and manipulated in the hope of identifying a finding that is statistically significant. If two or three such 'results' can be obtained then, from the wreckage of a failed study, something publishable might be obtained. The problem is that the publication has no meaning: the 'significant' findings could simply have arisen by chance. If 100 dredged comparisons are made, then several will be 'significant' at the conventional statistical level of $p = 0.05$.

Data Monitoring Committee (DMC)

The data monitoring committee (DMC) is part of the supervisory procedures for a clinical trial. It should be composed of individuals who have neither a direct involvement in the study, nor any vested interest in the outcome: its members should be intellectually and financially independent. The committee's role is primarily to ensure that the trial does not continue beyond a point at which a clear conclusion can be drawn from the data. Its duty is to protect patients from being pointlessly entered into a study that has already demonstrated clear superiority of one intervention over another. The implementation of the stopping rules, which should have been written into the trial protocol, is the responsibility of the DMC.

The DMC has several other responsibilities. The members should ensure that the data are gathered and analysed correctly and that the trial is conducted ethically. In other words, they are essential participants in the quality control procedures.

Database

A database is any structure, usually on a computer system, that stores data in a structured and retrievable fashion. Examples of relevance to oncology include: Medline and other databases containing medical literature; DNA sequence data available over the Internet; and patient-centred databases. One of the problems as information technology is introduced into healthcare is that databases that need to be linked turn out to be incompatible with each other. Data that is stored, but which, because of software incompatibilities, cannot be retrieved, is doubly wasted. First, because it cannot be used, and, second, because of the wasted effort entering it in the first place.

The basic structure of a database is that of a grid. The rows represent records and the columns are fields. A database for patients with colorectal cancer illustrates the practicalities (table 1).

The data can be presented in tabular form (as above) or as an index card for each record (patient) (figure 1).

Table 1

ID	No.	Name	Site	Date of diagnosis	Treatment
1	234567	Smith	rectum	03/08/99	chemotherapy
2	987654	Brown	hepatic fexure	31/07/99	XRT
3	286982	Jones	sigmoid	28/05/92	chemoradiotherapy

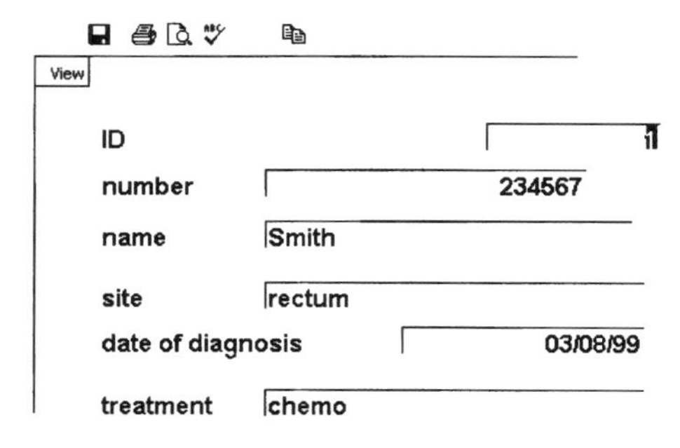

Figure 1 — Patient record.

There is much to be said for using generic commercial products that have the ability to export data into a universally accessible format rather than some sophisticated, but esoteric, self-designed system. Commercial databases, widely used for healthcare data include Access© (by Microsoft), etc.

A spreadsheet can be regarded as a form of database in which calculations involving the various fields can easily be performed. This facility can be used not only just for preparing summary statistics and for the visual presentation of data, but also, when used **iteratively** (**p. 169**), for modelling.

In the UK, the recent data protection legislation will have a profound impact on the design and use of patient-centered databases.

Daunorubicin

Daunorubicin is structurally similar to **doxorubicin** (**p. 107**). It differs only in the loss of an –OH group from a side-chain. Its clinical spectrum of activity, however, appears to be somewhat different: daunorubicin is predominantly used in the treatment of leukaemias and has played little role in the treatment of other tumours. In common with doxorubicin, it acts as an intercalating agent and is cardiotoxic.
A liposomal preparation, DaunoXome, has been used in the treatment of **HIV** (**p. 151**)-related Kaposi's sarcoma.

DCC

DCC is a tumour suppressor gene deleted in some human tumours: colon, rectum, prostate, breast. It is on chromosome 18q21.3. Loss of *DCC* is a relatively late event in colorectal carcinogenesis. The gene product has **sequence homology** (**p. 284**) with neural cell **adhesion molecules** (**p. 8**) such as N-CAM. It is a transmembrane protein of the immunoglobulin superfamily. Tumours with abnormalities in *DCC* may follow a more aggressive clinical course than tumours with intact *DCC* genes. Its normal **ligand** (**p. 180**) may be netrin-1, which acts as an attractant for axons in the developing spinal cord. When netrin-1 is absent, then *DCC* induces **apoptosis** (**p. 19**); if netrin-1 is present, then it will block apoptosis. Defective *DCC* genes could, therefore, block apoptosis and might lead to the inappropriate survival of abnormal cells. The *DCC* product is a substrate for **caspase** (**p. 47**) and *DCC* could be involved in inducing apoptosis if cells were in an environment unable to support them, e.g. if they had outgrown their blood supply. In this context, defective *DCC* might permit survival of cells in an inappropriate environment — this has important implications for **invasion** (**p. 167**) and **metastasis** (**p. 198**).

Decision Analysis

A repertoire of techniques used in an attempt to improve the quality of decisions. The methods are usually based on a decision tree, a branching structure in which a problem is structured as an initial choice between two or more options with subsequent branches governed by chance. The terminal branches correspond to all possible outcomes arising as a result of the original choice. Each outcome has a value, not necessarily monetary, associated with it. By a process known as folding back, the choices can be ranked

according to the expected value (as a product of likelihood times total value) of each choice. Folding back involves evaluating each choice in terms of the sum of the products of probability × value for each of the terminal branches associated with that choice.

These techniques were originally developed for use in business, but have been increasingly applied to medical problems and provide an interesting means for converting basic evidence into rational decisions. The assumption is that a rational decision is always the best decision, an assumption that is not always warranted.

In an oversimplified analysis (figure 1), the choice is between treatment and observation. The probabilities of survival will be different in the two groups, as will the utilities, based on a scale of 1–100. The toxicity of treatment is reflected in the −30 units assigned to the utility of death in a treated patient and the 80 units given to survival in a treated patient. In a real analysis the utility values would not be handled in this crude way but the principle would be similar. By substituting the probabilities of survival after treatment and observation in the tree, the more effective policy can be defined. **Sensitivity analyses** (**p. 283**) can define the critical values, those for which there would be an abrupt change in the decision, for these

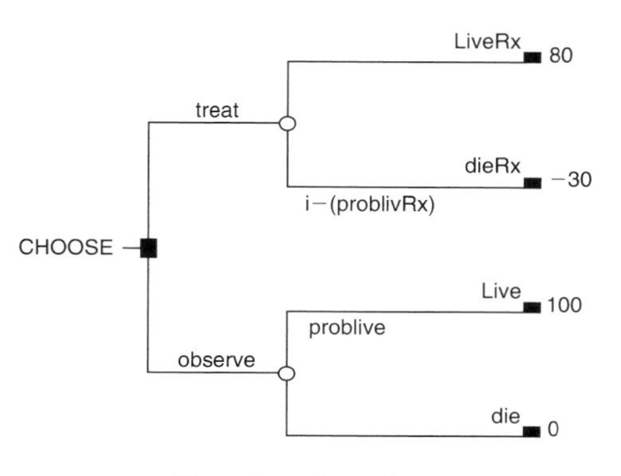

Figure 1 — An analysis.

probabilities. Evaluating the tree when the probability of survival after treatment is 0.9 and the survival probability after observation is 0.3 shows the expected utility of treatment to be 69 units and that of observation to be 30 units. The logical decision is, therefore, to treat. The decision will, however, change at some critical value for the probability of survival after treatment and this can be shown by **sensitivity analysis** (**p. 283**):

- problivR: = 0.00000 EU[treat] = −30.00000.

  EU[observe] = 30.00000 Choose: observe.
- problivR: = 0.10000 EU[treat] = −19.00000.

  EU[observe] = 30.00000 Choose: observe.
- problivR: = 0.20000 EU[treat] = −8.00000.

  EU[observe] = 30.00000 Choose: observe.
- problivR: = 0.30000 EU[treat] = 3.00000.

  EU[observe] = 30.00000 Choose: observe.
- problivR: = 0.40000 EU[treat] = 14.00000.

  EU[observe] = 30.00000 Choose: observe.
- problivR: = 0.50000 EU[treat] = 25.00000.

  EU[observe] = 30.00000 Choose: observe.
- problivR: = 0.60000 EU[treat] = 36.00000.

  EU[observe] = 30.00000 Choose: treat.
- problivR: = 0.70000 EU[treat] = 47.00000.

  EU[observe] = 30.00000 Choose: treat.
- problivR: = 0.80000 EU[treat] = 58.00000.

  EU[observe] = 30.00000 Choose: treat.
- problivR: = 0.90000 EU[treat] = 69.00000.

  EU[observe] = 30.00000 Choose: treat.
- problivR: = 1.00000 EU[treat] = 80.00000.

  EU[observe] = 30.00000 Choose: treat.

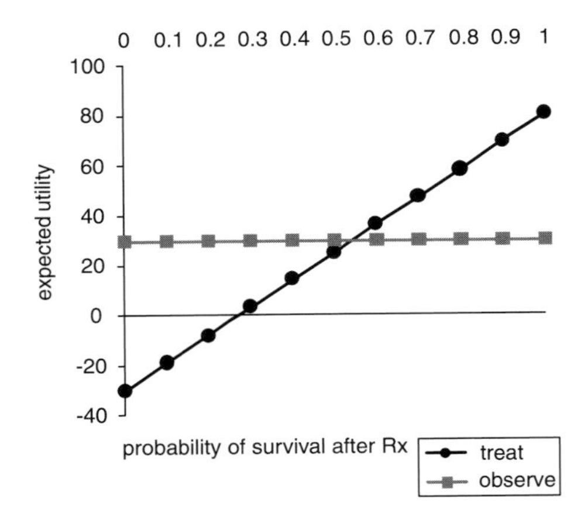

Figure 2 — Sensitivity analysis.

where problivR is the probability of survival after treatment and EU is the expected utility and indicates the value of the outcome, given the parameters used.

These data can be presented graphically (figure 2).

The **sensitivity analysis** (**p. 283**) shows that the decision flips from favouring treatment to favouring observation if the probability of survival after treatment falls to <0.55 (i.e. 55% cure rate).

Essential components of a decision analysis include: defining a formal structure for the problem; defining the probabilities associated with those events governed by chance; defining all possible outcomes for each choice analysed; assigning values, or utilities, to each outcome. A decision analysis is data-driven and a comprehensive review of the relevant literature is an essential first step. Through the use of a **Markov** (**p. 188**) process, essentially a reiterating series of decision cycles, the natural history of a disease or other time-dependent events can be modelled. This can be useful for assessing and comparing management policies for patients with cancer, whether for treatment or for follow-up.

Declaration of Helsinki

This statement on research ethics was first formulated in 1964. It arose out of concern, initially raised in the aftermath of the Second World War, about the legitimacy of experiments on human subjects. There was a desire to protect the rights of individuals from the abuses that had occurred during the Nazi era. The declaration has recently been criticised, mainly on the basis that it was devised to deal with the problems of medical research in the developed world and that, consequently, it did not address some of the issues of relevance to the developing world. A particular problem is the testing of drugs, e.g. treatments for **HIV** (**p. 151**), on patients in the developing world but then setting a price for the drug beyond the ability of Third World countries to afford. This relationship was seen as exploitative: with the developed world reaping all the benefit and the citizens of the developing world running all the risk. Similar concerns arose about the use of placebos in trials involving pregnant women who were known to be HIV-positive.

One version of the Declaration, and there have been several revisions and proposed revisions, is as follows:

I. Clinical research must conform to the moral and scientific principles that justify medical research and should be based on laboratory and animal experiments or other scientifically established facts.

 — Clinical research should be conducted only by scientifically qualified persons and under the supervision of a qualified medical person.

 — Clinical research cannot legitimately be carried out unless the importance of the objective is in proportion to the inherent risk to the subject.

 — Every clinical research project should be preceded by careful assessment of

inherent risks in comparison to foreseeable benefits to the subject or to others.

— Special caution should be exercised by the doctor in performing clinical research in which the personality of the subject is liable to be altered by drugs or experimental procedure.

II. Clinical Research Combined with Professional Care

In the treatment of the sick person, the doctor must be free to use a new therapeutic measure, if in his judgement it offers hope of saving life, re-establishing health, or alleviating suffering.

If at all possible, consistent with patient psychology, the doctor should obtain the patient's freely given consent after the patient has been given a full explanation. In case of legal incapacity, consent should also be procured from the legal guardian; in case of physical incapacity the permission of the legal guardian replaces that of the patient.

The doctor can combine clinical research with professional care, the objective being the acquisition of new medical knowledge, only to the extent that clinical research is justified by its therapeutic value for the patient.

III. Non-therapeutic Clinical Research

In the purely scientific application of clinical research carried out on a human being, it is the duty of the doctor to remain the protector of the life and health of that person on whom clinical research is being carried out.

The nature, the purpose and the risk of clinical research must be explained to the subject by the doctor.

Clinical research on a human being cannot be undertaken without his free consent after he has been informed; if he is legally incompetent, the consent of the legal guardian should be procured.

The subject of clinical research should be in such a mental, physical and legal state as to exercise fully his power of choice.

Consent should, as a rule, be obtained in writing. However, the responsibility for clinical research always remains with the research worker; it never falls on the subject even after consent is obtained.

The investigator must respect the right of each individual to safeguard his personal integrity, especially if the subject is in a dependent relationship to the investigator.

At any time during the course of clinical research the subject or his guardian should be free to withdraw permission for research to be continued.

The investigator or the investigating team should discontinue the research if, in her/his (or their) judgement, it may, if continued, be harmful to the individual.

Degrees of Freedom

This is not a simple concept. The origins of the concept of degrees of freedom (d.f.) lie in the idea that formulae used for comparative purposes require adjustment for the number of parameters within the equation that are estimated, as opposed to being directly measured.

Its use in practice is as follows:

- In an unpaired ***t*-test** (**p. 303**) it is the number of observations − 1.
- In a paired *t*-test it is the number of pairs of observations − 1.
- In a contingency table the degrees of freedom = (number of rows − 1) × (number of columns − 1). A four-way table, therefore, has 1 d.f.

For **chi-square** (**χ^2**) (**p. 62**) and *t*-testing the variability around the mean is an estimated parameter and is represented by the 1 in the −1 term in the calculation of degrees of freedom. We do not actually know the precise shape of the *t*-distribution or the χ^2 distribution for our particular data set — this depends on how much uncertainty there is in the measure of **standard deviation** (**SD**) (**p. 296**). We know the rough overall shape of the distributions and can correct for the uncertainty by using the concept of 'degrees of freedom'.

Delay in Starting Treatment

Delay in starting treatment for cancer is at best psychologically distressing for patients and at worst might seriously compromise cure. The natural history of a cancer can be divided into several phases. To some extent the boundaries between these phases are arbitrary and artefactual (**lead-time bias** (**p. 176**) is an example of this arbitrariness), but, nevertheless, the divisions are conceptually useful.

- Presymptomatic: **patient has malignant disease, but no symptoms**. This covers the period from the initial malignant transformation up to the point at which the disease produces sufficient physiological disturbance to be potentially recognisable. A symptomatic breast lump will contain at $\sim 10^9$ malignant cells. Occasionally, a tumour might present when there are fewer numbers of cells, often through non-metastatic manifestations, e.g. gynaecomastia due to HCG production by a germ cell tumour. **Screening** (**p. 278**) programmes are designed to detect cancer during this phase of its development: any failure in a screening programme, whether episodic or systematic, will cause delays in this phase. Often the delay is such that the patient presents with symptomatic disease — as an interval case. **Interval cases** (**p. 166**) will present even when a screening programme is meticulously carried out. This is because of **length–time bias** (**p. 177**): rapidly growing tumours cause symptoms in the interval between screening examinations.
- Symptomatic but preclinical: **patient has symptoms from malignant disease but has not sought medical advice**. This covers the period from the first subjectively detectable derangement due to the cancer up to the point at which patient seeks medical advice. The patient may dismiss bleeding caused by rectal cancer as due to piles and may not wish to trouble a busy GP with such a 'trivial' problem.
- Clinical but prediagnostic: **patient has symptoms, has sought medical advice but diagnosis has not been made**. The patient has brought significant symptoms to medical attention but the cancer has not been diagnosed. Delays can arise at this point for many reasons: symptoms are dismissed as 'trivial' by the primary care team; inadequate access to diagnostic facilities; wait for a hospital appointment; wait for a diagnostic test (endoscopy, X-ray, etc.); wait for hospital bed, operation, etc.
- Diagnosed but pretreatment: **diagnosis has been made but the patient has not started definitive treatment**. The cancer has been diagnosed, the patient knows the diagnosis but definitive treatment has not been carried out. Delays here involve problems of access to specialists (e.g. oncologists of whatever persuasion), problems of access to further investigations (CT scans MRI, etc.), problems of access to specialist treatment facilities (hospital bed, operating theatre time, radiotherapy treatment).

Implicit within the above scheme is the recognition that some delays in instituting cancer treatment are system-related, i.e. they reflect the organisation and delivery of healthcare services as a whole, and others are more personal, reflecting the behaviour and attitudes of individual doctors or patients. Any interventions designed to decrease delay in starting treatment for cancer need to address both these aspects.

A systematic approach can be taken to the problems associated with delay in the diagnosis and treatment of cancer:

Potential problems associated with delay in cancer treatment

- Tumour:
  - — decreased local control
  - — increased chance of distant spread
  - — increased complications
  - — destruction caused by tumour
  - — increased doses and treatment volumes.
- Patient:
  - — increased physical distress symptomatic for longer time with symptoms increasing in severity
  - — increased emotional and psychological distress.
- Society:
  - — shifts in practice, e.g. from XRT to surgery
  - — shifts in referral, e.g. to more distant centres
  - — cancer treatment in general perceived as futile and toxic — a vicious circle in which fear of toxic and ineffective treatment delays presentation
  - — economic costs of treating, but failing to cure, advanced disease are often greater than the costs of successfully treating early disease.

Table 1

Tumour doubling time (days)	Loss of local control (% day^{-1})
25	1.02
58	0.44
82	0.31

Data from Mackillop *et al.*: these are maximum figures (but how can we be sure that these maximal values might not apply to the patient we put on the waiting list 40 days ago?).

Table 1 shows a practical example drawn from a modelling study of potential loss of control associated with delay in starting radiotherapy for tonsillar cancer.

Dendritic Cells

Dendritic cells are a numerically inconspicuous, but functionally important, component of the immune system. They are of haemopoietic lineage and are involved in processing and presenting antigens to other components of the immune system. They may have a role in the immunotherapy of cancer since they might, by improving the presentation of tumour-related antigens, augment T-cell responses directed against tumours.

DerSimonian and Laird Method

A method used for combining trial results in a **meta-analysis (p. 196)**. The method is complex, both in principle and in calculation, but its essential feature is that it recognises that the treatment effect identified in each trial will vary randomly about a true mean representing the mean of the population of trials. This method is sometimes, therefore, referred to as a random-effects method, in contradistinction

to the fixed-effect assumption that underlies the **Mantel–Haenszel (p. 187)** procedure for pooling trial results. The DerSimonian and Laird method explicitly accepts a degree of heterogeneity among the results of the various trials — its conclusions, therefore, tend to be more conservative than the Mantel–Haenszel method.

Desmogleins

Desmogleins are **adhesion molecules (p. 8)** structurally related to desmosomes and which form part of the anchorage system for the intermediate filaments of the cytoskeleton.

Desmoplastic Small Round Cell Tumours (DSRCT)

DSRCT enter the differential diagnosis of **small round cell tumours (p. 289)**. They present, usually in males aged 15–25 years, as a painful intra-abdominal mass with peritoneal nodules. The tumours respond poorly to treatment and behave aggressively. There is a characteristic translocation t(11;22)(p13;q12) that generates a fusion gene which encodes for a protein in which the RNA-binding domain of the *EWS* gene product, an RNA polymerase, is replaced by the DNA-binding domain of the *WT*1 gene product. This abnormality is similar to, but distinct from, that associated with Ewing's sarcoma and PNET.

Details

'God dwells in the details', a remark attributed to Ludwig Mies van der Rohe (the modernist architect), among others, is as relevant to assessing science as it is to the scrutinisation of the blueprint of a building's plan and section. The remark serves as a reminder that the processes of fusion, condensation and agglomeration intrinsic to many statistical and experimental methods may obscure important details. For example, the data published in the breast overview suggested that **adjuvant treatment (p. 8)** with **tamoxifen (p. 304)** did not significantly improve survival in premenopausal women with breast cancer. This part of the analysis might, because of the **arithmetic fallacy (p. 21)**, be somewhat unreliable.
A re-analysis of the data for tamoxifen in premenopausal women did, in fact, suggest a survival benefit. Attention to detail is important, but it is also important simply not to dismiss **exceptions (p. 122)** or counterintuitive results as merely artefactual.

'The devil is in the details' — borrowed from the law — points to the difficulties of getting everything right: in science a lack of attention to detail could render a whole investigation worthless. The concept of PCR (**polymerase chain reaction**) **(p. 238)** took only a weekend for Kary Mullis to formulate, sometime in the spring of 1983. A convincing working version of the technique took a long time, it was March 1985 before a reproducible technique was developed: the time was spent getting the details right.

Determinism

In biology, the concept is that the fate of an organism is determined by a pre-existing set of instructions. In its crudest expression, the approach would assume that a person's destiny was entirely dictated by the nucleotide sequence in their DNA. The concept crops up in a variety of disguises: in E. O. Wilson's writings on socio-biology; in Richard Dawkins's notions concerning selfish genes and the origins of altruistic behaviour. It is a form of **reductionism (p. 263)** and is in direct opposition to concepts such as **vitalism (p. 325)**. The idea that all of cancer can be understood in terms of the molecular biology of genes and DNA is a deterministic belief that, so far, has produced little in the way of practical benefits.

Deterministic (Non-Stochastic) Effect

When the severity of an effect is directly proportional to dose, then it might be assumed that a direct causal relationship exists between dose and effect. There may be a threshold dose below which the effect is not observed, but thereafter increasing the dose increases the damage in a fairly predictable fashion. A **non-stochastic effect** (**p. 214**) is one for which severity is directly dependent on the dose of the causative agent. An example would be the dose-dependent cardiotoxicicty of **anthracyclines** (**p. 16**): there is a rapid increase in the severity of cardiomyopathy with total doses of **doxorubicin** (**p. 107**) >500 mg m^{-2}.

Dexrazoxane

This drug is used as a chemoprotective agent to mitigate the cardiotoxicity of anthracycline therapy. Chemically, it is a cyclic version of the chelating agent EDTA. Anthracyclines cause myocardial damage by forming superoxide **free radicals** (**p. 132**) that damage the mitochondria by peroxidating the lipid in the inner membrane. These superoxide radicals are more readily formed in the presence of iron and the dexrazoxane will inhibit the formation of superoxide radicals by chelating iron. To be effective, the dexrazoxane must be administered 15–30 min before the anthracycline. Dexrazoxane will potentiate any anthracycline-induced myelosuppression; otherwise it has little serious toxicity.

Diagnosis and Screening: The Algebra of Testing

The performance of a test used for diagnosis or screening can be put into a four-way table, expressed in a general algebraic form (table 1).

Table 1

Test result	Disease present	Disease absent	Total
Positive	a	b	a + b
Negative	c	d	c + d
Total	a + c	b + d	a + b + c + d

Note: by convention the disease present/test-positive cell is always placed at the top left of the table.

The percentage of patients who have the disease present and whose test result is positive is given by a/(a + c): this is the **sensitivity** of the test (other names include TPR, for true positive rate, and PiD, for positive in disease). The false-negative rate (FNR) = 1 − sensitivity: FNR indicates the proportion of patients whose test is negative, but who actually have disease.

The percentage of patients who do not have disease and who have a negative test is given by d/(b + d) and is termed the **specificity** of the test (other names include TNR, for true negative rate, and NiH, for normal in health). The false-positive rate (FPR) = 1 − specificity.

These concepts are computationally both simple and useful but they are difficult to reconcile with how we have to deal with the problem of diagnostic testing: calculating vertically, as for both sensitivity and specificity, implies that we already know who has, and who does not have, disease. But if we knew this we would not need to perform a diagnostic test: in the real world the data arrive horizontally, in terms of test results defined as either positive or negative. There are useful indices that adopt this orientation: the **positive predictive value** (**p. 242**) is calculated as a/(a + b), that it is the percentage of patients who have the disease who have a positive test result. The corollary is the proportion of patients who do not have disease who have a negative test, d/(c + d), the **negative predictive value**.

Table 2

Marker	Disease present	Disease absent
LDH 'positive'	82	35
LDH 'negative'	96	617

All this seems very simple and it would seem that all we need to do is to define thresholds in such a way that sensitivity and specificity are as high as possible. Unfortunately **PPV** (**p. 242**) and NPV both depend on the prevalence of the disease in the population under investigation. As prevalence falls, it is a simple consequence of the mathematics that PPV will fall and NPV will rise. This is most easily shown in a practical example (table 2), where real data on LDH as a tumour marker in testicular cancer are used.

Based on these data, the indices mentioned above can be calculated as:

- Sensitivity: 46%.
- Specificity: 94%.
- PPV: 70%.
- NPV: 86%.
- Prevalence: 21%.

Now imagine that LDH, with exactly the same specificity and sensitivity, is applied to a population in which only 2.1% of patients actually have disease.

- Sensitivity: 46%.
- Specificity: 94%.
- PPV: 14%.
- NPV: 99%.
- Prevalence: 2%.

PPV has fallen to only 14% (that is only 14% of patients with a test defined as positive actually have disease), and NPV has risen to 99% (only 1% of patients who have a negative test will actually have disease).

Apply LDH as a marker to a population with a high **prevalence** (**p. 244**) of disease, say 42%:

- Sensitivity: 46%.
- Specificity: 95%.
- PPV: 86%.
- NPV: 71%.
- Prevalence: 42%.

Now that the prevalence has risen the test is much better at identifying patients with disease: 86% of patients with a raised LDH actually have disease. The problem is that a negative test is no longer so good at excluding disease: nearly 30% of patients whose LDH is normal will turn out to have disease. This is because LDH has low sensitivity.

It is a function of the mathematical interrelationships between sensitivity, specificity and prevalence that a diagnostic test will be most useful in confirming or refuting the presence of a disorder when the prevalence of that disorder is between 40 and 60%. Here are data, using AFP, a somewhat more sensitive marker for NSGCT that LDH:

- Sensitivity: 71%.
- Specificity: 98%.
- PPV: 97%.
- NPV: 77%.
- Prevalence: 50%.

PPV and NPV are appreciably higher than for LDH but the low sensitivity means that NPV is still rather low.

Here is how a test that is 95% sensitive and 99% specific will behave when disease prevalence is 50%:

- Sensitivity: 95%.
- Specificity: 99%.

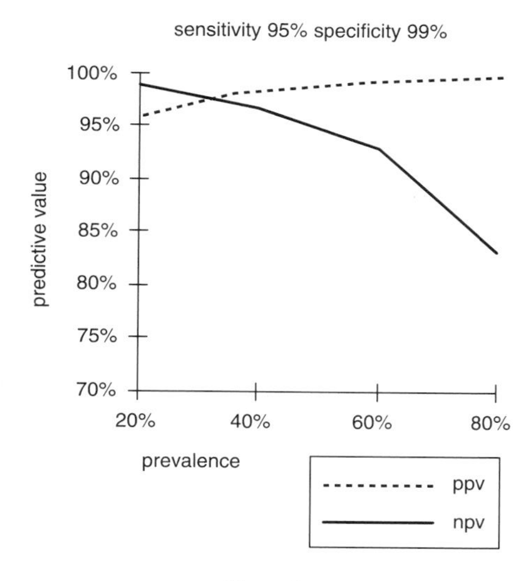

Figure 1.

- PPV: 99%.
- NPV: 95%.
- Prevalence: 50%.

When the prevalence is 50%:
PPV = specificity and NPV = sensitivity.

A graph showing the interrelationship between prevalence, PPV and NPV is given in figure 1.

There are two strategies that are useful in attempting to obtain as much predictive value as possible from a diagnostic test:

1. Use some preliminary procedure to select the population so that the prevalence of disease is likely to be between 40 and 60%.
2. Limit the scope of enquiry so that test performance will have sensitivity and specificity of ≥90%.

Strategy 1 is that used by experienced clinicians when narrowing down diagnostic possibilities so that one pathognomonic test will confirm or refute a particular diagnosis. It is rarely possible to do this with tests used for screening healthy populations.

Strategy 2 is less commonly employed but may be useful in certain circumstances. Although AFP is elevated in only ~70% of patients with active non-seminomatous germ cell tumours, there are certain subtypes, e.g. those containing yolk-sac elements, where AFP is almost always elevated. If the question about active disease were rephrased as 'does this patient have active disease containing yolk-sac elements?', then the sensitivity of AFP will be much improved. Unfortunately, any gains, in terms of improved sensitivity, may be offset by losses in terms of decreased prevalence, since only a subset of these tumours contain yolk-sac elements.

Although this discussion has dealt almost entirely with serum markers, the same logical process applies to any test (scan, blood test, functional test) used in diagnosis and screening: the origins of sensitivity and specificity can be traced to studies performed in the late 1940s, by Yerushalmy and others, on the interpretation and predictive power of plain radiographs.

How might some of the above considerations apply to the clinical practice of oncology?

1. Maintain a degree of scepticism about new marker tests or pathognomonic investigations: look carefully at the statistical evidence for such claims.
2. Consider using different cut-off values for different clinical circumstances, e.g. screening (where prevalence is low) versus staging (where prevalence is likely to be much higher). An example might be the use of Ca125 in screening for ovarian cancer as opposed to the interpretation of a Ca125 level in a patient who has just had an apparently curative resection.
3. In some circumstances change over time may be more useful than absolute values. Patients can be used as their own

controls, e.g. rising CEA levels in patients followed after surgery for colorectal cancer.

4. Combinations of markers may be more useful than values for individual markers, e.g. PLAP and LDH considered together in patients with seminoma.

5. We probably waste a lot of time and effort collecting data that are both meaningless and uninterpretable.

6. False-positive tests sometimes have an identifiable explanation: smoking (especially cigars) causes elevated PLAP; benign liver disease, e.g. binge drinking, can cause transient elevations of AFP.

Diagnostic Odds Ratio (DOR)

This statistic can be derived quite simply from a 2 × 2 table summarising the diagnostic performance of a test. DOR is, in words, the odds for a positive test result in persons known to have the disease divided by the odds of a positive test result in patients known to be disease free.

Algebraically, this is identical to:

$$\frac{(\text{sensitivity}/[1-\text{sensitivity}])}{([1-\text{sensitivity}]/\ \text{specificity})}.$$

It is an expression, like the **odds ratio** (**p. 217**) itself, that has benefits in terms of computational ease and statistical manipulation but which has major disadvantages in terms of comprehensibility either as a mathematical or as a verbally defined concept.

Diploid Cell

A diploid cell is one that contains two homologous sets of chromosomes (2n).

Disintegrins

A family of proteins that disrupt the activity of integrins. The family includes **contortostatin** (**p. 75**), extracted from the venom of the copperhead snake. By interfering with **invasion** (**p. 167**) and **metastasis** (**p. 198**), as well as **angiogenesis** (**p. 14**), the disintegrins have therapeutic potential.

DNA Damage

The DNA molecule is ~2 nm wide. Radiation characteristically distributes damage unevenly. One result is locally multiply damaged sites: the large blob contains 12 ion pairs and is 7 nm in diameter, the small cluster contains three ion pairs and is 4 nm in diameter. Other forms of damage include **intercalation** (**p. 163**) and cross-links between strands (interstrand) or within strands (intrastrand). Cross-links may also occur between the histones and the DNA, and between histones. All of these abnormal linkages will interfere with the function of the DNA.

There are three overlapping phases in the response of cells and tissues to radiation:

- Physical ($10^{-20} - 10^{-9}$ s).
- Chemical ($10^{-12} - 10^{2}$ s).
- Biological ($10^{0} - 10^{9}$ s).

The yard was, originally, the distance between the tip of the king's nose and the fingertips of his outstretched arm. Now the metre is used. If we takes a metre as equivalent to the human lifespan (~10^9 s) then the relative time intervals can be converted to length:

- Physical — smaller than a subatomic particle.
- Chemical — about the size of a red blood cell.
- Biological — the whole metre.

As in cosmology, the events of the first miniscule intervals of time have effects of comparatively immense duration.

Physical phase: produces fast electrons and ionisation

Type of ionization	No. Gy^{-1}	Energy (eV)	Target size (nm)
Sparse	1000	10–40	2
Moderate	20–100	100	2
Large	4–100	400	5–10
Very large	0–4	800	5–10

DNA damage Gy^{-1}/cell:

- 10 000 damaged bases.
- 1000 damaged sugars.
- 1000 single-strand breaks.
- 40 double-strand breaks.
- 150 DNA–protein cross-links.
- 30 DNA–DNA cross-links.

Imagine that a DNA molecule could be magnified so that a base pair occurred every yard:

- Human genome is 5.5 million miles long.
- DNA molecule is 6 yards wide.
- Smallest radiation cluster is the size of a large room.
- Largest radiation cluster is the size of a large house.
- Nucleosome is the size of a country mansion and contains 200 yards of DNA.

After 1 Gy radiation:

- Small or moderate cluster will occur every 5000 miles.
- Large or very large cluster will occur every 50 000 miles.
- Double-strand break will occur every 120 000 miles.
- Single-strand break will occur every 4800 miles.
- Damaged base will occur every 550 miles.
- Damaged sugar will occur every 5500 miles.
- DNA–protein cross-link will occur every 37 000 miles.
- DNA–DNA cross-link will occur every 180 000 miles.

Single-hit killing is easy to imagine; multiple hits adjacent to a single target are less easy to envisage.

Chemical phase: damage to critical molecules

- Direct damage to DNA via molecular excitation (1/3 of damage).
- Indirect damage to DNA via the radiolysis of water into **free radicals** (**p. 132**) and hydroxyl ions, etc. (2/3 of damage).

The ways in which radiation-induced DNA damage might be modified can be summarised as:

- Modification:
  - — mop up free radicals (scavengers), e.g. 10% EtOH, DMSO
  - — donate H to DNA, e.g. glutathione, cysteamine
  - — rapid enzymatic degradation of radicals, etc., e.g. catalase, peroxidase, **superoxide dismutase** (**p. 300**)
  - — increased vulnerability of certain sites (e.g. peripheral).
- Repair: fidelity is crucial — rejoining and repair are not equivalent

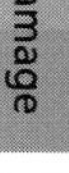

- — some sites are more rapidly repaired than others (e.g. preferential repair of transcribed regions)
- — time available: base damage, single-strand breaks need <1 h; double-strand breaks may need up to 4 h
- — genes (*AT*, *XP*).

- Inhibitors (e.g. 3 aminobenzamide).
- **Cell cycle** (**p. 52**) phase (peaks in resistance during G_1 and at S/G_2 border).
- Post-irradiation conditions (PLD repair).
- Effects of radiation mediated and/or modified by:
  - — direct physicochemical damage to DNA
  - — physicochemical effects on membranes
  - — effects on cytoplasmic receptors — NF-κB
  - — effects on signal transduction — kinase cascades
  - — effects on cell cycle control — *AT* gene product, *p53*, *GADD* genes
  - — effects on **growth factors** (**p. 144**)
  - — effects on cellular programmes, e.g. apoptosis
  - — microenvironmental factors — oxygen, nutrients, ebbs and flows
- Repair processes can result in
  - — complete repair with fidelity-clonogenic survival
  - — misrepair: mutation
  - — incomplete repair; change in clonogenicity
  - — no repair: loss of clonogenicity

DNA Library

A DNA library is a resource for the investigation of DNA sequences and genes. It consists of a large collection of DNA, from a given source (species, cell line or tissue), which has been cloned and preserved. Cloning uses the standard vectors, phage, **cosmids** (**p. 76**), etc. Larger fragments can be cloned using **yeast artificial chromosomes** (**YAC**) (**p. 330**). The DNA clones can be indexed, extracted and amplified for study. The disadvantage of this approach is that if one laboratory wishes to study a clone held in another laboratory, the vector and clone have to be physically transported between laboratories. The concept of **sequence-tagged sites** (**STS**) (**p. 284**) means that sequences that act as markers can be stored in computerised form and laboratories can share simply by communicating sequence data without the need to transport anything other than information. A cDNA library uses complementary DNA synthesised using mRNA as the template.

Interrogating and using a DNA library is not necessarily straightforward. There have to be procedures for examining the whole library for DNA sequences identical to that which is of interest and for ascertaining whether a newly entered sequence has any similarity to a previously known gene or sequence. Human DNA libraries, and their analysis, are essential components of the Human Genome Project. Availability and accessibility are essential if data held on DNA sequences are to have practical benefit. Problems of access and dissemination are primarily concerned with **database** (**p. 87**) design, and compatibility together with the practical application of information technology. There are at least 20 different databases of relevance to questions concerning DNA sequences, protein structures and functions. Until and unless all these resources can be combined and interrogated in real time, there will be a limiting bottleneck in the application of basic knowledge to practical questions.

The Human Genome Project illustrates the size of the problem in terms of handling information: there are between 30 000 and 40 000 genes in the human genome. Genes vary considerably in length, interrupted genes are common and although ~50% of human genes are between 2000 and 10 000 kb, some are >100 000 kb. Genes themselves account for only ~5% of the human genome, 95% of the DNA is apparently silent. There are 3×10^{12} nucleotides in the human genome. Now that 92% of the entire genome is fully sequenced it is no trivial task to index and match sequences for study to the genome as a whole.

DNA Polymerases

DNA polymerases are enzymes involved in the **replication** (**p. 267**) of DNA. They act by adding nucleotides, one at a time, to the 3′-end of the growing polynucleotide chain.

DNA Sequencing

The two main methods for ascertaining the sequence of nucleotides in a length of DNA were described almost simultaneously in 1977. The Maxam–Gilbert method, which is now less widely used, cleaves the DNA at specific bases in such a way that, on average, 1% of susceptible bases per DNA molecule are modified and cleaved. Dimethysulphate modifies the **purines** (**p. 252**); heating breaks the chains at guanine. Hydrazine treats the pyrimidines, exposure to high salt concentration causes cytosine, but not thymine, to react. The DNA is radiolabelled at one end before exposure to the chemical reaction. The cleavage fragments are run on an electrophoresis gel and the bands on the gel correspond to different lengths of DNA. The procedure is repeated for each of the bases and the ladders of fragments can be compared and the sequence in the original DNA can thus be reconstructed. This method allows ~300 nucleotides to be sequenced at a time.

The alternative method, described by Frederick Sanger, relies on synthesis rather than fragmentation. The DNA is first annealed with a labelled **primer** (**p. 246**). Dideoxy derivatives of the four nucleotides are prepared. The effect of the dideoxy group is to prevent further synthesis of DNA if the modified base is incorporated into the growing molecule. Reaction mixtures, each with a small amount of the dideoxy derivative of that base, are set up for all four bases. Four separate synthetic reactions are performed, one for each base. Each base-specific reaction will produce a series of DNA fragments of varying lengths all terminated by the dideoxy derivative of that base. The products of each of the four reactions are run on separate gels, which are then put into register with each other and the sequence can be read by zigzagging across the four parallel gels. Automated DNA sequencing equipment is now available and the speed of this equipment has dictated both the pace and cost of the Human Genome Project.

DNA Strand Breaks

There are two main types of DNA strand break following radiation: double- (DSB) and single- (SSB). These can be identified using a variety of assays:

- Electrophoretic methods:
  - **pulsed-field gel electrophoresis** (**p. 251**)
  - single cell gel electrophoresis.
- Sedimentation methods:
  - sucrose gradient sedimentation
  - nucleoid sedimentation.
  - Elution: neutral filter elution.

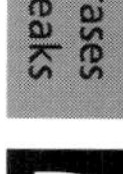

D_0 Dose

The D_0 dose is used in radiobiology to indicate the sensitivity of a cell, or cell type, to radiation. It is the dose that kills 63% of cells. After a D_0 dose the lesions per cell are:

- 1000 bases damaged.
- ~1000 single-strand breaks.
- 30–50 double-strand breaks.

Docetaxel

Docetaxel is a **taxane (p. 305)** with activity and toxicity similar to that of **paclitaxel (p. 226)**. Structurally, the differences between the two drugs are relatively minor. It is a more potent inhibitor of depolymerisation of the microtubules than paclitaxel, it is more active in S-phase and less active in M-phase. This is in contrast to paclitaxel, which is most active in M-phase. It causes more fluid retention than paclitaxel and skin eruptions are more common. It may also, particularly when combined with **vinorelbine (p. 325)**, produce colitis and typhlitis. It is active against a similar range of tumours, including non-small cell lung cancer.

Domain

A domain is a part of a molecule that is functionally or structurally distinct. In organisational terms it is one level of complexity up from a **motif (p. 207)**: if a protein were a neighbourhood then each house would be a domain and the various constituents of the houses (windows, doors, floors, drainpipes, etc.) would be motifs. A domain is a basic structural unit. Certain domains are associated with specific functions and, by domain shuffling, proteins may evolve through the joining of pre-existing domains. A knowledge of how to recognise particular domains, their structure and function, can save much time when attempting to deduce the structure and function of a protein from a knowledge of its amino acid sequence. Domains are typically 50–350 amino acids in length. Complex proteins contain many domains whereas simpler proteins may consist of but a single domain.

Dominant Negative Effect

This is the effect, classically shown by ***p53* (p. 225)**, whereby the expression of the mutant gene product overwhelms the activity of any wild-type protein present. When paired with a wild-type allele, the effect of the mutant gene is dominant. A mutation of this type is termed *trans*-dominant. In the case of *p53* the relative ineffectiveness of wild-type *p53* leads to a failure of tumour suppression. The dominant negative effect is a potential barrier to the **clinical effectiveness (p. 69)** of gene therapy for cancer. The presence of abundant quantities of the mutant gene will swamp the activity of the wild-type gene that has been so painstakingly reintroduced.

Dose

This is not a straightforward concept in oncology. Dose simply describes the quantity of an administered agent, drug or radiation. It is expressed in simple numerical terms: Gy radiation, mg drug. So far, so simple. The problem is that the numerical dose is not what really matters. It is the biological effects on the living organism that are important and, for any given dose, these will vary considerably between individuals and, over time, for any one individual. This is the problem of **heterogeneity (p. 150)**.

Dose-Distribution

This term is used in radiotherapy planning to indicate the spatial distribution of dose within the **irradiated volume** (**p. 169**). It is constructed by summing the **isodoses** (**p. 169**) for the beams used in the treatment. The ideal dose distribution is completely homogeneous across the tumour, with no variation in dose across the **planning target volume** (**p. 235**), and it completely excludes normal tissues from irradiation. This is an impossible ideal and all treatment plans are based on some compromise in terms of what is judged to be a clinically acceptable dose-distribution. The ICRU recommends that any dose variations across the treatment volume should be within the range −5 to +7%. Even this is hard to achieve: one study of radical treatments for lung cancer showed that, in the real world, the dose range was −50 to +18%. **Dose–volume histograms** (**p. 105**) are a useful means for quickly inspecting and comparing the performance of different treatment plans and dose-distributions.

Dose Intensity

This implies that it is not just total dose that produces biological effect but that the overall time over which a treatment is given is important. For any given dose, the shorter the time, the greater the treatment intensity. In radiotherapy this principle underlies the technique of **accelerated fractionation** (**p. 1**). In chemotherapy, dose intensity can be expressed as dose unit $time^{-1}$ (e.g. mg doxorubicin $week^{-1}$). The assumption is that greater dose intensity will produce more therapeutic benefit: the truth of this assumption has only infrequently been proven.

The concept of dose intensity dictates that as much treatment as possible should be given within as short a time as possible. The obvious implication is that any breaks in treatment to allow recovery of normal tissue should be kept to the absolute minimum — indeed such an approach implies going to the limits of acceptable toxicity: 'I will take you to edge of the cliff, I will push you off, but I will give you a hand as you try to claw your way back up.' The use of **growth factors** (**p. 144**) and other methods for ameliorating toxicity to normal tissues have enabled regimens of increasing dose intensity to be given safely.

Dose Modification Factor

A relatively crude method of assessing whether a particular manipulation improves the effectiveness of treatment. The concept is simple:

$$\text{Dose modifying factor} = \frac{\text{dose required for given effect in absence of modifier}}{\text{dose required for same effect in presence of the putative modifier}}$$

The problem with the approach is that it is only applicable to one level of effect, the extent to which dose modification occurs may well be dose-dependent and choosing one particular dose–effect combination at which to make the assessment could prove misleading. The OER (**oxygen enhancement ratio**) (**p. 221**) is an example of a dose-modifying factor — in this case the modifying factor is oxygen and radiation is the treatment with which it interacts. OER will not, however, be constant for all tissues and for all fraction sizes. Even *in vitro* there are problems: the **OER** (**p. 221**) for G_2 cells has been estimated to be ~2.3, for S-phase cells it is nearer to 2.9. In high-dose experiments an OER = 2.5–3.0 is assumed; for the doses used in clinical radiotherapy, 2.0 is probably more realistic.

Dose–Response

The relationship between dose and response is crucial to understanding both the origins

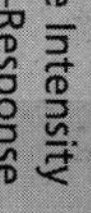

of cancer and its treatment. Environmental agents, such as ionising radiation, can cause cancer. Knowledge of the relationship between the dose (of the environmental toxin) and the response (probability of developing cancer) is necessary not only just to give sensible advice on minimising the risks to the population, but also in teasing out mechanisms of carcinogenesis. The distinction between **stochastic** (**p. 298**) and **deterministic** (**p. 214**) (non-stochastic) relationships is particularly important.

The concept of **selective toxicity** (**p. 282**), which is the basis of most treatment for cancer, can be expressed as a dose–response relationship. Curves can be constructed, both for tumour control and for normal tissue damage, in which increasing dose is plotted against probability. These curves can be generated mathematically using classical Poisson statistics.

At very low doses there is neither tumour control nor damage to normal tissues. At intermediate doses the probability of tumour control rises steeply but so also does the probability of damaging normal tissue. At very high doses all tumours are controlled, but there is also the absolute certainty of damage to normal tissues. The relative relationship between the two curves, that for tumour control and that for morbidity rate, will govern, in any given set of circumstances, an optimal dose. The most direct way to focus on this particular issue is to use an unequivocal endpoint for normal tissue damage: treatment-related death is fairly unequivocal. Using the dose–response formalism, one can then plot, for a range of doses, both tumour control and death from complications related to treatment (figure 1).

In the simplest situation there is clear separation of the two curves, the curve for the morbidity rate being parallel to that for tumour control but shifted to the right.

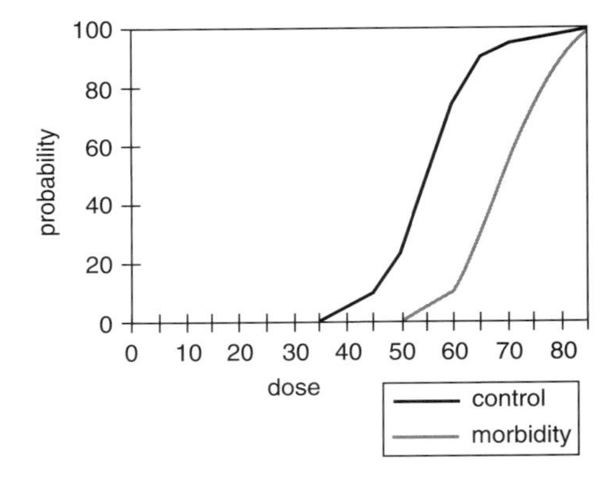

Figure 1 — Dose–response relationship that approaches the ideal for successful cancer treatment.

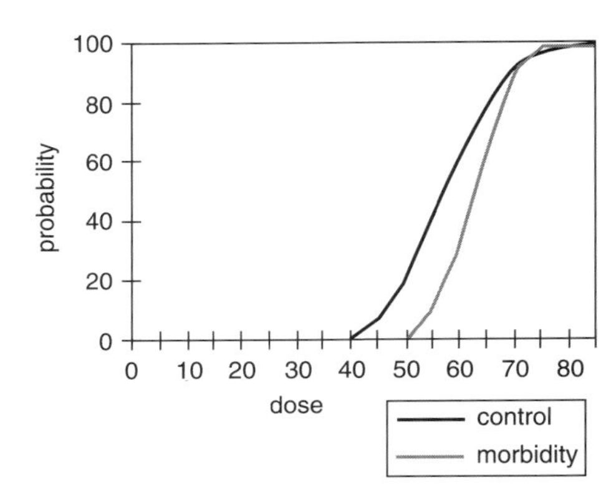

Figure 2 — Dose–response relationships for tumours and normal tissues are rarely parallel and are not often cleanly separated.

Choosing the dose becomes straightforward. A dose of ~55 will cure nearly 80% of patients and only ~5% of patients will die as a direct result of treatment.

Reality is, however, different. Curves are often not parallel and the curve for morbidity rate may not be as far to the right of the tumour control curve as one might wish. The decision about the optimal dose is, therefore, less straightforward (figure 2).

Matters are, in fact, even more complicated. We do not really have adequate data on dose

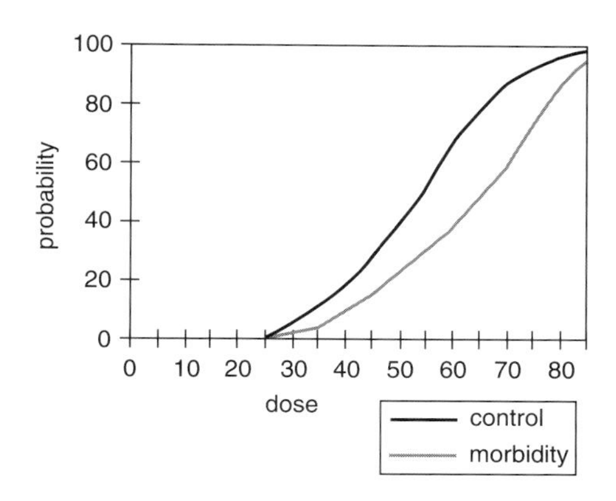

Figure 3 — Heterogeneity flattens dose–response curves.

and response: and it is unethical consciously to set out to obtain such information. We cannot knowingly under- or over-treat patients. We have to rely, therefore, on partial information and then try to project and extrapolate into a full dose–response relationship. The other problem is that of morbidity. An easy endpoint is chosen (treatment-related death) in the examples above, but this is way beyond where we would wish to operate. We are interested in more subtle, softer, endpoints: mucositis, diarrhoea, skin erythema, as examples of early reactions to radiotherapy; dry mouth, subcutaneous fibrosis, telangiectasia as examples of late effects. Once again, we have to consider the variability in patients, in tumours, and their responses to treatment. The effect of this variability is to flatten dose–response curves and this adds further to the difficulties in attempting to define a dose that will clearly separate the desirable effects of treatment (in terms of tumour control) and the undesirable effects (in terms of damage to normal tissues) (figure 3).

The question often asked is: what dose (of radiation or drug) is needed to cure cancer? There is no answer to this question. The appropriate dose will depend on many factors including the type of tumour, its site, and the general condition and age of the patient. The only possible answer to the original question is non-specific: the dose is that which will permanently eradicate the tumour without causing excessive damage, either in the short- or long-term, to the normal tissues. It is really the biological effect of a treatment that is important, rather than the dose itself.

Dose–Volume Histogram (DVH)

A technique used in radiotherapy treatment planning as a visual means of checking the **homogeneity (p. 152)** of dose across the target volume as well as of assessing the dose to critical normal tissues. When dose is assessed in three dimensions, then the term 'dose–volume histogram' is appropriate. If only one plane through the volume has been quantified then the correct term is 'dose–area histogram'.

There are two main methods for plotting DVH. The direct method, which is little used clinically, simply gives a frequency distribution of point doses within the tumour. The more useful technique is to use a cumulative DVH in which the percentage of volume (or area) receiving less than a given dose is plotted against dose (or dose as a percentage of mean target dose) (figure 1).

In the following example (figure 2), two plans, a and b, are compared using cumulative DVH firstly for the target volume.

And, second, for a dose-limiting normal tissue (figure 3).

The cumulative DVH show that Plan a is clearly superior both in terms of homogeneity of dose to the tumour, the whole volume is irradiated to within ±5% of the mean target dose, and in limiting the dose to the critical normal tissue.

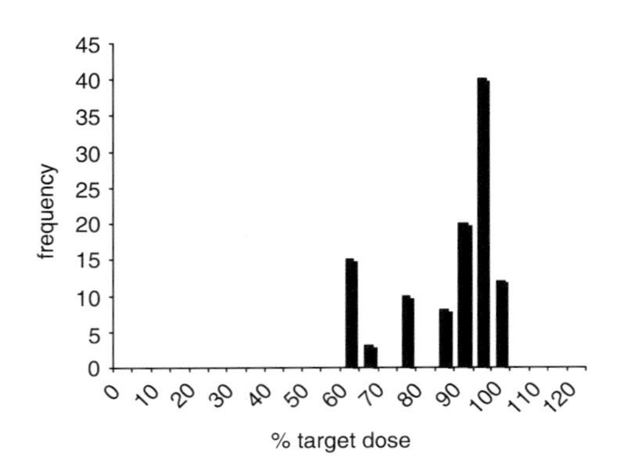

Figure 1 — Direct DVH for an inhomogeneously irradiated tumour (Plan b in figure 2).

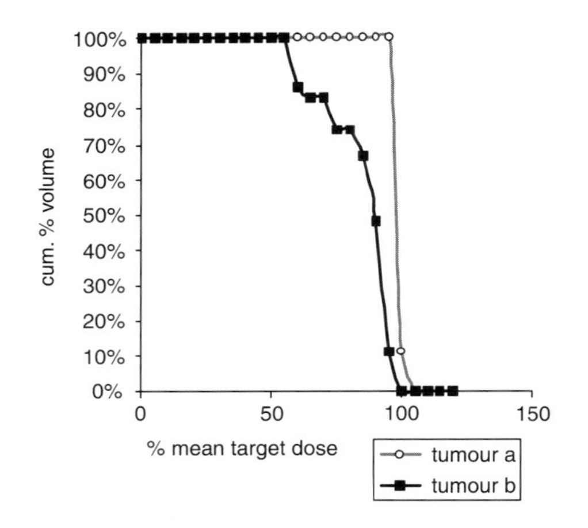

Figure 2.

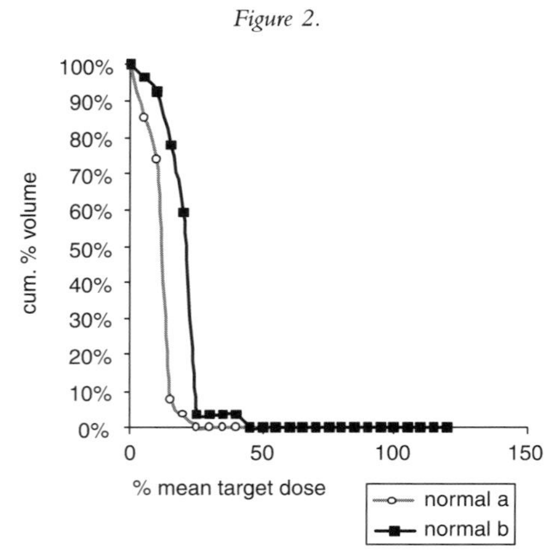

Figure 3.

With Plan b >50% of the critical volume receives 25% of mean target dose, whereas with Plan a, <5% receives a dose as high as this. All this is an oversimplification of course. In the real world the choice would be between a plan with the target DVH of Plan a, but the critical tissue DVH of Plan b, and a plan with the target DVH of Plan b but the rather better critical tissue DVH of Plan a. The art of clinical radiotherapy lies in intuitively making this type of trade-off correctly: the challenge for the future is to make such artistry reproducible, by putting it on an objective scientific footing.

Dot Blot

A rapid method for detecting specific sequences of nucleic acid. The dot of liquid containing the test sample is filtered, under suction, through a membrane. The relevant probe is then hybridised using the same technique as for **Southern blotting** (DNA) (**p. 292**) or **Northern blotting** (RNA) (**p. 215**). The technique can, by setting it up against standard samples of known concentration, be used quantitatively.

Double Minute Chromosomes (DMC)

These are not true chromosomes since they lack chromosomal structure. They are formed by gene amplification. The best example is amplification of the dihydrofolate reductase gene in cells exposed to **methotrexate** (**p. 199**). The multiple copies of the gene can confer resistance to the drug. When the drug is withdrawn the DMC disappear and the cells regain susceptibility.

Double-Time Gene

A gene that regulates **circadian rhythms** (**p. 65**) in *Drosophila*. Its product, the double-time protein, phosphorylates other clock proteins (e.g. PER).

Doubling Time

The time it takes for the volume, or number of cells, in a tumour to double in size or number. These parameters are not the same. The presence of stroma within a tumour means that the time taken for the volume to double will not be the same as the time taken for the number of cells to double.

Down-Regulation

The process by which continued exposure to a stimulus causes failure of response to that stimulus. Mechanistically, repeated exposure of a receptor to its agonist causes a decrease in the number of receptor molecules — the agonist is, therefore, no longer able to produce its customary effect. A clinically useful example is the down-regulation of gonadotrophin secretion that occurs following long-term administration of an **LHRH (p. 178)** analogue. This is exploited as chemical castration in the management of breast cancer and prostate cancer.

A variety of mechanisms can produce down-regulation: decreased **transcription (p. 313)** of genes coding for receptor proteins; increased endocytosis, and subsequent intracytoplasmic degradation, of receptors; formation of unstable mRNAs when the receptor gene is transcribed.

Doxorubicin

Doxorubicin is an **anthracycline (p. 16)** derived from *Streptomyces peucetius* var. *caesius*. It is used in the treatment of haematological malignancies and many solid tumours. It acts through a variety of mechanisms, the most important of which is probably **intercalation (p. 163)**. By inserting itself between the base pairs, and thereby changing the shape of the DNA molecule, it can inhibit both **DNA polymerase (p. 101)** and RNA polymerase. Doxorubicin also acts as a topoisomerase II inhibitor. By stabilising the complex between DNA and topoisomerase II, doxorubicin produces double-strand breaks.

Free radicals (p. 132) are produced as a result of the interaction between doxorubicin, iron and oxygen. These radicals may produce membrane damage as well as damage to DNA. Doxorubicin is cardiotoxic and the effect is related to the cumulative dose: total doses >550 mg m^{-2} incur an unacceptable rate of cardiac damage, tolerance is lower in patients who have radiotherapy to fields including the heart or who have been treated with **cyclophosphamide (p. 83)**. Idiosyncratic responses occur with some patients developing cardiac damage at cumulative doses of <250 mg m^{-2}. Clinically the cardiac damage is manifest as congestive cardiac failure due to cardiomyopathy. Arrhythmias are common but are not harbingers of cardiac failure and the development of SVT or PVC is not necessarily an indication for dose reduction. The cardiac damage is caused by the formation of free radicals, superoxide anions, which, due to relative deficiency of glutathione peroxidase in heart muscle, cause membrane damage to the mitochondria in the myocardial cells. The calcium binding of the myocardial cells is also impaired.

Prevention is the best management policy for anthracycline-induced cardiomyopathy. Total lifetime dose should be limited to 550 mg m^{-2}, less if radiotherapy to fields involving the heart has been given or is planned. Women and children are more susceptible than men to anthracycline cardiotoxicity. Myocardial function should be monitored before, during and after treatment. If left ventricle (LV) ejection fraction falls to $<50\%$ or falls by >15–20%

from baseline, then anthracycline therapy should be abandoned.

Doxorubicin given intravenously is heavily protein bound. The majority of the non-metabolised drug is excreted in the bile. Dose reductions may be advisable in patients with evidence of biliary obstruction. Elimination is triphasic with half-lives of 10 min, 3 and 24–48 h.

Drug Development

New drugs emerge in two main ways: by chance and by design. The chance discovery of the inhibition of bacterial growth by platinum electrodes led to the development of *cis*-platinum and its analogues. Cytopenia in sailors shipwrecked from a vessel carrying nitrogen mustard led to the development of alkylating agents.

Discovery through design rests upon several concepts:

- That certain classes of compound (e.g. antifolates) are likely to have anti-cancer effects.
- That promising compounds can be screened for activity using *in vitro* and *in vivo* tests.
- That a rational approach to the design of new compounds is to base them on the structure–activity relationships of existing anti-cancer drugs.
- That knowledge of key DNA sequences concerned with cancer development and propagation permits the prediction of the structure and function of the protein derived from that sequence and that appropriate drugs might then be designed specifically to block the production of function of the abnormal protein.

The testing of potential anti-cancer drugs is a structured process.

Preclinical testing

Preclinical testing involves both *in vitro* and *in vivo* tests. *In vitro* tests are based on the candidate drugs' effects on cell membrane function (dye exclusion tests) and DNA function (labelling, growth *in vitro*, etc.). *In vivo* tests are based on effects on tumours growing in animal models: either murine or human tumours growing as xenografts in immune-suppressed mice.

Phase I study

A Phase I study is designed to establish the **maximal tolerated dose** (**p. 190**) in the human. Volunteers, usually with disease that has failed all previous attempts at treatment, are treated in cohorts with gradually increasing doses of the candidate drug. The dose escalation scheme is based on a predetermined numerical sequence, e.g. a modified Fibonacci: n, $2 \times n$, $3.3 \times n$, $5 \times n$, $7 \times n$, $9 \times n$, ..., where n is a starting dose established on the basis of studies in animals. The toxicity level and pattern for each schedule and route of administration is then established. Pharmacokinetic and pharmacodynamic studies are also performed during Phase I studies.

End points for phase I studies

- **Subtoxic dose** (**p. 299**): one that causes consistent mild toxicity and which suggests that significant toxicity will be encountered at the next dose level.
- **Minimal toxic dose** (**p. 200**): one at which at least one of three patients shows reversible toxicity.
- **Recommended dose** (**p. 263**): one that causes moderate, reversible, toxicity in the majority of patients.
- **Maximum tolerated dose** (**p. 190**): the highest dose that can safely be tolerated.

This term has been used loosely — sometimes to describe it as the dose immediately below that which is usually lethal, sometimes to describe the dose at which any detectable toxicity appears.

Phase II study

A Phase II study employs the dose and schedule defined in the Phase I study to establish whether the drug has any meaningful clinical effect on human tumours. The convention is to treat 14–20 patients with a given tumour type and to look for clinical response. If there are no responses in a group of this size then the probability that the response rate is ≥20% is <5% (**Gehan's rule**, **p. 135**).

Phase III study

The purpose of a Phase III study is to compare drugs, defined as active in Phase II studies, with conventional treatment for that tumour type. Inevitably, today this implies conducting a **randomised controlled trial** (**p. 261**).

Phase IV study

This is really post-marketing surveillance. The pharmaceutical company responsible for a drug collects data prospectively on unexpected adverse events. There is always the possibility, no matter how carefully the initial studies have been carried out, that a drug has a rare but serious toxic effect. Only by actively seeking out such adverse effects can the risk be accurately estimated. This is less of a problem with drugs used for treating cancer, where the condition is often fatal, but can be a problem with drugs used in other contexts, e.g. the unexpected peritoneal fibrosis that occurred with the β-blocker practolol.

An enormous number, possibly ~500 000, compounds have been screened as potential anti-cancer agents. This contrasts with the ~50 drugs that are currently in routine clinical use. A survey in 1999 identified 354 agents of potential value in the treatment of cancer undergoing clinical testing in humans: only a minority will make the transition into routine clinical use. Recent data on the development of drugs for treating cancer suggest that for every drug eventually approved by the regulatory authorities, 5000 have undergone preclinical testing and five have entered clinical trials. It takes ~6.5 years to synthesise a compound and perform the preclinical testing; 1.5 years are spent in Phase I studies; 2 years in Phase II; Phase III will take ~3.5 years. In the USA, once the Phase III data are available, it will take the FDA ~1.5 years to approve the drug. This adds up to a total of 15 years from first synthesis to clinical licence.

This represents a huge investment of time and effort. In 1996, the average cost of bringing a drug from the laboratory to the clinic was estimated at US$500 million. Rosenberg's original discovery on the cytotoxic effects of *cis*-platinum was made during the mid-1960s; the early clinical tests were performed in 1970–71; the drug was licensed for more general use ~1979–80. This interval, of 15–20 years, is still fairly typical. A faster-track approach is possible — as has been shown in **AIDS** (**p. 3**) where, for some new drugs, the time from laboratory to clinic has been ~2 years. The plight of a patient with pancreatic cancer is every bit as desperate as someone with AIDS. Perhaps, for some patients, a less cautious approach to the introduction of new therapies might be appropriate. It is known that, when the outlook is deemed hopeless, patients will often accept enormous risks for very slender chances of benefit.

Drug Interactions

Drug interactions might be clinically important in cancer and **supportive care (p. 300)**, e.g. nifedipine might help reverse the development of **multiple drug resistance (p. 210)** (MDR) and thereby increase the effectiveness of cancer chemotherapy. Metoclopramide can act as a radiosensitiser. Anti-depressants, which act through selective serotonin re-uptake inhibition (SSRI), might interact with anti-emetics such as ondansetron and granisetron, which act through antagonism of the $5HT_3$ receptor. It is often hard to detect drug interactions — doctors looking after patients with cancer may not know their patients' full drug histories.

Dummy Variables

A dummy variable is one introduced into a **Cox Proportional Hazards (p. 79)** analysis to accommodate the use of categorical variables. If there are three categories of performance status, 0, 1 and 2, we can put in two variables g_1 and g_2 with the following codings:

- $g_1 = 1$ performance status is 0.
- $g_1 = 0$ performance status is not 0.
- $g_2 = 1$ performance status is 1.
- $g_2 = 0$ performance status is not 1.

A matrix can then be used to code the three categories of performance status:

Performance status	g_1	g_2
0	1	0
1	0	1
2	0	0

The use of only two dummy variables, g_1 and g_2, permits the coding of three categories of performance status.

Dynamic Wedge

A dynamic wedge is used in radiotherapy treatment. It is constant in shape but its dwell time within the beam is programmed by the appropriate software to mimic the characteristics of a fixed wedge. The advantage of the dynamic wedge, from the radiographers' point of view, is that there is no heavy piece of metal to be attached to the machine: it is a combination of hardware and software designed into a treatment machine and activated through the control panel.

E1A and *E1B* Genes

These genes are found in small DNA viruses and can inhibit the action of ***p53*** (**p. 224**) in the host cell. The evolutionary advantage is obvious: the virus can parasitise the cell without killing it through *p53*-mediated apoptosis. The **ONYX-015** (**p. 220**) mutant adenovirus has had its *E1B* gene altered so that the virus is specifically cytopathic for cells that lack wild-type *p53*.

E-Cadherin

An **adhesion molecule** (**p. 8**), loss of which enables cancer cells to become invasive. The molecule acts like a zip fastener, linking adjacent cells. Transfection of malignant cells with normal E-cadherin causes loss of the ability to invade adjacent cells.

Eddy's Principles

These principles, which have to do with the rational allocation of resources for healthcare, were first adumbrated by David Eddy:

1. The financial resources available to provide healthcare to the population are limited.
2. Because of this limitation, when deciding about the appropriate use of treatments, it is both valid and important to consider the financial costs of treatments.
3. It is necessary to set priorities.
4. It will not be possible to cover from shared resources every treatment that might have some benefit.
5. The objective is to maximise the health of the population as a whole, therefore, there is need to consider:
   - importance of outcome
   - probability of outcome
   - number of times service can be rendered per unit cost.
6. The priority a specific treatment receives should not depend on whether the patients under consideration are one's personal patients.
7. Determining the priority of a treatment will require estimating the magnitudes of its benefits, harms and costs.
8. Use, or obtain, empirical evidence. If empirical evidence contradicts subjective judgement then empirical evidence should dominate.
9. Criteria that should be satisfied before promoting a treatment for use:
   - evidence that, compared with no treatment, outcome is improved
   - its beneficial effects should outweigh its harmful effects
   - compared with the next best option it should represent a good use of resources (see Principle 5).
10. When making relative judgements, the judgements should, wherever possible, reflect the preferences of those who would receive the treatments.
11. The burden of proof that Principle 9 is satisfied falls on the promoters of the treatment.

Effective Doubling Time (T_{eff})

The effective doubling time defines the ability of a tumour to repopulate in the face of cell depletion — as a result, for example, of treatment with drugs or radiation. It can be estimated roughly as follows. Use clinical data to ascertain the extra dose required (compared with treatment without a gap) to achieve local control to compensate for a gap in a split course schedule. Assume that

10 logs of cell kill are required for local control. If, say, 10% extra dose is required then the regrowth was 10% of 10 logs, i.e. 1 log. One log = 3.32 doubling times. The effective doubling time, if the gap was 2 weeks, is 14/3.32 days, i.e. 4.2 days. An alternative, more accurate, method that can be used in animal studies is to compare clonogenic survival in tumours treated with 6-h interfraction interval, no time for repopulation, with those treated with daily fractions, where there is time for some repopulation. The ratio of the two surviving fractions can be plotted against overall treatment time and the slope of the line will give an indication of T_{eff}.

T_{eff} is distinct from T_{pot} since it will be influenced by several other factors including the intrinsic sensitivity of the cells to the cytotoxic treatment. In general, $T_{eff} > T_{pot}$: the data from experimental tumours typically show T_{eff} double T_{pot}. There is no clear **correlation (p. 75)** between the two parameters and thus it would be unwise to use T_{pot} as a proxy for T_{eff}. The position is further complicated by the differences in intrinsic **radiosensitivity (p. 260)** between different tumours of the same morphological appearance. In clinical practice values may range from 5 to 50–60 days, even though the corresponding T_{pot} may be similar. Data obtained from a population of patients will, therefore, give a poor indication of T_{eff} for an individual patient.

The mechanisms whereby a population of cells might accelerate repopulation in response to external stress include: shortening the overall **cell cycle (p. 52)** time; increasing the **growth fraction (p. 144)**; decreasing the cell loss factor. There is vigorous debate about which of these factors is the main mechanism whereby **accelerated repopulation (p. 2)** occurs during treatment with radiotherapy. Another debate concerns whether there is a lag period before accelerated repopulation occurs or whether it starts as soon as the first fraction has been delivered.

Eicosanoids

These are fatty acids used in signalling pathways involved in inflammatory responses. There are four main types:

- Prostaglandins.
- Prostacyclins.
- Thromboxanes.
- Leukotrienes.

Anti-inflammatory drugs affect eicosanoid synthesis: steroids inhibit phospholipase; aspirin blocks cyclooxygenase.

Electromagnetic Radiation

Electromagnetic radiation exhibits both wave and particle (quantum) behaviours. For everyday purposes it is the wave behaviour that dominates and is used to classify electromagnetic radiations. The radiation travels at the speed of light (3×10^8 m s^{-1}) and there is a simple relationship between frequency (ν), wavelength (λ) and velocity (c):

$$c = \nu\lambda.$$

Type of radiation	Wavelength	Frequency (Hz)
Radiowaves	5 m to 30 km	3×10^5 to 3×10^{10}
Microwaves	100 μm to 5 cm	6×10^9 to 3×10^{12}
Infrared	1–100 μm	3×10^{12} to 3×10^{14}
Visible light	400–700 μm	4.3×10^{14} to 7.5×10^{14}
Ultraviolet	10–400 nm	7.5×10^{14} to 3×10^{16}
X- and γ-rays	10 fm to 10 nm	3×10^{16} to 3×10^{22}

The quantum nature of electromagnetic radiation can be used to convert wavelength into energy. Planck's constant (h) gives the energy of the photon: 6.62×10^{-34} J s.

The energy of electromagnetic radiation is given by:

$E = hc/\lambda$.

This equation can be used to calculate the energy of X- and γ-rays, since 1 electron volt (eV) $= 1.602 \times 10^{-19}$ J: the range is from 124 eV to 124 MeV.

Electron

An electron is a negatively charged subatomic particle. Electrons orbit the nucleus as an electron cloud. Each electron has a mass of 9.108×10^{-31} kg and a charge of -1.6×10^{-19} Coulomb. The mass of an electron is 0.00055 amu.

Electron Volt (eV)

This is a unit of energy used to define the energy of a beam of radiation. It is defined as the kinetic energy acquired by a single electron as it is accelerated through a potential difference of 1 V: 1 eV $=$ 1.6×10^{-19} J; 1 keV $= 1000$ eV, 1 MeV$= 10^6$ eV. The relationship between energy and wavelength (λ) in metres can, by application of Planck's constant (h) and the speed of light (c) be simplified to:

Energy (eV) $= 1.24 \times 10^{-6}/\lambda$.

Electronic Equilibrium

Electronic equilibrium defines a state in which the number of electrons entering a volume of tissue is exactly equal to the number leaving it. As a beam enters a medium, the electrons have considerable forward velocity in the direction of the beam, and so more leave a volume than enter. The **absorbed dose** (**p. 1**), therefore, is less at the surface than is at depth of ~1–2 cm. This is the build-up region and accounts for the skin-sparing effect of megavoltage treatment.

Elimination Rate Constant

This is used to define the rapidity with which a drug is removed from the body. It can be calculated as:

$$\frac{\text{rate of drug elimination}}{\text{amount of drug in the body}},$$

or as:

$$\frac{\text{clearance}}{\text{volume of distribution}}.$$

E-mail

A means of sending information rapidly over the Internet. The requirements are a personal computer connected to the Internet via a modem, an e-mail address (typically of the form xxx@yyy.com) from which to send e-mails and an e-mail address for the intended recipient. The advantages of e-mail are its cost (contact anywhere in the world for the cost of a short local call), its rapidity of transmission and, since messages have to be typed in on a keyboard, prolixity is unusual. Its disadvantages include **spamming** (**p. 292**), the e-mail equivalent of junk mail, security may also be an issue.

End Replication Problem

The phenomenon whereby successive cell divisions cause progressive shortening of the chromosomes: a process implicated in **senescence** (**p. 282**). It ties in with concepts, such as the **Hayflick hypothesis** (**p. 148**), suggesting that a cell line might only give rise to a limited number of generations. In order for this limitation to apply, there has to be some sort of counting mechanism to log each cell division. Chromosome shortening might provide such a mechanism.

Endogenous Growth Theory

An economic theory that currently dominates many Western economies and is particularly relevant to the funding of scientific research. It has a mathematical base, which might be sound, based on such entities as returns to scale, marginal products, the Cobb–Douglas relationship, etc. In simple terms, a society should invest in science because science, by its very definition, brings knowledge. If we have more knowledge we can do more things for less money, if we can do more for less money, we can reinvest the surplus in the pursuit of more knowledge, and there is a quasi-exponential growth in wealth. A fundamental premise is that more knowledge produces a disproportionate increase in wealth: investing £10 in knowledge will make us £20 richer. The facts do not support this assumption, quite the reverse is true: to double GNP, investment in research has to be increased fourfold. But what are the costs of not investing in research?

Endowment Effect

The endowment effect describes circumstances in which we demand to be paid more to relinquish something than we would be prepared to pay to purchase it. As such, it is a variant of a **status quo bias** (**p. 297**). We may be using a treatment that cures 75% of patients, we would be unlikely to drop it in favour of a treatment that cures 76% of patients but would probably consider changing to a new treatment with an 80% cure rate. The origins of the endowment effect lie in our reluctance to embrace the unfamiliar — the treatment that we have been using is tried and tested, its strengths and limitations are known. The new treatment may look good on paper but we have no personal experience with it and so are reluctant to adopt it uncritically.

Endpoint

An endpoint defines a measurement or outcome used to assess the effect of an intervention. Endpoints have to be meaningful, and meaning is often in the eye of the beholder:

- A dead patient has zero quality of life.
- Fates worse than death can be defined.
- Does a patient really care about their PSA level?

> The assessment of roentgenographic tumor size, white blood count, and survival time does not indicate whether a patient is alive and vibrant or miserable and vegetating.
>
> (Feinstein, 1977)

The question of what constitutes successful treatment is crucial to the management of cancer. What matters to the individual patient is whether they are personally cured of disease. Those treating them may well be concerned with more subtle issues: reduction in odds of recurrence; rate of fall of tumour marker. Doctors and patients may well end up having entirely different expectations of treatment and this can cause major problems in communication and understanding.

Not all endpoints are equally valuable, some are more important than others and it is possible, crudely, to rank them according to strength.

Strength of endpoints

- Total **mortality rate** (**p. 207**).
- Cause-specific mortality rate.

- Quality of life.
- Indirect surrogates:
  - disease-free survival
  - **progression-free survival** (**p. 248**)
  - tumour response rate
  - **'proof of principle'** (**p. 248**).

There are well-defined criteria for assessing response rates to treatment, these totally ignore the considerations mentioned by Feinstein.

Complete response

Complete disappearance of all demonstrable malignant disease for at least 4 weeks: clinical (but ?radiological, ?pathological).

Partial response

Decrease in tumour size of at least 50%, lasting at least 4 weeks. No increase in sites or size (by >25%) of any other malignant disease. The 50% is either of an area, or of the product of two perpendicular diameters. There is a problem here in that the relationship between the product of two perpendicular diameters and volume is not always completely straightforward — even for simple spherical geometry the relationship is non-linear (figure 1).

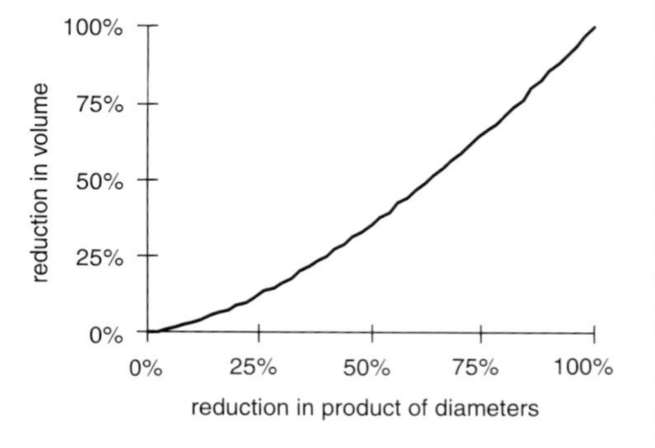

Figure 1 — Volume diameter for a sphere.

No change = stable disease

Anything that, without actual disease progression, is less than a **partial response** (**p. 228**).

Progression

An increase of >25% in tumour size, at any site.

Proof of principle

Proof of principle is a concept that is becoming more fashionable.

If a treatment ought to work it is because it is thought to have an effect, *x*, then proof of principle demands that we demonstrate that, indeed, the treatment truly does have effect *x*. Demonstration that *x* happens is, of course, no proof that the treatment will actually be beneficial. Satisfying the concept of proof of principle is an essential first step in the evaluation of a new approach — it is, by itself, not a sufficient justification for adopting a new treatment. The demonstration that a given vector-construct system was capable of restoring expression of normal *p53* to a bronchogenic carcinoma would provide important proof of principle: it would not, however, establish such an approach as therapeutically effective.

Problem of time

The problem with the main endpoints used in evaluating cancer treatment is that there is a considerable delay between the intervention and in obtaining an interpretable result. A new form of **adjuvant treatment** (**p. 8**) for breast cancer would be primarily evaluated in terms of long-term survival and would require assessment in a **randomised controlled trial** (**p. 261**). The trial itself would take at least 2 years and 10- and 15-year survival data would be required for definitive conclusions. It would, therefore, take 12 years from the treatment of the first patient until unequivocal data could be obtained — for a treatment of marginal benefit it could be as long as 20 years before the true worth of the treatment were known.

Dissatisfaction with this state of affairs has led to the introduction of surrogate, or proxy, endpoints. Examples would be the use of PSA fall after radical radiotherapy to the prostate, or the fall in GH level as a result of treating a pituitary tumour, as a predictor of ultimate outcome of treatment. The effectiveness of changes in management can be assessed in a more timely fashion using this approach: the key assumption, and one which is often unproven, is that the **surrogate endpoint** (**p. 300**) is an infallible predictor of the ultimate endpoints: survival and tumour control. These surrogate endpoints have the advantage that they may detect important changes early, without having to wait several years for the full expression of the natural history of the disease. There still needs to be proof that they correlate with the real outcome of interest, survival: a fall in PSA is really only fully meaningful if it indicates cure or long-term remission.

Eniluracil

Eniluracil is an inhibitor of dyhydropyrimidine dehydrogenase (DPD), the rate-limiting enzyme responsible for **5-fluorouracil (5FU)** (**p. 130**) catabolism. It is a uracil analogue with an ethynyl group substituted at the 5′-position. The **half-life** (**p. 147**) of 5FU is normally ~10 min, when Eniluracil is given with 5FU the half-life increases 10-fold. By combining Eniluracil with low doses of oral 5FU, it is possible to mimic continuously infused 5FU without the need for intravenous therapy.

Epidemiology of Cancer: Some Facts, Figures and Trends

The EU data, projected for 2000, emphasises the important effect of lung cancer on mortality rate. Although lung cancer causes the largest number of deaths, colorectal tumours and breast cancer are the most frequently occurring tumours. It is a sad irony that those malignancies that can be cured by chemotherapy (lymphoma, leukaemia, testicular cancer) are relatively uncommon. In comparison with the rest of the EU, the UK has higher that expected mortality rates for both oesophageal and lung cancers, and, in women, for bladder cancer. The rates in the UK are lower for oral cavity, liver and leukaemia in men, and for thyroid and uterine corpus in women.

The most dramatic time-related trends in cancer mortality rate over the past 50 years have been the rise in lung cancer and the decline in stomach cancer. Recently, there has been a decline in the death rate for men with lung cancer, but the rate in women has continued to rise. In both the USA and UK there has been a recent fall in the mortality rate for breast cancer. The death rate for colorectal cancer has fallen over the past 15 years.

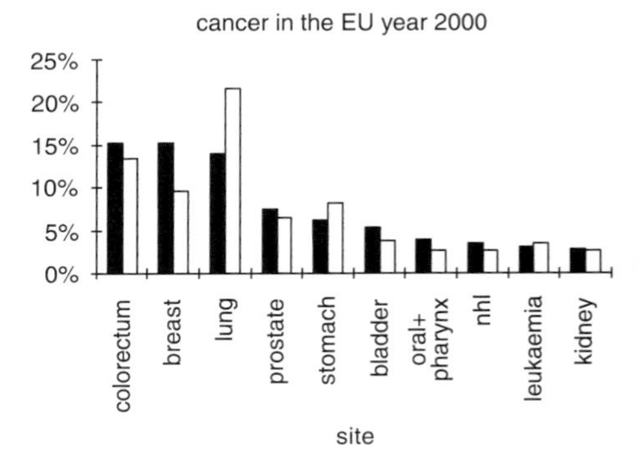

Figure 1 — Estimated EU figures for 2000. Solid bars, new cases as percentage of all new cancers; white bars, deaths as percentage of all deaths due to cancer (data adapted from Eur J Cancer *1997;* ***33:*** *991).*

This fall is more in women than in men. It is difficult to interpret trends in **incidence** (**p. 160**) and mortality rate for prostate cancer because of ascertainment **bias** (**p. 22**) introduced, particularly in the USA, as a result of the sporadic use of PSA as a screening test.

This, predominantly Western, perspective tells only part of the story. Patterns of cancer incidence and mortality rate are very different in the developing world but, because of difficulties in gathering accurate data, it is not easy to obtain a true picture of events. For example, the current world-wide epidemic of **HIV** (**p. 151**) infection will, through its association with lymphomas and Kaposi's sarcoma, affect the incidence of malignant disease. It will, however, be difficult to detect this, given the paucity of medical care in those communities most affected.

There are some striking geographical variations in cancer incidence and the pattern of variation provides interesting clues as to the relationship of cancer both to the environmental and to genetic factors (figures 2 and 3).

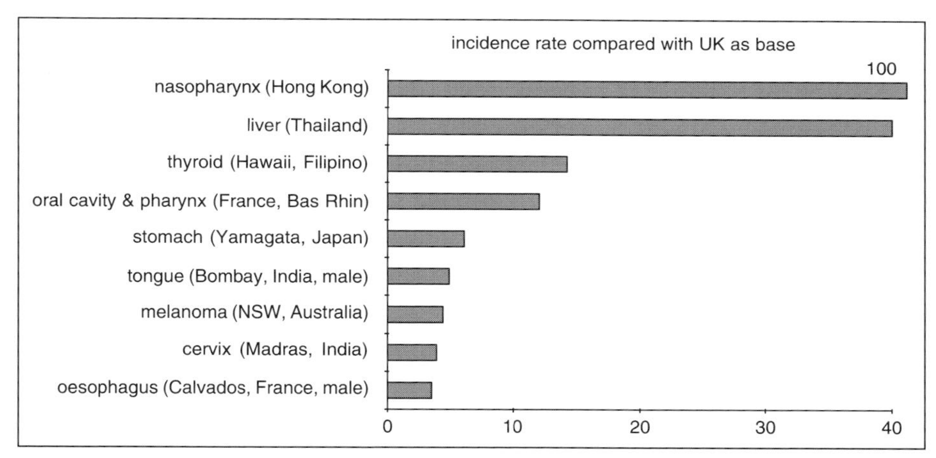

Figure 2 — Incidence rates, using age-standardised rates for the world population, compared with rate in the UK for those tumours whose geographical incidence is higher than that in the UK.

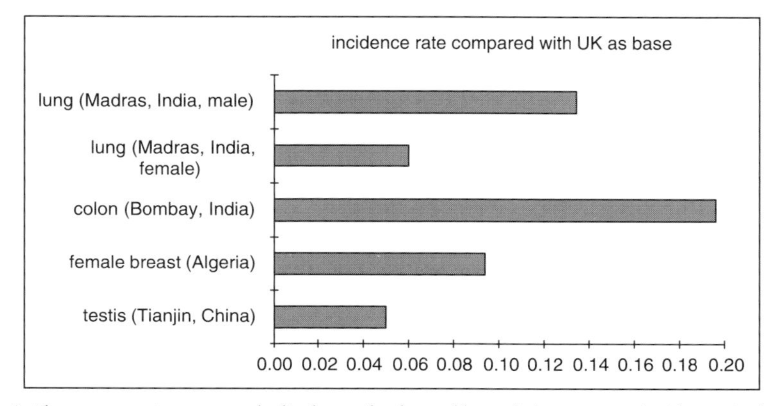

Figure 3 — Incidence rates, using age-standardised rates for the world population, compared with rate in the UK for those tumours whose geographical incidence is lower than that in the UK.

These data, as with all comparative data, must be interpreted with caution. **Biases** (**p. 28**) associated with case finding and reporting can distort figures but large differences probably reflect genuine effects.

Descriptive and geographical epidemiology can yield important insights into the cause and prevention of malignant disease.

Comparison of data concerning cancer incidence and mortality rates, over time and between centres, can be both practically useful and intellectually provocative.

Practical usefulness

- Allocating resources for cancer treatment — anticipating demand.
- Allows estimation of economic burden of cancer upon a community.
- Devising strategies for cancer prevention.
- Mortality rate : incidence ratios can provide a crude measure of the effectiveness of treatment.
- Identification of trends, both welcome (e.g. declining incidence of gastric cancer in UK) and unwelcome (rise in lung cancer in females in the UK).

Intellectual provocation

- There is an intellectual train of thought, rooted in epidemiological observation, that can be traced from Percival Potts's observations on the scrotal cancers of chimney sweeps to the recognition of the role of glutathione reductase polymorphisms in cancer susceptibility.
- Comparison of cancer incidence between countries can highlight dietary influences upon carcinogenesis, e.g. a high-fat diet in New Zealand and its association with breast cancer. It is not just dietary content that is important but also methods of food preparation, e.g. smoked fish and the incidence of oesophageal cancer in South China.
- Comparison of cancer mortality rate data between and within countries may draw attention to remediable deficiencies and disparities in care.

Ultimately, all successful strategies for managing cancer have their roots firmly in epidemiology.

Epitope (syn. Antigenic Determinant)

That part of an antigen that provokes the specific immune response is called the epitope. Antigens are usually large molecules and, since an epitope is only about five-to-six amino acids in size, one antigen may posses several epitopes. The epitopes will vary in their ability to provoke an immune response, binding properties, etc.: some epitopes are more antigenic than others.

Equipoise

This term is used in medical decision-making to describe the state of mind of a decision-maker who genuinely does not know which of two competing options is the better. The scales are balanced, there is no tilting towards one choice. The mind of a clinician planning to offer a patient **randomisation** (**p. 260**) into a controlled clinical trial should be in a state of equipoise about the interventions under trial. Otherwise, there are ethical problems: the patient could be randomised to receive a treatment that, in the clinician's opinion, was inferior. The term **'uncertainty principle'** (**p. 319**) is sometimes used as a synonym for equipoise: it adds nothing to the analysis of the problem, except confusion.

Erythropoietin

The erythropoietins are haemopoietic growth factors that act on erythroid precursors and stimulate the production of red cells. They are produced primarily in the peritubular interstitial cells of the kidney and their production is influenced by tissue oxygenation: induced by hypoxia, inhibited by high pO_2 levels. There are two main types, α and β. The erythropoietin gene is at chromosome 7q22. The gene codes for a peptide 193 amino acids in length. During secretion, the terminal 27 amino acids are removed leaving the active polypeptide: 166 amino acids long with a molecular weight of ~18 000 kDa. Both types of erythropoietin are, thanks to recombinant DNA technology, available commercially.

Erythropoietin interacts with a receptor expressed on the surface of all committed erythroid precursor cells. This receptor belongs to the **haemopoietin receptor superfamily** (**p. 147**). It is a transmembrane protein 507 amino acids long. The gene is on chromosome 19. Binding of erythropoietin to its receptor blocks apoptosis in the colony-forming unit-erythrocyte (CFU-E) pool.

Patients with cancer are often anaemic and this anaemia is of complex origin: blood loss, haemolysis (both within and outwith the marrow), nutritional deficiencies, myelosuppression induced by disease and treatment, impaired responsiveness of marrow cells to growth factors and nutrients, etc. Treatment with erythropoietin is currently being assessed in randomised controlled trials as a means of dealing with the problem of hypoxic cells potentially limiting the effectiveness of radiotherapy. Erythropoietins are also being used to treat fatigue, and the consequent impaired quality of life, in patients treated with cytotoxic chemotherapy.

Etoposide

This is an epipodophyllotoxin derivative that acts as a **cell cycle** (**p. 52**) phase-specific cytotoxic drug. It preferentially kills cells in G_2-phase. It is derived from the mandrake root. It inhibits topoisomerase II by binding to the enzyme–DNA complex and by preventing repair of **DNA strand breaks** (**p. 101**). It can be used orally but is usually given intravenously. Its main toxicity is myelosuppression. It can cause **second malignancies** (**p. 280**), particularly leukaemias: characteristically M4 or M5 in the FAB classification with no preceding myelodysplasia; there is a characteristic 11q23 chromosome abnormality.

Euthanasia

A process whereby, with or without the consent of the patient or subject, a lethal injection is administered. Euthanasia, with consent, is an option that may be considered by individuals with incurable disease and a **quality of life** (**p. 253**) that is so poor as to be unendurable. Euthanasia means, literally, a good death: something to which, given the inevitability of a conclusion to life, is desirable. Assisted suicide is a variation on euthanasia with consent: the patient is provided with the materials and apparatus to take their own life. Although the debate about euthanasia is primarily couched in moral and ethical terms, it is entirely possible that its ultimate acceptance may rest on less exalted grounds.

Evaluation of Claims

As consumers of scientific 'product', we are continually exposed to claims about the value and importance of the work presented. These claims can be made concerning any aspect of oncology, from prevention to terminal care. Examples could include:

- Prevention: using cancer patients to exemplify to friends and relatives the dangers of smoking will lower the incidence of lung cancer.
- Cause: electric power lines cause leukaemia.
- Screening: routine mammography for women <50 years of age is worthwhile.
- Diagnosis and classification: certain subtypes of DCIS are worth recognising.
- Prognosis: EGF receptor status on axillary nodes is an important **prognostic factor** (**p. 248**).
- Treatment efficacy: **gemcitabine** (**p. 136**) is better than 5FU for treating metastatic pancreatic cancer.
- Treatment economics: prophylactic **G-CSF** (**p. 135**) saves money in patients treated with high-dose therapy for breast cancer.
- Treatment morbidity: epirubicin is less toxic than doxorubicin.
- Follow-up: PLAP is a useful marker after treatment for seminoma.
- Terminal care: the majority of people accept the need for **physician-assisted suicide** (**p. 235**).

The disciplines of statistics and clinical epidemiology provide the framework within which claims can be evaluated:

- Could the observation have arisen by chance?
- Has **bias** (**p. 28**), both conscious and unconscious, been eliminated?
- Have the data been appropriately presented?
- Are the **endpoints** (**p. 114**) appropriate?
- Was the experimental design appropriate?
- Has the study been carried out correctly?
- Are the results consistent with other similar studies?
- Are the results relevant?
- Was the study worth doing in the first place?

The burden of proof lies with those who make claims but, all too often, research is presented in a way that is aimed at inflating the relevance and strength of the conclusions. It is up to the consumer, therefore, to maintain a healthy degree of scepticism and to use the tools provided by statistics and clinical epidemiology to make their own judgements. Just because something is in print does not mean that it is true: you are, at this moment, reading print — *caveat lector.*

Events per Independent Variable (EPV)

This is a simple and useful method for assessing whether the results of a multivariate analysis are likely to be robust. EPV is simply the number of events (e.g. relapses) observed in the study divided by the number of variables that have been included in the prognostic or predictive model. A proportional hazards or **logistic regression** (**p. 183**) analysis may become unreliable when the number of variables tested is high relative to the number of events. As a rule of thumb, generally EPV > 10, even this is, perhaps, too small a figure and only those studies with EPV > 20 are likely to be completely reliable.

In practical terms: if we have a group of 200 patients, of whom 40 have died, and wish to identify factors of potential prognostic significance, we should certainly look at no more than four factors and would be better advised to assess no more than two.

Evidence, Levels of Evidence

Evidence is material brought forward in support of a claim or contention. Its most familiar role is in the court room from which a simple lesson emerges: the value of evidence varies considerably, from the worthless and misleading, devoid of truth, to that which is robust, imbued with truth and totally compelling.

PDQ ranking (1997) of levels of evidence

- **Randomised controlled trials** (**p. 261**):
  - — double blinded
  - — non-blinded
  - — (**meta-analyses**) (**p. 196**).
- Non-randomised clinical trials.
- Case series.
  - — population-based, consecutive
  - — consecutive, non-population-based
  - — non-consecutive.

Appraisal Instrument for Clinical Guidelines (Cluzeau *et al.*)

I. Evidence from well-designed **randomised controlled trials** (**p. 261**), meta-analyses or systematic reviews.

II. Evidence from well-designed cohort or case-control studies.

III. Evidence from well-designed non-experimental descriptive studies, such as comparative studies, **correlation** (**p. 75**) studies and case control studies.

IV. Evidence from expert committee reports or opinions and/or clinical experience of respected authorities.

Category IV evidence leaves many questions unanswered: who is an 'expert'; what constitutes 'authority' and when should it be 'respected'.

Rules of evidence and clinical recommendations on the use of anti-thrombotic agents (Sackett, *Chest* 1989; 95: 2S–4S)

- Level I evidence:
  - — randomised controlled trials that are big enough to be either:
    - positive, with small risk of false-positive conclusions
    - negative, with small risk of false-negative conclusions
  - — meta-analysis.
- Level II evidence: randomised controlled trials that are too small so that they show either:
  - — positive trends that are not statistically significant
  - — no trend, but large risk of false-negative conclusions.
- Level III evidence: formal comparison with non-randomised contemporaneous controls.
- Level IV evidence: formal comparison with historic controls.
- Level V evidence: case series.

The NHS R&D grading system for therapeutic interventions is shown in Table 1.

Table 1

Grade	Level	Type of evidence
A	1a	**systematic review** (**p. 302**) of a homogeneous set of randomised trials
	1b	individual randomised trial with a narrow **confidence interval** (**p. 70**)
	1c	all or none phenomenon
B	2a	systematic review of a homogeneous set of cohort studies
	2b	individual cohort study or a randomised trial of low quality
	2c	'outcomes research'
	3a	systematic review of a homogeneous set of case control studies
	3b	an individual case-control study
C	4	case series
D	5	expert opinion without any systematic review or a recommendation based only on biological principles

The all or none phenomenon (A1c) is as follows: before the intervention no-one survived (or suffered the adverse event) after the intervention everyone survived (or suffered the adverse event).

The message that emerges from these three systems is that the **randomised controlled trial** (**p. 261**) is accepted as the **Gold Standard** (**p. 141**) against which all evidence should be judged. The difficulty is that all not problems can be addressed in such a fashion: a randomised controlled trial is simply not feasible for many rare conditions, and yet we still need evidence on which to base the management of these diseases. There is less agreement, once we descend below the level of the randomised controlled trial, about how we should rank the available evidence. To make rational judgements we needs meta-evidence, evidence about evidence, and we simply don't have it.

Exceptions

Exceptions, findings that break the accepted rules, are an important source of progress in science. The classic example from astronomy is the anomalous perihelion of Mars, the accurate observation of which was, for many, the final physical proof that Einstein's Theory of Relativity was correct. The question here is, of course, that of chickens and eggs. Do exceptions, like grit in oysters, produce an intellectual irritation that finally yields the pearl of theory; or does the theory arise as a result of pure abstract speculation which, only later, is applied to concrete anomalies?

An example from the recent history of oncology: for years it was known that long-lived T-lymphocytes died an intermitotic death after irradiation. This was anomalous, since radiation was supposed to exert its cytotoxic effect through damaging DNA and thereby interfering with mitosis. The rediscovery of **apoptosis** (**p. 19**), now known to be an important mechanism for radiation-induced cell killing, provided the explanation. Perhaps if more attention had been paid to the anomaly, knowledge might have advanced more rapidly. The usefulness of exceptions was well known to the statistician R. A. Fisher, one of whose maxims was 'treasure your exceptions'.

Experimental Design

The design of experiments or clinical studies is crucial to the success of the endeavour: a poorly designed experiment can only give an approximate answer. No matter how brilliant the insight or the hypothesis being tested, if the design is flawed the results will be unconvincing. The primary purpose of an experiment is to discover truth, and truth is no respecter of persons, dogmas, prejudices, fashion or any of the other factors that might influence why or how a particular experiment or study is performed. It follows that an essential feature of experimental design is that not only should the experiment elucidate truth, but also that it should do so in a convincing and persuasive manner. There are, therefore, a few general points to be considered in designing an experiment:

- Be sceptical.
- Anticipate criticism.
- Aim to convince.
- Eliminate **bias** (**p. 28**).
- Statistics are integral to the design, they are not simply bolted on as an afterthought.

Beware of the pitfalls:

- Data dredging.
- Multiple comparisons.
- Subgroup analyses.
- Asking questions not addressed in the study design.
- **Recursive logic** (**p. 263**).
- Bias.
- Add-ons.
- **Statistical launderette** (**p. 297**).

Expert

A person deemed to have especial knowledge and experience of a particular subject. The term is decreasing in value and is now virtually interchangeable with the slightly pejorative term 'pundit'. Experts are rarely selected by any objective process and, too often, are simply self-defining. I feel expert, therefore, I am an expert. Self-defined experts are particularly pernicious: they may know very little more than those whom they presume to advise but dangerously, have little insight into the depths of their ignorance. Expertise and wisdom have a relationship analogous to that between precision and accuracy: it is possible to become an expert simply by repeating the same mistakes, 'experts, text spurts'. I am, myself, an expert on experts. We are back with an old riddle — you meet a man who says he comes from Crete and tells you that all Cretans are liars.

Expert System

An expert system is a computer-based system, usually incorporating programming techniques developed in the investigation of artificial intelligence, designed to replicate the behaviour and decision-making of an expert. The computer asks questions, formulates the problem, frames choices and provides a hierarchy of preferred solutions just as if the real expert were present. The aim is to make expert advice portable and consistently available, as it were, at point of sale. In spite of >20 years of research, the impact of expert systems on medicine has been small: perhaps because we lack the expert knowledge to programme into the systems, perhaps because human 'experts' feel threatened by something they would be better to regard as an opportunity.

Explanatory Trials

Randomised controlled trials (**p. 261**) can be used to learn about the way in which potential treatments might work or to measure the extent to which they are clinically useful. The former approach uses an explanatory trial, which is in contrast to the latter aim, that associated with a pragmatic trial. The distinction between explanatory and pragmatic trials is conceptually useful but, in real terms, the boundaries are often blurred.

The requirements for the two types of study are rather different.

Explanatory	Pragmatic
Restrictive entry criteria to maximise **homogeneity** (**p. 152**) of study population	Broad entry criteria to embrace all-comers: since these form the population to which the trial results will be applied
Careful data collection an analysis, study intermediate outcomes as well as simply assessing survival, pristine approach to measurement	Simple data collection, e.g. survival or relapse and not much else, non-obsessional approach to data gathering
Ensure that subjects comply with allocated treatment	Relaxed attitude to compliance — what will happen in the real world when we try to use this intervention?
Small sample size	Large sample size

Exposure

A technical term used in radiation dosimetry. It indicates the amount of ionisation produced. Exposure is defined as: the absolute value of the total charge of all the ions of one sign produced in air (δq) when all the electrons liberated by photons in a volume of air are completely stopped in air:

$$\text{Exposure} = \mathrm{d}q/\mathrm{d}m,$$

where dm is a volume element of air with mass dm.

Exposure is measured as Coulombs kg^{-1} — it has no other name. In the early days of clinical radiotherapy, exposure was used as a proxy for dose. The unit was the Roentgen (R):

$$1\ \mathrm{R} = 1\ \mathrm{esu\ cm^{-3}} = 2.58 \times 10^{-4}\ \mathrm{C\ kg^{-1}}.$$

Extrapolation Number (*n*)

The extrapolation number is derived from the **cell-survival curve** (**p. 57**) for irradiated cells and is an indication of the number of targets per cell that need to be inactivated before the cell is, in the reproductive sense, killed. It can be estimated by (unsurprisingly) extrapolating the linear part of the curve to the y-axis.

There is a mathematical relationship between D_o, D_q and n:

$$D_q = D_o \log_e n,$$

where D_q is the **quasi-threshold dose** (**p. 254**) and D_o is the slope of the linear portion of the cell survival curve.

Fab Fragment

An antibody is comprised of both heavy and light chains, two of each. The molecule is 'Y'-shaped. The heavy chains constitute both the stem and the branches of the Y, the light chains contribute only to the branches. The Fab fragment comprises the branches without the stem. It contains the variable portion of the light chain (V_λ), the variable portion of the heavy chain (V_H), the constant portion of the light chain (C_λ) and the first part of the constant portion of the heavy chain (C_H1). The Fab fragment, therefore, contains the antigen-binding site of the immunoglobulin molecule.

Factor Analysis

A repertoire of techniques primarily used in psychometric studies and which have found their way into oncology, primarily by way of **quality-of-life (p. 253)** assessment. The purpose of factor analysis is to reduce the number of variables for consideration by identifying groups of variables, which, through having some factor in common, cleave together. For example, in a quality-of-life questionnaire there may be items asking about:

- Able to climb stairs.
- Able to walk on the level.
- Able to dress oneself.
- Do friends visit frequently?
- Do family visit frequently?
- Do you feel lonely?

The first three questions could cluster around a factor to do with physical disability; the last three questions could coalesce around a factor about social support. Factor analysis is a complex business and has spawned its own vocabulary: rotation; Eigen values; principal components analysis; varimax rotation. **Maximum likelihood estimation (p. 189)** may be used as a means of finding appropriate values to use within a factor analysis. The origins of factor analysis are found in attempts to answer the question: what is intelligence? The success of the approach can, in part, be measured in terms of the well-known litany: what is intelligence? That which is measured by intelligence tests.

Factorial Design

A method for designing a clinical trial so that two or more interventions can be considered simultaneously. A practical example will illustrate: the UKCCCR trial of postoperative management for patients who had surgical excision of DCIS (ductal carcinoma *in situ*) of the breast. Since this disease was 'manufactured' by the breast-screening programme, no one knew what to do with it. Consequently, there was an urgent need to assess a number of competing options, the most important being no further treatment, adjuvant radiotherapy or adjuvant **tamoxifen (p. 304)**. The factorial design was used to define the groups for randomisation in the DCIS study:

Allocation group	Adjuvant XRT	Adjuvant tamoxifen
A	yes	no
B	yes	yes
C	no	yes
D	no	no

All possible combinations of **tamoxifen (p. 304)** and radiotherapy (including neither) are investigated using the factorial approach.

Familial Adenomatous Polyp (FAP) Syndrome

This autosomal dominant condition has an incidence in Western populations of ~1 : 7500. It is due to mutations, usually deletions, in the ***APC* gene** (**p. 18**). The gene itself is at 5q21–q22, the gene product is probably involved in cellular adhesion. Antibodies to *APC* will precipitate catenins, proteins known to interact with **cadherins** (**p. 39**). *APC* might be a mediator of contact inhibition and is probably one of the earliest steps in the multistep process that leads from normal mucosa to invasive cancer.

The classical clinical feature is the presence of multiple, at least 100, polyps in the large bowel. Unless prophylactic colectomy is performed, the risk of bowel cancer is, by 40 years of age, almost 100%. Other tumours are also found: stomach cancer, tumours of the ampulla of Vater, thyroid carcinomas and CNS tumours. Retinal pigment abnormalities are often associated with FAP, congenital hypertrophy of the retinal pigment (CHRPE). The coexistence of FAP and a cerebral tumour is known as **Turcot's syndrome** (**p. 317**). To confuse matters: Turcot's syndrome is either medulloblastoma + FAP or HNPCC + glioblastoma.

Fanconi's Anaemia

This condition, which is inherited as an autosomal dominant, is characterised by pancytopenia and chromosome fragility. In many patients there are also other associated abnormalities: absent radii, skin pigmentation with café au lait spots and renal abnormalities. There is an increased risk of malignancy as well as increased toxicity following exposure to ionising radiation or alkylating agents. There are several different forms of the condition each with a specific mutated site: Fanconis A, 16q24.3; C, 9q22.3; and D, 3p26−p22. Fanconi B and E are unmapped.

The associated tumours are leukaemia, head and neck cancer, cancer of the cervix, vulvar carcinoma, and oesophageal cancer.

Fas (syn. CD95, APO-1)

A member of the TNF receptor family expressed by a wide variety of cells. When Fas ligand (FasL) binds to the Fas receptor the cell undergoes apoptosis. The Fas–FasL interaction unleashes a cascade of events, primarily mediated by **caspases** (**p. 47**), that results in programmed **cell death** (**p. 55**). In a **paradigm** (**p. 226**) of neatness, justice and **parsimony** (**p. 228**) the role of the Fas–FasL system in lymphocyte-mediated cell killing is not only that lymphocyte-mediated cell death is accomplished using this system, but also that, once the cytotoxic lymphocytes are no longer required, they themselves commit suicide through apoptosis activated by the Fas–FasL interaction. Inherited deficiency of Fas expression causes a non-malignant lymphoproliferative syndrome with autoimmune manifestations directed primarily against peripheral blood cells.

Fas Ligand (FasL)

Fas ligand is expressed by activated cytotoxic T-cells and, by binding to the Fas expressed on target cells, triggers **apoptosis** (**p. 19**). Tumour cells, by expressing Fas ligand, may evade immunological attack. FasL on the tumour binds to Fas on the, no longer cytotoxic, lymphocyte and it is the lymphocyte rather than the tumour cell that undergoes apoptosis. It is a case of the biter bit.

Fertility and Cancer Treatment

The germ cells of the **testis** (**p. 306**) and ovary are very different in their organisation: the testicular germ cells are a classical self-renewing population, with stem cells amplification, differentiation, etc. The ovarian germ cells have all been formed by the time a girl is 1 year old.

Ovary

By puberty there are only ~4000 ovarian germ cells per ovary, compared with 10^6 at the fifth month of intrauterine development. They rest, as primary oocytes, in the **prophase** (**p. 249**) of meiosis until shortly before ovulation. The final meiotic division eventually produces a mature ovum and three polar cells. These go on to participate in the formation of ripening ovarian follicles. The intermediate stages of follicular development are more radiosensitive than the mature or immature stages. Thus, after ovarian irradiation, there can be temporary fertility, then temporary sterility, followed by a regain of fertility. This pattern reflects the differing radiosensitivity of differing stages of follicular development and, in contrast to the pattern observed in the testis, has nothing to do with recruitment of stem cells. The number of primary oocytes declines with age; hence, the increased sensitivity to irradiation of the ovary with age. The ovary may be surprisingly resistant to irradiation in the younger woman: pregnancy can occur after doses of 8 Gy. Radiotherapy was once used to treat infertility: 341/796 women treated with ~2 Gy in 3f fell pregnant; 27/61 women treated with ~2.3 Gy in 3f subsequently conceived. These women, remember, had previously been defined as 'infertile'. Ovarian ablation is inevitable with fractionated doses >24 Gy and, in women >40 years of age, nearly all will experience ovarian failure with doses >6.5 Gy. There are major (×7-fold) species differences in radiosensitivity. This is probably due to the arrest occurring in different stages of prophase.

Testis

The germinal epithelium requires the support of the Sertoli cells for normal function. The time to expression of radiation-induced damage is dictated by the time-course of normal **spermatogenesis** (**p. 292**) (usually ~75 days from primordial cell to mature sperm). The spermatogonia are the most radiosensitive cells; it takes ~2–4 weeks after irradiation for the effect of depletion of spermatogonia to be made manifest as falling sperm count. Doses of <2 Gy will produce temporary infertility, most men will recover fertility after ~12 months. Doses of 5 Gy almost invariably produce permanent sterility. At intermediate doses there is a dose-dependent prolongation of the period of infertility.

Many **cytotoxic drugs** (**p. 84**) can produce infertility, but Procarbazine is probably the most potent.

The crucial difference, in terms of response to cytotoxic therapy, between the ovary and the testis is that, in the latter, provided a few stem cells survive treatment, spermatogenesis can be reconstituted. In the mature ovary, however, there are no stem cells and there is no mechanism for compensatory replacement of lost oocytes.

Advances in reproductive technology — *in vitro* fertilisation, semen cryopreservation and, more recently, oocyte harvest and preservation — have improved the prospects for restoring or preserving fertility in patients treated for cancer.

Fibonacci Sequence

The original numerical sequence described by Fibonacci was as follows: n, $n+1$, $n+n+1$, $n+n+1+n+1$,... Starting at 1, the sequence goes: 1, 2, 3, 5, 8, 13, 21,... This type of sequence is ideally suited to dose-finding studies in Phase I clinical trials. In practice, the sequence is altered slightly, i.e. a 'modified Fibonacci sequence'. If n is the initial dose, then escalations are: $2.0n$, $3.3n$, $5.0n$, $7.0n$, $9.0n$, $12n$, $16n$. The proportional increments decrease as the numbers become larger. This strategy is to avoid excess toxicity as doses rise. Three

patients are entered at each dose level, with at least 3–4-week intervals between the first and second patient. Once toxicity is observed, then six patients are entered at each dose.

Fibonacci means 'blockhead' in Italian and was the nickname given to Leonardo of Pisa. He brought numbers from the Arab mathematicians to Europe — published as the *Liber Abaci*. The Fibonacci sequence has its origin in a calculation of the number of pairs of rabbits that would be produced from an original breeding pair. The answer was 233 pairs after 12 months; 1, 2, 3, 5, 8, 13, 21, 34, 55, 89, 144, 233.

Fibre FISH

A technique that facilitates the examination of multiple **DNA sequences (p. 101)** simultaneously. By shining ultraviolet light at a cell, the DNA can be persuaded to extrude itself out onto a slide. Different labels, with colour coordinated fluorescence, can then be used for **FISH (p. 129)** . The colour sequences from different tumours can then be compared.

Fibroblast Growth Factor-Binding Protein (FGF-BP)

FGF-BP is a 17 kD protein that mobilises and activates fibroblast growth factors (FGFS). It might be an important component of the **angiogenic switch (p. 15)** and, as such, could be a potential therapeutic target in cancer treatment.

Fibronectin

An **adhesion molecule (p. 8)** that functions as a matrix protein. Decreased production enables cancer cells to detach themselves from the tumour mass. Fibronectin, therefore, plays an important role in the formation of metastases.

Fine Needle Aspiration (FNA)

A technique used to obtain small quantities of cellular material for diagnosis. A needle is introduced into the area of interest: tumour, lymph node, etc., and suction is applied to a syringe attached to the needle. The aspirated cells enter the lumen of the needle and then can be used for preparing smears for light microscopy or cell suspensions for more sophisticated analysis (PCR surface markers, etc.).

Techniques such as **PCR (p. 238)** and **FISH (p. 129)** can identify genetic abnormalities in cells obtained by FNA. Imaging, e.g. ultrasound or CT, can direct FNA of less accessible lesions such as pulmonary nodules or retroperitoneal nodes. FNA can often provide a tissue diagnosis without recourse to more invasive and time-consuming procedures such as incisional biopsy or laparotomy. The results of morphological examination of FNA specimens can be available within 1 h of the specimen being taken. The time advantage, compared with conventional histology, may be offset by the disadvantage that FNA can only provide information about cellular morphology. It can tell nothing of tissue architecture. This can pose problems, particularly in the assessment of the lymphomas. Sampling error is another potential difficulty with FNA — the small sample may be untypical of the tissue as a whole. A negative FNA does not rule out malignancy — the technique may be specific but it is not necessarily sensitive.

First-Pass Metabolism

This occurs when a drug is metabolised, at least in part, before it can be measured in the systemic circulation. Oral testosterone is extensively metabolised in the liver by first-pass metabolism, which is why it is ineffective when administered orally.

FISH (Fluorescent *In Situ* Hybridisation, Chromosome Painting)

A method for identifying specific sequences of DNA or RNA in chromosomes or tissues. A fluorescently labelled probe, complementary to the sequence of interest, is applied to either fixed tissue or a chromosome preparation. The specific complementary sequences will bind to each other. Any probe left unbound will be removed by washing. The sequences of interest, if present, will fluoresce when viewed under light of the appropriate wavelength. The position of these sequences can thus be assessed by direct vision. FISH is, therefore, an extremely useful tool in mapping the genome. It is also very useful in identifying chromosomal abnormalities, particularly translocations. An alternative technique, but one which uses radioactivity, is to label the complementary sequences with radioisotope and then to localise any bound probe by using autoradiography.

The advantages of FISH over conventional cytogenetic methods for identifying abnormalities in human tumours are: it is technically easier; more rapid; more sensitive; it can identify hidden segments that cannot be found using traditional methods; and it can be used on frozen sections, or paraffin-embedded, archival, tissue.

Flucytosine

Flucytosine is a fluorinated **pyrimidine** (**p. 252**) analogue used in the systemic treatment of fungal infections. It is deaminated in fungal cells, but not in mammalian cells, to fluorouracil. Deamination is due to the presence of cytosine deaminase, an enzyme present in fungal cells, but not in mammalian cells. The fluorouracil disrupts fungal synthesis of RNA and protein. The main toxicity is myelosuppression. Therapeutic levels are achieved in the cerebrospinal fluid following oral administration. Resistance can, however, develop rapidly and for severe infections it should be combined with another agent such as amphotericin B.

Fludarabine

Fludarabine is an **antimetabolite** (**p. 17**) that is structurally similar to cytosine arabinoside. Its active metabolite, 2-fluoro-ara-A, inhibits **DNA polymerase** (**p. 101**) and other enzymes involved in DNA synthesis. It causes premature termination of synthesis of the DNA chains. It is preferentially transported into malignant cells. Its main clinical use is in the treatment of chronic lymphocytic leukaemia (CLL), which has become unresponsive to therapy. It is myelosuppressive and neurotoxic and may cause **interstitial pneumonitis** (**p. 166**).

Fluence

A means of assessing the amount of radiation when the particles produced are not necessarily travelling in parallel (for parallel beams the intensity can be used). Fluence is defined as:

$$\phi = \mathrm{d}N/\mathrm{d}a,$$

where $\mathrm{d}N$ is the number of particles that enter a sphere of cross-sectional area $\mathrm{d}a$.

FACS (Fluorescence-Activated Cell Sorting)

This is an invaluable tool with many applications in cancer research: in conjunction with monoclonal antibodies, it can quantify subpopulations of lymphocytes and other cells; it can be used to analyse and define the kinetic properties of human tumours and normal tissues; and it can be used to sort and analyse chromosomes.

The principle is to use a dye, which fluoresces at a specified wavelength, to label

a particular type of cell or intracellular constituent. The tagged cells are passed in suspension, one at time, through a system comprising a light beam and detector. The light used is of a wavelength that will cause the dye to fluoresce. The labelled cells are counted by the light detector and the total number of cells passing through the apparatus is also counted. The proportion of labelled cells can be easily calculated.

The method can be used to estimate cell cycle parameters. If DNA is stained with a fluorescent dye, then cells in G_2 will have twice the intensity of fluorescence compared with cells in G_1. Since the DNA content of the cell doubles during S-phase, cells in S-phase will have a spectrum of values varying between these two limits. Propidium iodide can stain the DNA of non-viable cells, cells stained with Hoechst 33342 can maintain viability.

Bromodeoxyuridine (BrdU) can be a substitute for thymidine during DNA synthesis. It can be detected using a monoclonal antibody that is itself labelled so that cells that contain BrdU will fluoresce green. Giving BrdU and then performing FACS analysis on a cell suspension from the tissue or tumour of interest provides a straightforward way of assessing the labelling index without using radioisotopes. The approach can be further extended using bivariate staining. If cells are simultaneously stained with propidium iodide (DNA fluoresces red) and BrdU (S-phase cells fluoresce green) then the potential doubling time can be calculated from a single sample obtained 4–8 h after BrdU administration. T_s can be estimated by plotting the relative DNA content (red colour) of S-phase cells (green cells) against time after BrdU administration. The intensity of DNA fluorescence must double during S-phase and this provides the basis for a graphical calculation of S-phase duration (Figure 1).

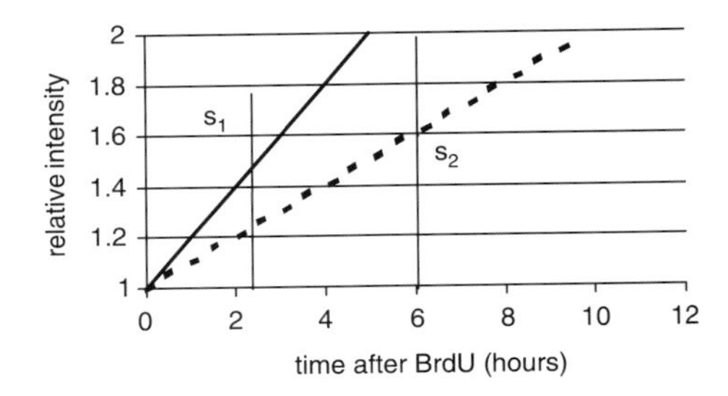

Figure 1.

The observation s_1 corresponds to a cell population with an average S-phase duration of ~5 h; the observation s_2 comes from a population where the duration of S-phase is ~10 h.

5-Fluorouracil

5-Fluorouracil (5FU) is a cytotoxic agent that inhibits thymidylate synthetase (TS). It has a short plasma half-life of 10 min and its oral absorption is unpredictable. It is a cell cycle phase-specific agent, acting only in S-phase. Although from first principles it would seem, therefore, to be an unpromising drug for cancer treatment, it is in fact widely used in the treatment of gastrointestinal cancers and breast cancer. It is also used topically in the treatment of malignant and premalignant disease of the skin.

It is a **prodrug** (**p. 246**) that is converted in the tissues to the active metabolites 5-fluorouridine-5′-triphosphate (FUTP) and 5-fluoro-deoxyuridine-5′-monophosphate (FdUMP). These intracellular metabolites, unlike the parent compound, have prolonged half-lives. TS is inhibited when a ternary complex forms with FdUMP and the reduced folate mTHF. FUTP acts as a false substrate in RNA synthesis, replacing uridine triphosphate (UTP). The antitumour effects of 5FU are increased by the administration of folinic acid. The mechanism is mainly by stabilisation of the ternary complex of TS/FdUMP/mTHF.

It is metabolised by dihydropyrimidine dehydrogenase, an enzyme which can be inhibited by **eniluracil** (**p. 116**). About 80% of an intravenous dose is metabolised in the liver.

Flutamide

This drug is a non-steroidal anti-androgen used in the treatment of prostate cancer. It antagonises the uptake of testosterone and its metabolites by prostate tissue. Its main side-effects are gastrointestinal discomfort, diarrhoea and nausea. Rarely, it may cause hepatic damage. It is usually used in combination with an **LHRH** (**p. 178**) agonist such as **goserelin** (**p. 143**) or buserelin.

Focus Skin Distance (FSD)

The distance between the source of X-ray production and the skin of the patient. When the source is very small, as in a linear accelerator, then the term 'focus skin distance' (FSD) is appropriate. When the source has finite dimensions, as in a cobalt unit, then **source skin distance** (**SSD**) (**p. 292**) is the better term. Linear accelerators typically have FSD = 100 cm, cobalt units of 80 cm and for orthovoltage units FSD ~50 cm. When a treatment unit is being used isocentrically, then FSD (e.g. 100 cm for a linear accelerator) for a parallel opposed pair is:

fixed FSD − (separation/2).

Folinic Acid (Leucovorin, Citrovorum Factor)

Folinic acid is a formyl derivative of terahydrofolic acid. The biologically active isomer is the L-isomer. The drug is usually provided as a racemic mixture of the inactive D-isomer as well as the active L form. Its primary use in oncology was to rescue cells from the toxic effects of folate antagonists, in particular methotrexate. Planned rescue with folinic acid is part of the regimens using high-dose **methotrexate** (**p. 199**) that has been used to treat sarcomas and lymphomas. The rescue is usually started 24 h after the methotrexate has been given and is maintained until the serum methotrexate level has fallen to <0.05 μm. In contrast to folic acid, it does not require reduction by dihydrofolate reductase, being rapidly metabolised to a series of active reduced folates (MTHF, mTHF). These reduced folates are polyglutamated by foly-polyglutamate synthetase. The polyglutamated folates are larger than ordinary folates and are also negatively charged. These factors increase the cellular retention of folate.

Folinic acid will increase the anti-cancer effect of **5-fluorouracil** (**5FU**) (**p. 130**), which acts primarily as an inhibitor of thymidylate synthetase (TS) through the covalent binding of mTHF and 5-fluorodeoxyuridine. Folinic acid increases the intracellular levels of mTHF available for binding to TS.

Founder Effect

The founder effect describes circumstances in which a population contains individuals with identical portions of chromosome compatible with an origin from a single ancestor — the founder. Classically, the founder effect is identified when an individual with a particular genetic abnormality emigrates to a community, which, in breeding terms, is relatively closed. Recent data suggest that the founder for a BRCA mutation originated in south-west Scotland.

Fractionation: Novel Fractionation Schemes

Fractionation has been used since the earliest days of clinical radiotherapy. Its introduction was, in part, due to the unreliability of the apparatus and the need for frequent rests, rather than based on any firm biological

principles. Fractionation can now be put on a more sound biological basis.

Smaller fractions, for any given total dose, will selectively spare late, as opposed to acute, effects. The shorter the overall treatment time, then the less opportunity there will be for tumour cells to proliferate during treatment. These two considerations lead to different, but not exclusive, approaches:

- **Hyperfractionation** (**p. 156**): defines any treatment in which fraction sizes are lower than those conventionally used. Fraction sizes in hyperfractionated radiotherapy are typically <1.7 Gy. Pure hyperfractionation is possible and may permit escalation of the total dose without any untoward increase in late effects: 70 Gy in 7 weeks at 2 Gy per fraction/day (35 fractions) is equivalent to 80.5 Gy in 7 weeks using two fractions of 1.15 Gy/day (70 fractions).
- Acceleration: for obscure historical reasons, it is accepted that the conventional overall treatment time for clinical radiotherapy is 6–8 weeks. Any treatment given in an overall time <6 weeks would be considered to be accelerated. This ignores the fact that in many departments 3–4 weeks would be regarded as the standard time over which to give **radical treatment** (**p. 259**). Pure acceleration, by the conventional definition, would involve giving 60 Gy in 30 fractions in 3 weeks: this is not tolerated because the rate of dose–accumulation exceeds the tolerance of the acutely responding tissues. In practice, a combination of acceleration and hyper-fractionation is used, e.g. 70 Gy at 2 Gy per fraction/day in 7 weeks might be comparable with 72 Gy at 1.6 Gy (t.i.d.) in 5 weeks.
- **CHART** (**p. 62**): continuous hyperfractionated accelerated radiotherapy is an attempt to obtain the best of both worlds: hyperfractionation and acceleration. A total dose of 54 Gy is given continuously, there are no rests at weekends, as thrice daily fractions, each of 1.5 Gy for a total of 12 days: 54 Gy in 36 fractions in 12 days. Recently completed randomised trials suggest that **CHART** (**p. 62**) has little advantage in terms of local control or survival for cancers of the head and neck. There is, however, evidence of benefit for patients with lung cancer. A revised version of CHART, more easily accommodated within existing departmental practices, is currently under investigation (CHARTWEL).

Frameshift Mutation

This type of mutation involves the deletion or insertion of a small number of nucleotides (one, two or four) so that the sequence of nucleotide triplets is altered. This means that the reading frame for the genetic code will be shifted, or displaced, and so a completely incorrect sequence will be read over a long length of DNA — even though only an apparently trivial alteration has taken place. The triplet sequence of the genetic code means that deletions or insertions of three consecutive nucleotides cannot produce frameshift mutations.

Free Radical

Free radicals are highly reactive, usually short-lived, chemical species. They are characterised by the presence of an unpaired electron, which confers the reactivity. Since water is the predominant molecule in mammalian cells, it is the free radicals formed from water that are, biologically, the most important. The radicals are not ions; they are electrically neutral with equal numbers of electrons and protons. It is simply that the electrons are arranged in such a way that the normal valency requirements are left unsatisfied. A radical is notated as '·'. The three most important free radicals are the

hydrated electron ($e_{aq}^{\bullet}$) the hydroxyl radical ($OH^{\bullet}$) and the hydrogen radical ($H^{\bullet}$).

Funnel Plot

This is a graphical method for assessing whether the results of a **meta-analysis** (**p. 195**) have been affected by publication bias. The funnel plot is obtained by plotting the size of the trial against the effect observed. The plot should be broad at its base, small trials bracketing widely around the pooled estimate of effect. The results of the larger trials should converge around the pooled estimate, the scatter plot narrows towards an apex. The results should be symmetrical about the pooled estimate and, therefore, resemble a funnel lying on its side. If publication **bias** (**p. 28**) has occurred, then there will a relative under-representation of small negative trials and the funnel will have an asymmetrical base, indicating where the 'suppressed' data would have been found (figures 1 and 2).

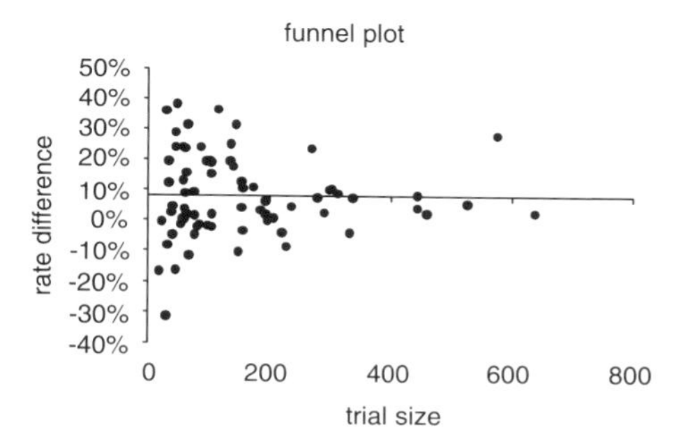

Figure 1 — Data from 74 randomised studies of chemotherapy in head and neck cancer, the pooled estimate of treatment benefit was 8% and the funnel plot is reasonably symmetrical around this — suggesting that the conclusions of the analysis have not been weakened by publication bias.

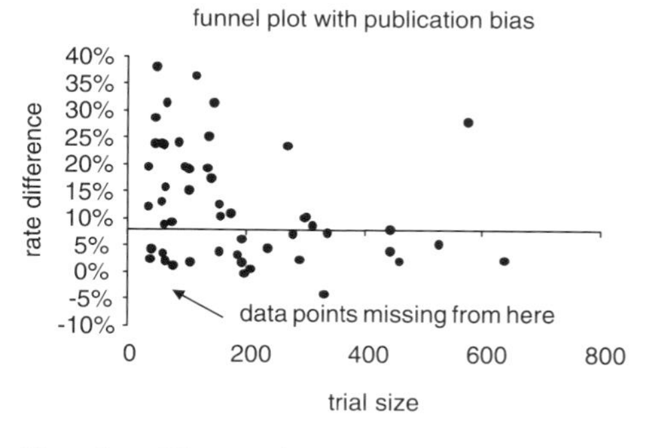

Figure 2 — The same data as in figure 1, but with the smaller negative trials omitted. The base of the funnel is now asymmetrical — consistent with publication bias.

Fusion Protein

A protein formed when the translocation of DNA from one chromosome to another results in the production of an abnormal protein, which consists of amino acid sequences from both the host chromosome and the translocated chromosome. Leukaemic cells in patients with acute promyelocytic leukaemia (APL) have a t(15:17) chromosomal translocation and the abnormal protein product (PML/RAR α) contains the retinoic acid binding **domain** (**p. 102**) of the **retinoic acid receptor** (**RAR**) (**p. 271**) α. This offers therapeutic potential: treatment with all-*trans*-retinoic acid can induce differentiation and improve survival in patients with APL. The PML moiety is less well characterised but appears to act as a **transcription factor** (**p. 313**).

Fv Fragment

A construct derived from an immunoglobulin molecule; it consists of the V_H chain linked to the V_λ chain by a flexible peptide bridge that prevents dissociation of the two short polypeptide chains. This relatively simple single polypeptide chain can be fused with a phage vector and then be inserted into *E. coli*. This permits the production of large quantities of specific antibody, which is effectively of human origin and, therefore, suffers few of the disadvantages associated with murine monoclonal antibodies.

G Protein-Linked Receptor

This is a receptor protein formed as a single polypeptide chain. The free $-NH_2$ terminal projects into the extracellular space; a transmembrane portion is comprised of a series of loops, the ends of which project into both the extracellular space and the cytosol. The free –COOH terminal projects into the cytoplasm. This versatile class of receptor is ideally constructed for conveying messages from the extracellular environment into the interior of the cell. There are >100 types of G protein-linked receptor, their DNA sequences being highly conserved. The binding site for the **ligand** (**p. 180**) is near the $-NH_2$ terminal, in the extracellular space. Sequences in the loop area, just within the plasma membrane, bind to a class of GTP-binding proteins, the trimeric G proteins. Towards the –COOH end are regions that, when phosphorylated, cause the receptor to become desensitised.

The trimeric G proteins will inactivate themselves by hydrolysing their bound GTP if the extracellular signal is turned off. If the trimeric G proteins remain activated, their cellular effects are mediated by two main pathways: **cyclic AMP** (**p. 82**) or Ca^{2+}. A crucial feature of the G protein receptors is that once the ligand has been bound, continued action of adenylyl cyclase/cAMP is possible. The effect can persist for a short while even after the ligand has dissociated from the receptor. The overall effect is that G protein receptors can amplify an extracellular signal in terms of its effects on intracellular adenylyl cyclase and cAMP production.

Trimeric G proteins can also act via a pathway involving inositol phospholipids. The products are inositol triphosphate and diacylglycerol. Inositol triphosphate acts as an intracellular messenger and can cause release of Ca^{2+} and cause events such as muscle contraction. Diacylglycerol can activate the Ca^{2+}-dependent enzyme protein kinase C. This in turn can activate gene transcription by at least two separate mechanisms: direct activation by a cascade of protein kinases; phosphorylation of a inhibitor of gene transcription — activation by inactivating the inactivator. Cleavage of diacylglycerol produces arachnidonic acid, an intracellular messenger in its own right, which can stimulate the production of **eicosanoids** (**p. 112**).

Gadolinium (Gd)

A contrast agent used in **magnetic resonance imaging** (**MRI**) (**p. 208**). It is a strongly ferromagnetic silvery white solid. It is this property that is exploited in its use as an MRI contrast agent. It increases the contrast between the $\mathbf{T_1}$ (**p. 303**) and $\mathbf{T_2}$ (**p. 303**) of the tissues. Tumours will often preferentially take up gadolinium (Gd) and this shortens T_1 and produces a larger signal. This can be used to differentiate tumours from normal tissues.

Galbraith Plot

This provides a graphical method for assessing the degree of heterogeneity among **odds ratios** (**p. 217**) in a **meta-analysis** (**p. 195**). The z statistic for each trial (the log of odds ratio divided by **standard error** (**p. 296**)) is plotted against the reciprocal of the standard error. The slope through the origin corresponds to the overall log odds ratio, 95% confidence bounds are plotted above and below and delineate an area within which, in the absence of **heterogeneity** (**p. 150**), all trials would be expected to lie. Any outliers should be looked at carefully since, on the treasure your **exceptions** (**p. 122**) principle, they might provide important information.

Gap Junction

A specialised cell–cell junction in which the plasma membranes are 2–4 nm apart and there is a water-filled channel that allows molecules to flow directly from one cell to another. The size limit for such transfer is ~1000 daltons. The channels are formed from six transmembrane subunits, connexins. Nearly all embryonic cells have gap junctions. Physiological changes in Ca^{2+} or pH can influence the function of gap junctions. When a cell is damaged, its gap junctions seal off, isolating the cell from its neighbours so that the metabolic changes associated with injury are not transmitted to neighbouring cells.

Gardner's Syndrome

This is a variant of **familial adenomatous polyposis** (FAP (**p. 125**), syn. familial polyposis of the colon, FPC). It is inherited as an autosomal dominant and is associated with mutations in the ***APC* gene** (**p. 18**). Clinically, there are multiple polyps of the gastrointestinal tract, particularly the colon, associated with jaw cysts, osteomas, abnormal teeth, other tumours outwith the bowel, retinal pigment deposits (congenital hypertrophy of the retinal pigment, CHRPE), carcinomas of the colon developing in early adult life, and desmoid tumours. Gardner's syndrome, which was originally described in a Mormon kindred, probably arises as a form of FAP with a particular cluster of phenotypic abnormalities.

GATA1 and GATA2

These **zinc finger proteins** (**p. 331**) (together with a cofactor FOG, friend of GATA) are important in differentiation and maturation during erythropoiesis and thrombocytopoiesis. The complex, of transcription factor and cofactor, is crucial for determining the tissue specificity of gene expression: as such, it might be a model for explaining why, although all cells have the same genetic material, some become red cells and others neurones.

G-CSF, Filgrastim

G-CSF is a haemopoietic colony-stimulating factor. It is a polypeptide chain of 175 amino acids, which, using recombinant technology, is produced in *E. coli*. It selectively stimulates the production of neutrophils as well as mobilising peripheral blood stem cells. It is used to prevent or treat neutropenia and is part of the conditioning regimen used before harvesting peripheral blood stem cells (PBSC) for marrow rescue after high-dose therapy. It binds to receptors on the surface of granulocytes that have already committed themselves to neutrophil differentiation and it stimulates cell division, differentiation and activation. The natural growth factor is normally produced by fibroblasts, endothelial cells and monocytes.

Gehan's Number; Gehan's Rule

These techniques are used to estimate the number of patients that need to be treated before we can be reasonably certain that a treatment does not work. Their main use is in Phase II studies of new drugs. If a new drug is ineffective, there is no point in continuing a study past the point at which it can be shown, beyond reasonable doubt, to have no significant activity. The number usually used is 14. This assumes that we would reject as ineffective any drug that produced a response rate of $<20\%$ and that we also would accept the risk that, 5% of the time, a false-negative conclusion might be drawn.

The mathematics behind this are simply those of probability and lotteries. Imagine a bag containing 100 balls of different colours, 20 of the balls being red. The probability of not drawing a red ball first time is 0.8, i.e. $(1-0.2)$. The probability of not drawing a red ball at the first or second draw is 0.8×0.8, i.e. 0.64 the probability of not having drawn a red ball by the *n*th draw is $(0.8)^n$. By the 14th draw the probability of not having drawn a red ball is 0.044. Thus, if there were, in fact, <20 red balls in the bag, we could be 95% certain of this after 14 draws had not produced any red balls.

The general relationship between the response rate that is defined as useful, *r*, and the risk we are prepared to take of drawing a false-negative conclusion, α, is:

$$\text{number of patients required} = \frac{\log \alpha}{\log (1 - r)}.$$

This can be plotted graphically: the percentage rates attached to each line refer to the false-negative rate (figure 1).

An important point that is often forgotten is that achieving one response within 14 patients simply indicates that it is worth continuing the study, not that the effectiveness of the new drug has been proved. Gehan's method is a method for excluding ineffective drugs. It is not designed to test the hypothesis that a new drug has significant activity.

There are many other methods for estimating the sample size appropriate for a Phase I or II study. In modified Play the Winner, 20 patients are treated and we add five for each response observed, up to a maximum of 20. Other approaches have used Bayesian

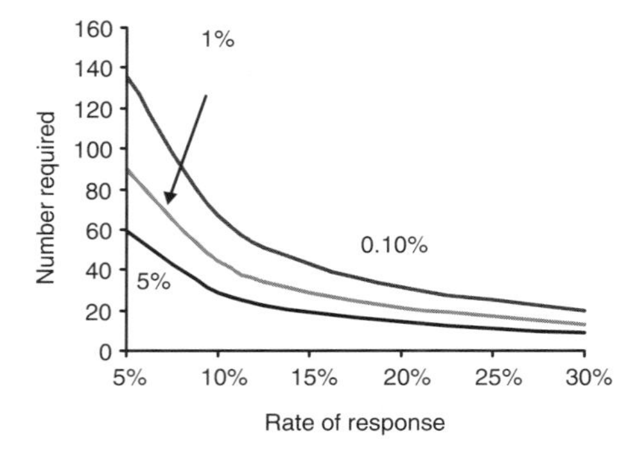

Figure 1 — Gehan's method.

inference or decision theory. Blackwelder described a method based on hypothesis testing and the binomial distribution: this has been used in Phase I/II dose-finding studies in radiotherapy.

Gemcitabine

Gemcitabine is an antimetabolite with activity against several tumour types, including non-small cell lung cancer and pancreatic cancer. It is **cell cycle (p. 52)** S-phase-specific, with cells accumulating at the G_1/S boundary after treatment with gemcitabine. It was the first **cytotoxic (p. 84)** drug to be licensed in the USA on the basis of an effect on **quality of life (p. 253)** as opposed to the more traditional criteria of tumour response. It is structurally similar to cytosine arabinoside. The triphosphate derivative of gemcitabine is a competitive inhibitor of DNA polymerase. Before the polymerase is halted a further base pair is inserted, a phenomenon known as 'masked termination'. This makes repair of the defect more difficult and adds to the effectiveness of the drug. Gemcitabine can, by inducing the activity of deoxycytidine kinase, stimulate its own activation. This prolongs the duration of its intracellular activity. Its main toxicity is haematological.

Gene Gun

A method for delivering plasmid DNA to human cells so that expression of the plasmid genes can be used to generate therapeutically useful proteins. The approach has been used in the delivery of **cancer vaccines (p. 45)**. The plasmids are coated onto gold beads 1 μm in diameter. These are then accelerated, using compressed helium, to the speed of sound and injected into the cytoplasm of skin epidermal cells. The plasmid DNA then expresses the appropriate protein, which then reaches the general circulation. The gene gun is about the size of a fountain pen and injections are entirely painless. An advantage of this delivery system is that it requires much less DNA, perhaps 100 times less, than intramuscular injection for successful gene expression.

Gene Therapy

Gene therapy is based on the simple premise that if bad genes cause or perpetuate cancer, then replacing the bad genes with good versions should be beneficial. There are major ethical and practical problems associated with this approach to managing cancer.

The main ethical problem is whether, for diseases with a predictable pattern of inheritance, such as **retinoblastoma (p. 271)**, it is ethical to attempt to correct the genetic defect in germline cells. Any genetic manipulation will be passed on from generation to generation. Somatic cell gene therapy for an individual patient is a less difficult ethical issue: the therapeutic episode begins and ends with that individual.

The practical problems associated with gene therapy are those of delivery. How do we get a functional version of the gene exactly where we want it to go? What is the best vector to carry the construct to its destination and, once the destination has been reached, how do we ensure that the whole system functions appropriately?

There are two main approaches to delivery. In *ex vivo* gene delivery cells growing in culture are exposed to the construct and then infused into the host. In *in vivo* gene delivery the construct is infused or injected directly into the host.

There are three main vector systems that have been used for gene therapy:

- Plasmid DNA: this is relatively simple to prepare but the efficiency of delivery is low. It cannot be used for *ex vivo* transfer and, *in vivo*, it is most effective when given by intratumoral injection, a procedure that is not always feasible.
- Viral delivery systems:
  - — Recombinant retroviruses: a retrovirus is an obvious vector for gene therapy since its natural behaviour is to insert **DNA sequences (p. 101)** into the host genome. The technology to produce appropriate constructs is well developed and gene transfer, both *ex vivo* and *in vivo*, has been successfully achieved in a variety of systems. Retroviruses will insert sequences only into the genome of dividing cells: this may be both an advantage, in terms of protection of normal tissues, and a disadvantage, in terms of leaving resting but clonogenic cells behind. The effective concentration of infective particles achieved by the production systems used for retrovirally based gene therapy is, unfortunately, low. This may limit the effectiveness of the procedure.
  - — Adenoviruses: adenoviruses do not insert DNA into the host genome and, therefore, can deliver gene therapy to non-dividing cells. It is easy to produce high concentrations of virus particles and so the sheer weight of numbers can compensate for any inefficiencies in delivery. The adenovirus is recognised as foreign by the host. Repeated administration is, therefore, limited by the development of an immune response to the vector.

There is a variety of therapeutic approaches using gene therapy:

Suppress expression of oncogene (e.g. antisense)

Block the gene product (mRNA) and stop it directing protein synthesis. If the amino acid

sequence of the gene product is known then, using the genetic code, the **DNA sequence** (**p. 101**) can be predicted. A complementary oligonucleotide can be synthesised that is complementary to an RNA sequence arising from that gene. If the oligonucleotide is introduced into the cell it will, through the complementary binding of base pairs, block any transcription of the mRNA by the ribosomes. The synthesis of the gene product will be specifically abolished by an oligonucleotide that, in theory at least, should be completely non-toxic to normal cells. This approach has been used in non-Hodgkin's lymphomas — an **antisense oligonucleotide** (**p. 17**) is constructed to a sequence of mRNA coded by the *bcl* gene. *bcl* protects cells from **apoptosis** (**p. 19**) and overexpression of *bcl* contributes to the propagation of the **malignant process** (**p. 185**). Antisense treatment will produce a fall in the amount of *bcl* gene product produced but, unfortunately, tumour responses are unpredictable and the therapy is more toxic than might be expected. The findings suggest that antisense therapy may be less selective than the theory suggests.

Restore defective function of tumour suppressor gene

Over 2000 mutations of ***p53*** (**p. 224**) have been recorded in association with human malignancy. Mutations in *p53* are almost certainly the commonest genetic abnormality associated with malignant disease in humans. Restoration of normal *p53* function is, therefore, a logical target for gene therapy. Early trials have already taken place. In one study a *p53*/adenoviral CHECK construct was introduced into lung cancers by intratumoral injection. Subsequent biopsies showed expression of normal *p53* (**proof of principle** (**p. 248**)), and some tumour responses were observed in the patients so treated.

Enhance the immunogenicity of a tumour

A construct containing a cytokine (e.g. **IL-2**) (**p. 164**) or immunostimulatory molecule (B7) can be used to target tumour cells. The construct will, by activation of effector cells, produce immunologically specific killing of tumour cells. A cascade of events will be triggered, which will deal not only with the original tumour, but also with any metastases. The beauty of this approach is that because of its self-propagating nature, only a few tumour cells need to be infected to unleash an army of cells specifically toxic to tumour cells. Gene transfer need not be particularly efficient. The approach is suitable for *ex vivo* use.

Insert cytotoxic genes into cells that will specifically target the tumour

This approach has treated melanoma with **TNF** (**tumour necrosis factor**) (**p. 317**). Tumour infiltrating lymphocytes (TIL) are obtained from a biopsy of the patient's tumour. These cells will specifically interact with tumour-related antigens. Inserting the gene for TNF then modifies the TIL. The modified cells are grown in culture and then infused into the patient. The TIL target the melanoma and release TNF at precisely the site at which it might have most effect.

Therapy using modified virus

ONYX-015 (**p. 220**) is a mutant adenovirus with activity against malignant disease. The virus has been designed specifically as an anti-cancer agent. The reasoning is as follows. Virus infection can activate *p53*, which in turn might activate apoptosis. Small DNA viruses, however, contain genes, *E1A* and ***E1B*** (**p. 111**), that can inhibit the activation of *p53* and any consequent apoptosis. In ONYX-015, a part of *E1B* is deleted. If the ONYX-015 infects a cell with normal *p53*, then the *p53*, in the absence of normal viral

E1B gene product, can inhibit viral replication: virally induced cell killing is, therefore, minimal in cells with normal ***p53*** (**p. 224**). The mechanism for this inhibition is, as yet, unknown. Many cells in human tumours contain mutant *p53*. This mutant *p53* cannot inhibit the replication and cytopathic effects of ONYX-015. The net result is selective killing by the virus of tumour cells with their mutant *p53* and selective sparing of normal cells with their wild-type *p53*. The experimental data so far bear out this hypothesis. **ONYX-015** (**p. 220**) appears to be selectively toxic to human tumour cell lines, either grown in culture or as xenografts. It is encouraging that several of these lines were derived from tumours resistant to conventional chemotherapy.

Mark tumour cells

If a specific genetic marker is inserted into tumour cells, then any cells remaining after treatment can be easily identified and the therapy altered accordingly. The marker most commonly used for this purpose is neoR (neomycin phosphotransferase).

Suicide gene therapy

If a gene that activates a prodrug is inserted specifically into malignant cells, then when the prodrug is subsequently administered only the cancer cells will activate it. Normal cells will lack the enzyme to activate the drug and so will be unaffected by it. Cancer treatment can be made specific by persuading cancer cells to incorporate a gene that will, given subsequent manipulation, cause their own destruction. Several prodrug/enzyme systems are available for this type of therapy:

- Thymidine kinase will activate Ganciclovir.
- Linamarase will convert amygdalin (non-toxic) to cyanide (highly toxic).
- Cytosine deaminase will convert **5-fluorocytosine** (**p. 130**) to 5-fluorouracil (5FU).

This approach has been used to treat gliomas. The herpes simplex virus thymidine kinase (*HSVtk*) gene is delivered to the tumour. The patient is then treated with Ganciclovir, a drug already approved for the treatment of cytomegalovirus (CMV) infection. Ganciclovir is activated only where there is thymidine kinase activity, predominantly within the tumour, and the normal tissues are left relatively unharmed.

Bystander killing is an important collateral advantage of this approach to therapy. Treatment may kill even those cells that have not incorporated the construct. The explanation for this probably lies in cell–cell transfer of toxic metabolites: via either gap junctions or apoptotic vesicles.

The concept of target-specific activation of **prodrug** (**p. 246**) for cancer therapy appears in a variety of guises: VDEPT, GPAT, **ADEPT** (**p. 7**). In virus-directed enzyme prodrug therapy (VDEPT), the molecular switch, which is known to control a cancer-specific gene, is (using a viral vector) introduced into the tumour. This then activates the gene and, in turn, the gene activates the enzyme, which converts the prodrug to active drug. Since only the cancer cells contain the specific gene, only the cancer cells will activate the drug. This approach is an example of a broader strategy termed GPAT — genetic prodrug activation therapy: other vectors, such as **liposomes** (**p. 182**), might be used to introduce the molecular switch into the tumour. ADEPT is antibody-directed prodrug therapy. An antibody, directed against a tumour-specific antigen is linked to an enzyme. The enzyme then converts an inactive precursor (prodrug) to an active cytotoxic agent. Since this activation will only occur where enzyme is present, and since, thanks to the antibody, there is only enzyme present where there is tumour, the approach offers a means of improving the **selective toxicity** (**p. 282**) of cancer therapy. Examples of this approach include: MAb BW431/26 conjugated to alkaline **phosphatase** (**p. 233**), the antibody binds to CEA, the enzyme activates **etoposide** (**p. 119**) phosphate (prodrug) and the locally active cytotoxic agent is etoposide; and MAb 323/A3 (antibody to a membrane glycoprotein found in carcinomas) conjugated to β-glucuronidase, epirubicin glucuronide is given as prodrug, epirubicin is the active agent. This approach is also sometimes called ADC (antibody-directed catalysis).

Risks associated with gene therapy

- **Insertional mutagenesis** (**p. 161**): a genetic sequence inserted into the host genome might itself act as a mutagen and is, therefore, potentially oncogenic.
- Oncogene activation: the inserted sequence might activate a proto-oncogene and, thereby, produce malignant transformation.
- Infection: when viral vectors are used there is always the possibility that a passenger virus might accompany the transfer of the vector. This is quite apart from any infective properties the vector itself might possess.
- Recombination: as part of the general genetic shuffling occurring *in vivo* as part of gene therapy, genes might combine to produce a novel sequence with properties that are unknown, unknowable and, possibly, deleterious.

Genomics

Genomics is the study of gene sequences: it does not deal with whether a protein is actually made by the gene.

Geometric Mean

For data whose distribution is skewed, the standard calculation of the mean (arithmetic mean) maybe misleading. The geometric mean reflects more accurately the average: its calculation, and that of its **standard deviation** (**SD**) (**p. 296**), is as follows:

- Transform observations by taking the $\log_{10}$ of each value.
- Calculate the SD of transformed observations.
- Calculate the mean of transformed observations.
- Construct a 95% **confidence interval** (**p. 70**) (CI) for the transformed observations as: mean $\pm 1.96 \times$ SD.
- Convert these values back to untransformed values by using antilogs.

Figure 1 shows data (hypothetical) on tumour size distribution in patients with screen-detected breast cancer.

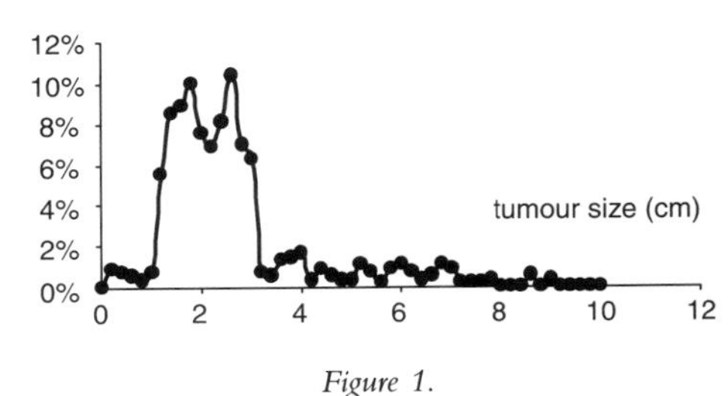

Figure 1.

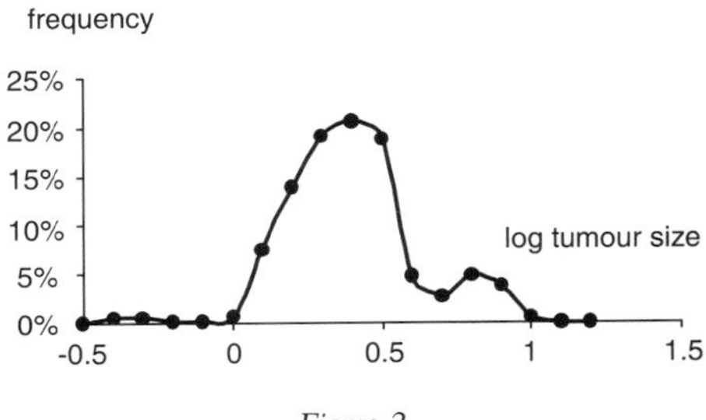

Figure 2.

After log transformation (figure 2).

The arithmetic mean is 2.50 (SD 1.52) with 95% CI = 2.41 − 2.65. The geometric mean is 2.16 (SD 1.79) with 95% CI = 2.03 − 2.24. The geometric mean is significantly lower than the arithmetic mean — it is less influenced by the few large tumours (>5 cm) in the series. The clue to the skew in the data, and its correction by transformation, lies in the comparison of means and medians (table 1).

Table 1

	Arithmetic	**Geometric**
Mean	2.50	2.16
Median	2.16	2.17

Germline Mutation

The abnormality is present in the ovum and is, therefore, present in all cells derived from that ovum. Since this will include the germinal epithelium, germline mutations, in contrast to somatic mutations, can be transmitted down the generations.

Global Positioning

This involves gaining an approximate idea of the location of a gene within the genome. There are several useful methods:

- SCM (**sequence-tagged sites** (**STS**) (**p. 284**) content mapping): mapped STS are compared against the clone by **polymerase chain reaction** (**PCR**) (**p. 238**).
- Hybridisation: mapped clones probed against clones mounted on membranes.
- RH (radiation hybrid) mapping: check to see whether a given STS resides in one of many RH chromosome fragments. By using 100 RH fragments, a signature of the chromosomal location of the STS can be derived. Inner product mapping (IPM) is a derivative of this approach that offers the potential ability to sequence extensive lengths of chromosome rapidly and inexpensively.

Glucagonoma

A rare endocrine tumour associated with excess production of glucagon. It usually arises in the endocrine pancreas. Its most spectacular clinical manifestation, other than hyperglycaemia, is necrolytic migratory erythema: an unpleasant skin condition characterised by episodic dispersed areas of erythema, resembling thrombophlebitis, but with conspicuous tendency to undergo necrosis. It can arise as part of a **multiple endocrine neoplasia** (**MEN**) **syndrome** (**p. 194**).

GM-CSF (Sargramostim)

GM-CSF, granulocyte–macrophage colony-stimulating factor, is a growth factor for haemopoietic cells. It stimulates the production and differentiation of macrophages and granulocytes; it also stimulates the function of the mature cells, improving their ability to migrate into the tissues and combat fungal and bacterial infections. Physiologically, GM-CSF is produced by activated T-cells, fibroblasts, endothelial cells and monocyte/macrophages. It is available as genetically engineered product: as a polypeptide ~125 amino acids long produced in yeast and available as a subcutaneous injection. It is used in treating or preventing neutropenia in patients treated with high doses of **cytotoxic drugs** (**p. 84**). It can also help mobilise peripheral blood stem cell (PBSC) for harvesting by cell separation.

Gold Standard

An overused expression describing an intervention or test considered to be the best available. A test regarded as a Gold Standard gives results closest to the possible approximation to absolute truth. The performance of potential new interventions or tests is judged against the standard. Gold Standards are often expensive or cumbersome. A new procedure may not perform as well as the Gold Standard but, because of cost or practicability, it might be preferred to the Gold Standard procedure for clinical use. Having been calibrated against the Gold Standard, its deficiencies will be well defined and a rational interpretation of results will be possible.

Gompertzian Growth

The calculation of the commercial rates for annuities was, and is, an important aspect of the insurance business. Benjamin Gompertz described in mathematical terms, the relationship between age and **mortality rate (p. 207)** to put the calculation of the price of an annuity on a more rational basis. He postulated that, as people aged, their resistance to death decreased and he derived an equation defining the relationship between age and the risk of dying. It so happens that the equation describes, fairly accurately, observed patterns of tumour growth. This pattern is sometimes described as Gompertzian, which (in not so plain English) describes a pattern of exponential growth in which the **doubling time (p. 107)** doubles exponentially. In the early stages of tumour growth there are few constraints on proliferation and the tumour cells increase in number exponentially. The cells progress inexorably round the cell cycle and the kinetics of the tumour are simply those of the cell cycle for that particular cell type. The time taken for the tumour cells to double in number (the **doubling time** t_d, **p. 107**) is simply the **cell cycle (p. 52)** time. However, as the tumour grows, oxygen and other nutrients have to diffuse

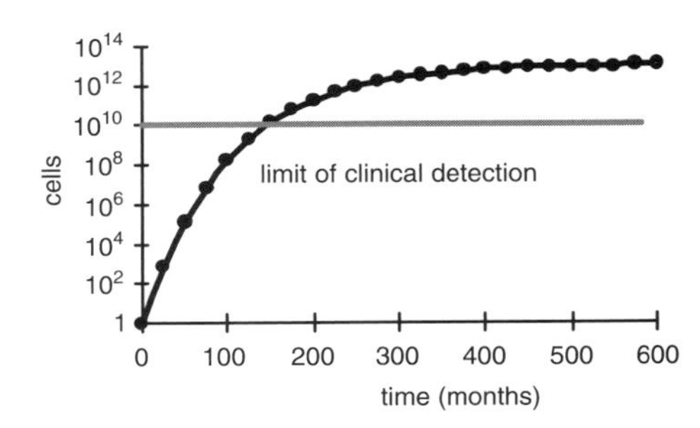

Figure 1 — Gompertz's equation applied to the growth of a tumour. Time is plotted in months. A lump 1 cm in diameter contains ~10^9 cancer cells and is at the threshold of detectability using modern X-ray and scanning techniques. A cancer containing 10^{13} cells is larger than a human being. In this example, it takes ~12 years of proliferation before the cancer crosses the threshold of clinical detectability, within the next 5 years the cancer grows to a size that will kill the patient.

over increasing distances to reach the cells at the centre of the tumour mass. This imposes constraints on proliferation since the various cell cycle checkpoints ensure that poorly nourished cells do not divide. The effect on the tumour as a whole is that, as it increases in size, the rate of growth slows. Since many cancer treatments are most active against proliferating cells, it is ironic that proliferation is slowing just as tumours become detectable (figure 1).

Goodness of Fit

The extent to which observed data correspond to a theoretical or mathematically derived distribution. The expression is usually, but not exclusively, used in connection with data that can be presented graphically.

Gorlin Syndrome

A recessively inherited syndrome conferring susceptibility to basal cell carcinomas and primary CNS tumours. ***PTCH* (p. 251)** is the gene responsible.

Goserelin

Goserelin is an analogue of GnRH (gonadotrophin releasing hormone): it is a GnRH agonist. It stimulates the release of the gonadotrophins, luteinising hormone (LH) and follicle-stimulating hormone (FSH), from the anterior pituitary. The normal control of gonadotrophins is by episodic rises and fall in GnRH. The sustained administration of a GnRH analogue, such as goserelin, will lead to the **down-regulation** (**p. 107**) (i.e. suppression) of the release of LH and FSH. The reduction in FSH and LH levels in turn causes a fall in levels of oestrogen in women and testosterone in men. Goserelin can therefore be used in the systemic treatment of hormone-responsive malignancies such as breast cancer and prostate cancer. It provides a form of medical hypophysectomy. Because it is an agonist, goserelin will cause an initial rise in levels of FSH and LH. This could cause an initial flare of tumour activity, which should be blocked by treatment with an anti-androgen (**flutamide** (**p. 131**)) or cyproterone acetate) or anti-oestrogen (**tamoxifen**) (**p. 304**). It is also used in the treatment of non-malignant conditions such as endometriosis or dysfunctional uterine bleeding. More controversially, it has been used as a form of chemical castration for men with a history of repeated sexual offences.

Graft versus Host Disease

A syndrome encountered after allogeneic **bone marrow transplantation** (**p. 34**) in which the engrafted immunocompetent cells attempt to reject their, immunologically distinct, new host. Clinically, it resembles an autoimmune syndrome with skin rashes, lethargy, renal and hepatic impairment. It can be treated, at least in part, by increasing the dose of immunosuppressive therapy. Some degree of graft versus host disease may be beneficial — the graft may also try to reject any residual leukaemic cells (graft versus leukaemia effect).

Graft versus Leukaemia Effect

A phenomenon seen after allogeneic bone marrow transplantation in which the engrafted immunocompetent cells destroy residual leukaemic cells of host origin. The effect may explain why, in some circumstances, allogeneic transplants might be associated with a lower leukaemic relapse rates than autologous transplantation since in the latter approach there is no self/non-self distinction to be made. There is a price potentially to be paid, however: **graft versus host disease**.

Gray

A unit of **absorbed dose** (**p. 1**) of radiation named after Hal Gray who was an early leader in British radiation biology and physics. Following political upheavals in his laboratory, he was, at one point, reduced to writing papers in a car parked between Hammersmith Hospital and Wormwood Scrubs Prison. Denied access to his laboratories, assistants scurried back and forth bearing data and questions.

$$1 \text{ gray (Gy)} = \begin{cases} \text{dose absorbed when} \\ 1\text{ J is deposited} \\ \text{in 1 kg medium.} \end{cases}$$

The older unit of absorbed dose was the rad:

$$1 \text{ rad} = 100 \text{ ergs g}^{-1} \text{ irradiated medium}$$
$$= 10^{-2}\text{ J kg}^{-1}$$

$$100 \text{ rad} = 1 \text{ Gy}$$

$$1 \text{ rad} = 1 \text{ cGy (centigray).}$$

Grey Literature

That part of the scientific literature that has not been published in widely circulated **peer-reviewed (p. 229)** journals has been termed the grey literature. Its importance emerges when an all-inclusive approach to the synthesis of evidence, such as **meta-analysis (p. 195)**, is used. Around one-quarter of **randomised controlled trials (p. 261)** are never published. Some of these, usually negative, studies can be traced through meeting abstracts, dissertations, policy documents, research reports, consensus statements — the grey literature. The European Association for Grey Literature has compiled a database, SIGLE, containing many such non-publications.

Gross Tumour Volume (GTV)

A term used in radiotherapy treatment planning that defines the known extent of the tumour, but does not include possible microscopic extension. In simple terms, it would correspond to that which is palpable, or demonstrable on scanning. It would not include impalpable disease or any potentially involved regional nodes. It is the smallest volume within the definitions proposed in ICRU Report 50.

Grounded Theory

A form of induction in which a theory is based on a series of observations: 'we have observed this pattern of behaviour in this group of people, now let us concoct a theory to explain why they do the things they do'. It is more often encountered in sociology than in mainstream science.

Growth Factor

Growth factors are proteins whose presence is essential for the proliferation of the cells on which they act. The colony-stimulating factors (CSF) of the marrow are growth factors, and each factor stimulates proliferation of a particular lineage: **erythropoietin (p. 119)** and interleukin 3 (IL-3) for erythrocyte production; GM-CSF for granulocyte–macrophage precursors; **G-CSF (p. 135)** for granulocytes. Other growth factors include epidermal growth factor, platelet-derived growth factor, nerve growth factor, etc.

Growth Fraction (GF)

A simple four-compartment model can describe the cells within a tumour or tissue:

- Actively dividing (cycling) cells.
- Cells that have differentiated and are incapable of further division.
- Cells capable of division but which are resting.
- Dead cells.

GF is the ratio of the number of proliferating cells to the total number of viable cells: that fraction of cells within a tumour or tissue that is actively proceeding around the **cell cycle (p. 52)** is the GF. It can be estimated using continuous labelling with tritiated thymidine (proportion of labelled cells equals growth fraction) or by extrapolation from the **labelling index (p. 173)**, assessed using a pulse of tritiated thymidine ($GF \sim LI \times 3$, since cell cycle duration is about three times as long as S-phase). GF can be measured more accurately using pulsed injection of tritiated thymidine and then, after several cell cycle times, by counting labelled cells and labelled mitoses:

$$GF = \text{fraction of labelled cells} \div \text{fraction of labelled mitoses.}$$

Estimates of GF vary widely between tumours: from >90% for acute leukaemias and high grade lymphomas to <10% for some adenocarcinomas: most human tumours have GF ~40%. GF can be used, in combination with an estimate of the cell cycle time, to estimate the maximum rate at which tumour cell number could double: T_d, the **potential doubling time** (**p. 241**).

GSPT1 Genes (π Class Glutathione S-transferase)

GST genes in general are protective against carcinogens, whether endogenous or environmental. The *GSPT1* system is particularly important in defence against **oxidative stress** (**p. 221**), e.g. hydrogen peroxide and its organified derivatives. Defective or deficient *GSPT1* might be an important early step in carcinogenesis within the prostate.

Guidelines

There is now quite an industry that has arisen around the formulation and dissemination of guidelines. Guidelines come in two main categories: clinical, designed to help a physician decide on, and implement, the best management for an individual patient; and those designed to ensure that those commissioning or purchasing healthcare know what to ask for. The two categories of guideline will, to some extent, overlap, but the formulation of the conclusions will be somewhat different. The essential basis of any guideline is a structured and comprehensive review of the evidence relevant to the matter in hand. One definition of a clinical guideline is: 'systematically developed statements to assist practitioner and patient decisions about appropriate healthcare for specific clinical circumstances'. The term 'practice parameters' is sometimes used to describe guidelines and it brings with it, appropriately enough, connotations of restriction, control and the setting of boundaries. Guidelines provide a mechanism whereby attempts to contain costs might be disguised as methods for improving standards. The way the evidence is summarised and presented will depend on the uses to which the guideline will be put. Poorly constructed guidelines and those driven by an economic agenda will do little to improve healthcare: it is vital, therefore, to have a set of critical standards for appraising guidelines. Guidelines cannot be taken as holy writ. The process by which guidelines are developed and implemented is one amenable to scrutiny. No guideline can be better than the evidence that went into it and it is not entirely frivolous to point out that there should be, and are, meta-guidelines: guidelines about guidelines. Any clinical guideline should contain, with adequate explanation, the following features:

- Specification of purpose, rationale and importance.
- An account of those involved in developing the guidelines, and why they were selected for the task.
- Clear definition of the issues or problem that the guidelines are to address.
- Definition of the subjects (patients, population) to whom the guidelines should apply.
- Definition of the potential users of the guidelines.
- All potential options for the particular problem are explored and defined.
- There is a defined method for obtaining and analysing evidence, the evidence is clearly identified and referenced, criteria for accepting and rejecting evidence are defined in advance of any assessment of the evidence.
- Evidence should be systematically assessed and graded according to its quality and its applicability to the problem in hand.

- Risks and costs of interventions are specified and, where possible, quantified. This is in addition to a similar analysis of the benefits.
- Any subjective judgements by the guideline developers should be clearly identified as such.
- The views and preferences of patients should be solicited, acknowledged and incorporated into any recommendations.
- Any recommendations should be graded according to the strength of the underlying evidence.
- A time limit, or date for a full review, should be placed on the use of the guidelines.

Guidelines can directly address process, they can only influence outcome indirectly. The assumption that good outcome always follows from good process is not entirely warranted. We can be right for the wrong reasons and, conversely, disasters can occur despite impeccable performance. Reviews of guidelines tend to concentrate on process not outcome: a review of 59 evaluations of guidelines showed that their effect on outcome was only assessed in 11 studies. The effect was beneficial in nine of the 11, but what was the effect in the remaining 48 studies? The overall quality of published guidelines is not good: in a recent review only about one-third of guidelines contained an adequate account of how the evidence had been assessed and summarised.

Haemopoietin Receptor Superfamily

This is a family of transmembrane proteins involved in the regulation of haemopoiesis. The common feature is a cysteine-rich common binding site at the N-terminal end. Members of the family include the receptors for interleukins 2–4, 6 and 7, **erythropoietin** (**p. 119**), prolactin, **G-CSF** (**p. 135**), and GM-CSF.

Half-life

The half-life of an isotope or drug defines the time required for the activity, or concentration, to decrease to 50% of the initial value. Each radioactive isotope has a half-life characteristic of that isotope and which is impossible to change by any means whatsoever. It is this constancy of half-life that enables estimates of geological time to be made on the basis of the constant decay of carbon isotopes. The mirror image of half-life is exponential growth. Half-life equations can describe radioactive decay as well as changes in plasma drug concentration over time.

The equations are of the general form:

$$t = \log_e 2\, \lambda^{-1},$$

where t is the half-life, $\log_e 2 = 0.693$ and λ is a constant representing the change in activity unit time^{-1}. For a radioactive isotope, λ is the **transformation constant** (**p. 313**), $-(\Delta N/\Delta t)/N$, where $\Delta N/\Delta t$ is the rate of decay, also termed the **activity** (**p. 5**) of the sample, and N is the number of radioactive atoms present in the sample.

The biological half-life of a drug is calculated as:

$$\frac{0.693}{\text{elimination rate constant}},$$

where the **elimination rate constant** (**p. 113**) is calculated as:

$$\frac{\text{rate of drug elimination}}{\text{amount of drug in the body}},$$

or as:

$$\frac{\text{clearance}}{\text{volume of distribution}}.$$

Halo Effect

A phenomenon encountered in psychometric testing in which satisfaction in one item under scrutiny diffuses to influence satisfaction in other, totally unrelated, items. An effective analgesic might, through a general glow of wellbeing produced in a patient who has been suffering from severe pain, diminish chemotherapy-induced nausea. Both pain and nausea scores would improve, but this does not imply that the analgesic is also an anti-emetic.

HAMA (Human Anti-Mouse Antibodies)

The HAMA response can occur when mouse monoclonal antibodies, or other antibodies of non-human origin, are administered to humans. The administered immunoglobulin is recognised as a foreign protein, which it is, and an immunological response is mounted. Some of the antibodies so produced will be the mirror image of the monoclonal antibody itself. In this world through the looking glass, some HAMA will be proteins of human origin that exhibit tumour antigens. Since the original monoclonal antibody was directed against a tumour-associated antigen, and an antibody to an **epitope** (**p. 118**) that is itself an antibody to an antigen will itself closely resemble the original antigen, HAMA can provide the equivalent of a tumour vaccine

and could thereby have therapeutic potential. The disadvantages of the HAMA response are the obvious ones associated with any allergic response: anaphylaxis, serum sickness, myalgias, etc.

Hand–Foot Syndrome

An unusual toxicity associated with chemotherapy for cancer. The hands and feet become shiny, red, swollen and tender. The skin of the palms and soles may desquamate and there is often fissuring in the web spaces and around the nails. The syndrome is most commonly associated with protracted inhibition of dyhydrofolate reductase (e.g. with continuously infused 5-fluorouracil or **capecitabine**) (**p. 46**). It has also been reported in patients treated with liposomal anthracyclines such as Caelyx. The problem usually resolves when treatment is interrupted. Pyridoxine may ameliorate the incidence and severity of the syndrome.

Haploid

A haploid cell is one, such as the mature ovum or sperm, that contains only one set of chromosomes (n). Most other mammalian cells are diploid (2n). Bacteria, however, are normally haploid.

Hayflick Hypothesis

Hayflick, in 1965, showed that human cells grown in culture had a limited overall life span: that the maximum number of population doublings is ~50. This suggested that the limitations on human life span may simply reflect the finite number of generations permissible in critical cell populations. This hypothesis ties in neatly with the **end-replication problem** (**p. 113**) and **telomere** (**p. 306**) shortening as a mechanism of cellular **senescence** (**p. 282**).

Hazard Rate

A theoretical measure of the risk of the occurrence of an event occurring within a specific time interval. Plotting the daily hazard rate can be instructive: in a trial of chemotherapy for lung cancer the daily hazard rate for death was relatively constant except for between 7 and 20 days after starting treatment. During this interval, the daily hazard rate was increased fourfold. The likely interpretation is that there was an excess of early deaths related to toxicity of chemotherapy in this study.

Hazard rates can also be used in the rational design of screening programmes or follow-up strategies after treatment. Screening examinations or clinic visits can be scheduled when the hazard rate for the event of interest is at its maximum.

Hazard Ratio

This is used to measure the relative survival experience in two groups. The ratio is calculated as the observed : expected ratio of events in one group divided by the observed : expected ratio in the other group. The 'expected' values are calculated using the **log-rank procedure** (**p. 183**). A hazard ratio = 1.0 indicates that the survival in the two groups is identical. If the hazard ratio in group A relative to group B is 0.75, then this suggests that over the duration of the study there have been fewer events in group A than would be expected. The hazard ratio gives a clear indication of just how different the survival in the two groups has been; the **log-rank test** (**p. 183**) can test the hypothesis that any observed difference has arisen by chance.

The calculation of a **confidence interval** (**CI**) (**p. 70**) for the hazard ratio involves log transformation, since the hazard ratio itself

will not be Normally distributed. It is very important to look at CI on the hazard ratio for any study that is negative by the **log-rank test (p. 183)**: a small study may be spuriously negative and the clue to this will lie in the wide CI around the hazard ratio.

Health Economics

Pukka economists rather look down upon health economists: 'some of the most useful work employs only elementary economic concepts'; 'health economics is an applied field in which empirical research predominates'. The 'real' economists see health economists as operating without a sound theoretical base, simply adopting pragmatic approaches to problems as they occur, and publishing their findings in journals rarely read by economists and for which there is no adequate peer review by trained economists. This relationship may be similar to that between medicine and pure science: when we realize that we will never cut the mustard as scientists, we apply for medical school.

There are major species differences among health economists, particularly when the USA is compared with the UK. These differences may bamboozle the unwary particularly since, as mentioned above, a sound base in theory may be lacking. All of which means that health economics should be treated sceptically — beware of credulity: if health economists cannot totally convince other economists why should they convince us?

Heat Shock Proteins (HSP)

The genes controlling these proteins are expressed following environmental insult to a cell. The response to thermal disturbance can be viewed as a model of the response to stress in general. Repression of previously active genes together with expression of the genes for the heat shock proteins is a typical response.

There are six main categories of HSP in higher organisms and their sequences are highly conserved: this suggests that the heat shock response has had an important role throughout evolution. Many of the HSP function as molecular **chaperones (p. 62)** and, presumably, help a cell to rebuild itself after potentially lethal insult. HSP also mediate the phenomenon of thermotolerance, which may have some relevance for therapeutic hyperthermia.

HER-2/neu (c-*erbB*2)

This is an oncogene that might be overexpressed in several different tumour types including breast cancer, ovarian cancer, gastric cancer, lung cancer and stomach cancer. Overexpression correlates with increased metastatic potential, resistance to **tumour necrosis factor (TNF) (p. 317)** and increased tendency to form tumours. It may also be associated with resistance to chemotherapy. Not surprisingly, tumours that overexpress the gene are often carry a poor prognosis. Downregulation of the gene will decrease malignant transformation. This can be accomplished experimentally by a DNA-binding protein, PEA3, which is encoded by a gene of the *ets* family. PEA3 produces downregulation of HER, etc. by binding to the promoter region of the gene. An alternative approach is to attempt to target the protein product of the *HER-2/neu* (c-*erb*B2) gene. The gene product is a membrane protein, p185, and a humanised monoclonal antibody (marketed as Herceptin) to this protein has already shown some benefit in the treatment of breast cancer. The approach has now entered Phase III trials.

Hereditary Non-Polyposis Colorectal Cancer (HNPCC)

A dominantly inherited disorder in which, within affected kindreds, there is an increased incidence of colorectal cancer as well as other tumours (endometrium, stomach, small bowel, ovary). About 1%, or less, of colorectal cancer arises as a result of this syndrome. The frequency of the syndrome is probably ~1:10 000. Four separate **DNA mismatch repair genes** (**p. 26**) have been implicated in the syndrome (*hMSH2*, *hMLH1*, *hPMS1*, *hPMS2*).

Heritage Drugs

Heritage Drugs are drugs that are, for historical reasons, within a pharmaceutical company's portfolio but which are considered unexciting or are out of patent or are, for other reasons, unlikely to be marketed enthusiastically: e.g. **cyclophosphamide** (**p. 83**), chlorambucil, melphalan, etc. When did you last see an advertisement for **methotrexate** (**p. 199**)?

Heterogeneity

Heterogeneity is an indicator of the extent to which entities differ. As such it is the direct opposite of **homogeneity** (**p. 152**). Although the term 'heterogeneity' is used widely in biology and statistics, in physics the preferred term is 'inhomogeneity'. The meanings are, however, identical. A full discussion of heterogeneity and its assessment is not necessary since much of it would simply be a negation of the discussion of homogeneity. A specific test for heterogeneity of use for enumerated data, such as colony counts, is the **Poisson heterogeneity test** (**p. 237**). Treating any patient with cancer involves, of necessity, dealing with heterogeneity: tumours and patients are heterogeneous.

Table 1 shows the wide variation in sensitivity to radiation both between and within different types of tumour. The higher the SF_2, the less the radiosensitivity.

Table 1

Site	**No.**	**SF_2**	**Coefficient of variation (%)**[a]
Cervix	52	0.47	38
SCC H&N	140	0.32	47
SCC H&N	34	0.45	26
High-grade glioma	16	0.52	42
High-grade glioma	21	0.51	28

[a](SD/mean) × 100.
SCC, squamous cell cancer; SF2, surviving fraction after 2 Gy.

The effect of this heterogeneity is to flatten dose–response curves, which in turn will obscure the effects of treatment upon susceptible subpopulations. Chromosomal heterogeneity is a specialised term used to describe a situation in which a gene that causes disease may be on more than one chromosome. Allelic heterogeneity occurs when a disease may be caused by more than one type of mutation within a specific gene.

Heteroscedasticity

The term is used in **regression** (**p. 264**) analysis to indicate systematic (either increasing or decreasing) variation in the plot of residuals versus fitted values. This is in contrast to homoscedasticity, in which the **residuals** (**p. 269**) vary only randomly from the fitted values. The presence of **heteroscedasticity** (**p. 154**) implies that the assumptions within the regression model are incorrect and call in to question the validity of the model.

Heuristic

An approach that uses trial and error to define a solution to a problem. In its original

sense it had to with a system of pedagogy in which the pupil was encouraged to find out things for themselves. It thus has resonances with the concept of the bootstrap theories of particle physics, in which things appear to pull themselves up by their own bootlaces. Examples of heuristic approaches in oncology include **neural networks** (**p. 212**), maximum likelihood methods of multivariate analysis and, perhaps, much of the clinical practice of **cancer management** (**p. 40**). After all, only rarely do we know what an intervention is going to achieve and so all that we do is, to some extent, trial and error.

Hevin

Hevin is a gene, on chromosome 4, that modulates interactions between endothelial cells and lymphocytes: its expression enables lymphocytes to migrate through the tissues. The *hevin* gene product is a calcium-binding protein with some similarities to a molecule, SPARC, which prevents cell adhesion. There is some evidence that *hevin* expression is decreased in prostate cancer and in non-small cell lung cancer. Both SPARC and *hevin* can inhibit the progression of cells from G_1 to S-phases. *Hevin* may function as a tumour suppressor gene.

High LET Radiation

High LET (linear energy transfer) radiation is densely ionising and **cell survival curves** (**p. 57**) for high LET radiations (e.g. neutrons, pions, neon nuclei, α-particles) are very different from the typical survival curves obtained with X-rays. The survival curves for high LET radiations have little or no shoulder and steep slopes. Cell killing is virtually independent of oxygen levels at the time of irradiation. These radiobiological advantages are, for clinical applications, eroded by the practical difficulties in obtaining reliable, well-calibrated beams.

HIV (Human Immunodeficiency Virus)

HIV is a lentivirus of primates that, in humans, is found in two forms: HIV-1 and -2. Other members of the family include slow viruses that attack various species of monkey as well as chimpanzee. Evaluation of **sequence homologies** (**p. 284**) among the immunodeficiency viruses suggests that HIV split off from an ancestral virus, found in *cercopithecus* species, between 600 and 1200 years ago. The current hypothesis is that at some subsequent time the virus spread from ape to man, and that there were human reservoirs of infection in Africa by the mid-twentieth century. Tourism, air travel and human behaviour then combined to disseminate the virus rapidly over the past 30 years.

HIV are RNA containing retroviruses. Their possession of reverse transcriptase means that they can produce DNA from their RNA. This DNA can be incorporated into the host genome and causes virus production and cell death. HIV have a particular affinity for $CD4^+$ lymphocytes, but some variants have a predilection for macrophages. The structure of HIV is relatively simple: a lipid bilayer membrane from the host cell surrounds the virion. This comprises an envelope–protein complex (gp120 is outer, gp41 is transmembrane), the production of which is controlled by the virus; a matrix protein (p17) is within the lipid bilayer but outwith the capsid itself. The major capsid (p24) surrounds the viral RNA and the nucleocapsid protein (p7/p9) is in intimate contact with the viral RNA. The virus produces three enzymes: reverse transcriptase, protease and integrase.

Viral RNA contains several genes: *gag* produces p17, p24 and p7/9; *pol* produces the three enzymes; *env* codes for the envelope precursor protein gp160

(which protease splits to gp120 and gp41). There are also other genes: *tat, rev, vif, nef, vpr.* HIV-1 contains *vpu*, HIV-2 contains *vpx*. The tat protein up-regulates viral transcription, the *nef* product is needed for efficient replication. *In vivo*, the rev protein stimulates export of viral RNA from the host nucleus. Reverse transcriptase converts the viral RNA into a DNA sequence that can, through the action of integrase, be inserted into the host genome. The protease is essential for the construction of the viral envelope. The main targets for treatment of HIV infection have been **reverse transcriptase** (**p. 272**) (AZT) and protease (saquinovir).

HIV infection is the cause of the acquired immunodeficiency syndrome (**AIDS**) (**p. 3**).

Homeobox

This sequence of DNA is 180 nucleotides long and is, in evolutionary terms, highly conserved. It was originally described in *Drosophila*. It is found in many other eukaryotic genes and is concerned in development, directing which part of the embryo becomes which part of the mature form. The associated protein, the homeodomain, is 60 amino acids long. Homeodomain proteins function as **transcription factors** (**p. 313**). In mammals, there are >38 genes in the homeobox family that group into four main clusters: Hox A, B, C and D.

Homeotic

A term introduced by the geneticist William Bateson to describe genes that act as switches and regulators in embryological development. For many years, homeosis was a metaphysical concept lacking any concrete basis. Today, with the description and analysis of the homeobox, it has a firm foundation in direct observation.

Homogeneity

Homogeneity is an indication of the extent to which entities are similar to each other. The entities can be patients, clonogenic cells within a tumour or clinical trials. However, individual variation is conspicuous in biological science: Gertrude Stein was wrong: a rose is not a rose is not a rose. A clonogenic cell within a tumour is not necessarily identical, in terms of its radiosensitivity, to every other clonogenic cell within that tumour. There will certainly be many similarities, assessed as homogeneity, but there will also be differences between cells, expressed as **heterogeneity** (**p. 150**).

Many parametric statistical tests, e.g. the ***t*-test** (**p. 303**), rest on the assumption of **homoscedasticity** (**p. 154**). This unwieldy term is just another way of stating that the variance in the two groups of observations is similar, that is homogeneous. The *F*-test is a statistical measure of this assumption: the null hypothesis is that the variances are homogeneous.

In tumour biology the steepness of the **dose–response curve** (**p. 103**) is an indication of the homogeneity of the clonogenic cells in terms of their response to the cytotoxic agent, be it drug or radiation: the steeper the curve, the more homogeneous the population. TCP is the tumour control probability plotted against dose of cytotoxic agent, in this case radiation (dose expressed as BED in Gy). The tumour contains multiple homogeneous subpopulations of cells, each with a different sensitivity. The curves for each subpopulation are steep. The composite curve is, however, much flatter, reflecting the heterogeneity of the total population (figure 1).

Homogeneity of the trials analysed is a critical prerequisite for a **meta-analysis** (**p. 195**). The statistical methods for

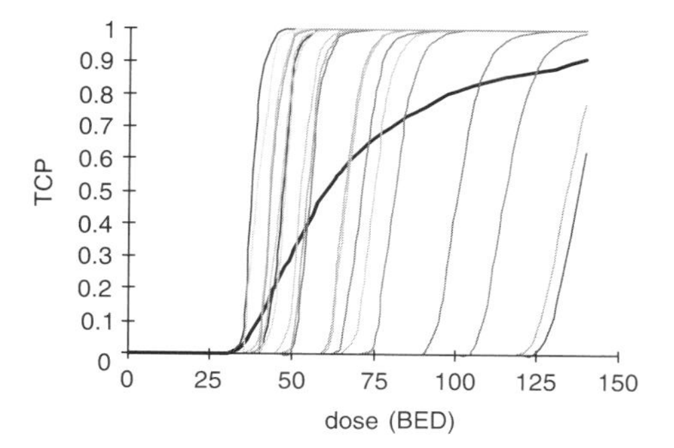

Figure 1.

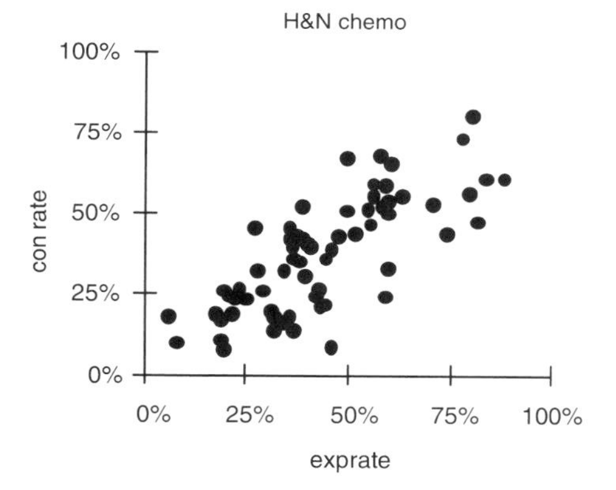

Figure 2.

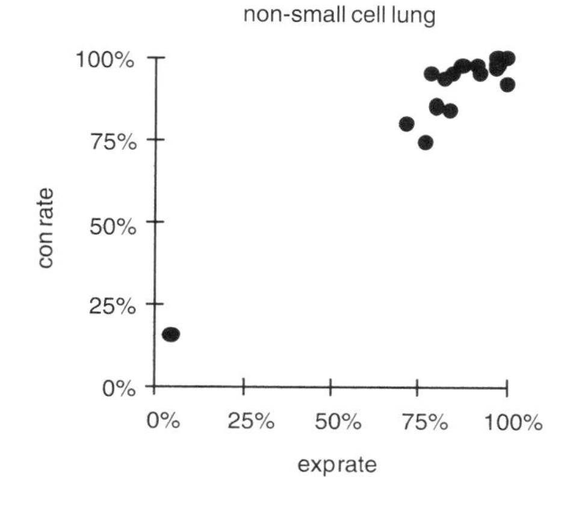

Figure 3.

combining trial data (see Meta-Analysis — Combining Trial Data), are most reliable when this criterion is fulfilled-indeed the **Mantel–Haenszel (p. 187)** method is based on the assumption of a fixed treatment effect in each trial, that is homogeneity. Homogeneity of trial results can be easily assessed visually by plotting, for each individual trial, the event rate in the control arm versus the event rate in the treatment arm. This type of plot is sometimes called a **L'Abbe plot (p. 173)**. Homogeneous trials will cluster together; heterogeneous trials will show no such clustering. Figures 2 and 3 demonstrate heterogeneity.

The head and neck trials show only moderate homogeneity whereas the non-small cell lung cancer trials are, with one obvious exception, much more homogeneous. The exception in this case is so exceptional that it demands explanation. The explanation is that all these trials compared death rates in patients treated with radiotherapy alone with the rates in patients treated with both radiotherapy and chemotherapy. The outlying trial was a trial of **adjuvant treatment (p. 8)** after surgery. All the other trials were non-surgical studies: as such, it would be inappropriate to include it with the others. The **L'Abbe plot (p. 173)** gives a clear indication of a problem with homogeneity that, without visual inspection of the data, might have been overlooked.

The number of trials in a **meta-analysis (p. 196)** is often small and so simple regression analysis or χ^2 testing is not adequate for assessing homogeneity. The null hypothesis would assume that the trials were homogeneous. If the sample size were small then, even if trials were in fact heterogeneous, the null hypothesis might be accepted inappropriately. The method of DerSimonian and Laird offers a means of calculating a statistical index of trial homogeneity, the Q statistic.

The **DerSimonian and Laird method** (**p. 93**) for pooling trial results is, partly because it explicitly accepts **heterogeneity** (**p. 150**), inherently more conservative than the Mantel–Haenszel method.

Homogeneity also has a place in radiation dosimetry. For most simple calculations of dose at depth, homogeneity of the irradiated medium is assumed. In fact human tissues are rarely homogeneous. In treating a lung tumour the radiation beam will traverse in sequence: soft tissue; bone (rib); air (in alveoli); tumour. It will then exit through an equally heterogeneous mixture of tissues. Each tissue will attenuate the beam differently: bone will attenuate more than air. Allowance must be made for these various inhomogeneities and appropriate correction factors applied.

Homology/Homologous

The extent to which sequences of nucleotides in DNA or RNA, from different sources, are similar is a measure of their homology. The extent to which sequences are homologous is indicated statistically — a very low probability that the observed sequences could have arisen by chance indicates a high degree of homology. The presence of **sequence homologies** (**p. 284**) in species that are, in evolutionary terms, widely separated suggests that the coded protein is likely to be of fundamental importance. Homology is also useful in assessing gene function: homologous sequences are likely to be associated with similar functions.

Homoscedasticity

The assumption, which is implicit within many regression techniques, that the residuals are randomly distributed around the predicted values, is termed homoscedasticity. It is the opposite of **heteroscedasticity** (**p. 150**),

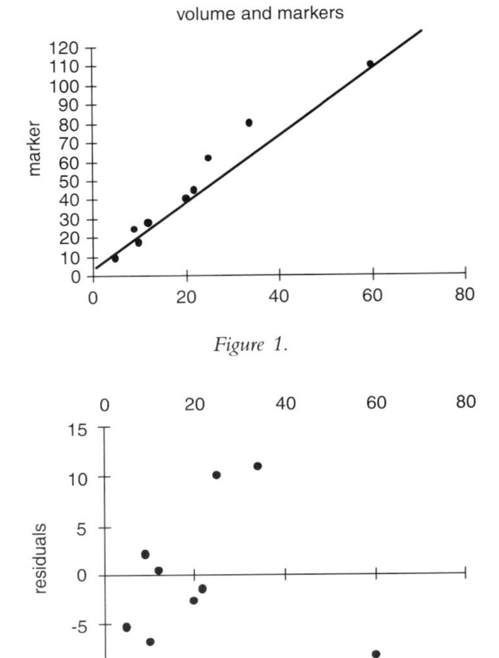

Figure 1.

Figure 2.

which implies that there is some pattern to the residuals. Plotting the residuals against the predictor variable can easily assess scedasticity. Figures 1 and 2 show that the tumour marker data are homoscedastic and that it is legitimate to use linear regression to predict marker level from a knowledge of tumour volume. There is no particular pattern or trend in the distribution of the data points in this plot.

5HT Receptors

$5HT_3$ receptors are concerned with the control of vomiting, and antagonists such as Ondansetron and Granisetron are potent anti-emetics used widely in oncology and general medicine. $5HT_3$ receptors are found both in the chemoreceptor trigger zone and at the terminal branches of the vagus nerve in the intestinal wall. Since 5HT is a

fairly ubiquitous neurotransmitter, important interactions are possible both in theory and practice. The selective serotonin re-uptake inhibition (SSRI) antidepressants act by potentiating 5HT-induced synaptic transmission. There is, therefore, the possibility of mutual antagonism between those anti-depressants acting through potentiation of 5HT and those anti-emetics whose action is to block the $5HT_3$ receptor. The side-effects of the $5HT_3$ antagonists include headache, vomiting, constipation and disturbing dreams. The introduction of these agents has made it possible to treat many patients with chemotherapy regimens that, previously, would have been considered intolerable.

HRAS Oncogene

This oncogene was first described in the Harvey murine sarcoma (hence its acronym). Along with *KRAS* and *NRAS*, it forms the *RAS* family of **oncogenes** (**p. 218**). These oncogenes encode a protein, p21. *HRAS*1 and *HRAS*2 have now been described and located: *HRAS*1 is on chromosome 11p15.5; *HRAS*2 is probably on the X chromosome.

The *p21 RAS* gene product appears to stabilise the product of the ***MYC* oncogene** (**p. 211**). The *MYC* product has a short half-life but, when a pathway acting via *RAS/RAF/MAPK* is activated, its half-life is prolonged. The *RAS* gene product inhibits *MYC* degradation by blocking a metabolic pathway dependent on the 26S **proteasome** (**p. 249**). *RAS* and *MYC* can, therefore, potentially at least cooperate in producing and sustaining malignant transformation.

Human Papilloma Viruses (HPV)

The human papilloma viruses are small DNA viruses that have been implicated in the causation of several different types of cancer in humans: cancer of the cervix, anal cancer and certain cancers of the head and neck. Using the polymerase chain reaction (PCR) (**p. 238**), between 15 and 35% of squamous carcinomas of the head and neck can be shown to contain HPV sequences, >90% of cancers of the cervix are associated with evidence of HPV infection. The commonest HPV is type 16, although other types (18, 33, 31) are occasionally associated. HPV is particularly frequent in tumours of the oropharynx and less commonly found in tumours of the oral cavity and larynx.

The role of HPV in carcinogenesis is complex and is mediated by oncoproteins E6 and E7, which are produced by virally transformed cells within the epithelium. These oncoproteins, both of which posses a zinc-binding **motif** (**p. 207**), inhibit the function of tumour-suppressor genes such as *p53* and the *Rb* gene. The E7 protein may also have a more direct effect on the **cell cycle** (**p. 52**) by increasing the activity of cyclin A and E. The effect is to abolish the G_1/S transition checkpoint. These functional abnormalities correlate with histological evidence of premalignant change. Further abnormalities, loss of the viral regulatory proteins E1 and E2, together with acquired mutations (especially on chromosome 3), correlate with the appearance of invasive squamous carcinoma.

Humanised Antibodies

Humanised antibodies are antibodies that have been engineered, often using recombinant DNA technology. The variable region is usually of murine origin, whereas the constant region is of human type. The chimeric nature of the construct means that it is possible to immunise the mouse to rear antibodies to specific antigens; the constant region, being of human type, is not recognised as foreign and, therefore, the problems of allergic reactions

and human anti-mouse responses (HAMA) can be avoided.

Hyperfractionation

Hyperfractionation implies that a larger number of smaller fractions are used in a course of radiotherapy treatment, assuming that total dose and overall time are kept constant. As with **accelerated fractionation** (**p. 1**), it is difficult to provide an unambiguous definition because standard practices vary: in some departments, the standard fraction size is 1.8–2.0 Gy, in others 2.75–3.0 Gy. A pragmatic definition would be to define any regimen using fraction sizes of ≤1.7 Gy as hyperfractionated.

Hyperfractionation is often given as multiple daily fractions so that overall treatment time will not be unduly prolonged. The rationale behind pure hyperfractionation is that, because of the relative sparing of late effects as fraction size is reduced, it might be possible to give a higher dose to the tumour without increasing the dose-limiting late effects. Pure hyperfractionation is possible and may permit escalation of the total dose without any untoward increase in late effects: 70 Gy in 7 weeks at 2 Gy per fraction day^{-1} (35 fractions) is equivalent to 80.5 Gy in 7 weeks using two fractions of 1.15 Gy day^{-1} (70 fractions).

Since acute reactions in normal tissues may respond similarly to tumours there is, implicit within this approach, the risk of increased acute damage to normal tissues. There would also be the risk of so-called consequential late effects — where long-term damage results from the acute reaction never having healed properly.

Hyperplasia Adenoma Polyp Cancer Sequence

This hypothesis was first proposed by Vogelstein *et al.* as a description of the morphological and genetic changes involved in the formation of cancer of the colon. A simplified account of this sequence is as follows (figure 1). An initial series of mutations takes place as a result of which

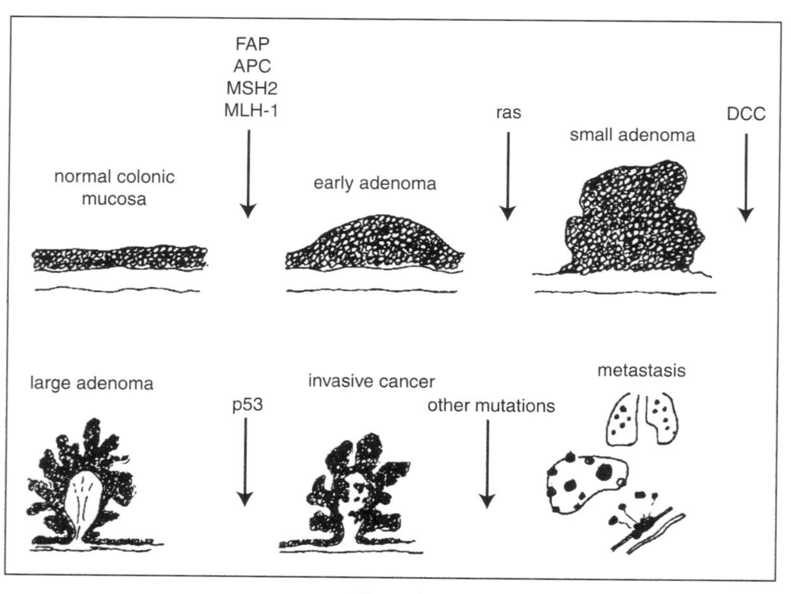

Figure 1.

the lining of the crypt cell changes from normal to hyperplastic: at least three genes (*APC*, *MLH*-1, *MSH*-2) are involved. The genes are concerned with DNA repair and cell–cell adhesion. A sequence of changes, involving *MCC*, *k-ras* and the gene controlling expression of the adhesion molecule N-CAM is associated with the change from hyperplasia to premalignant polyp. The change from premalignant polyp to carcinoma is associated with loss of function of normal (wild-type) *p53*. *p53* is a tumour suppressor gene that, in response to the detection of **DNA damage** (**p. 98**), triggers either repair of the damage or the death of the cell. The hypothesis is that cells with mutant *p53* do not die appropriately and thereby a further series of genetic abnormalities associated with malignant transformation might be propagated. As a result of the sequence of genetic changes, the carcinoma exhibits the full malignant phenotype: uncontrolled proliferation; **invasion** (**p. 167**); metastasis.

Hypothetico-Deductive Approach

An approach to science that first involves the framing of a hypothesis: a conjecture about the true state of things. This is then subjected to a series of tests designed to test its validity. An essential feature is that the deductions and hypotheses are mutually modifying: experience tempers theory and theory directs investigation. The earliest origins of the hypothetico-deductive approach are probably found in the work of Parmenides.

According to Karl Popper, refutability is a key property of a hypothesis: the most useful tests of a hypothesis are those aimed at disproving it. The hypothetico-deductive approach is in direct contrast to the inductive approach: in which observations are assembled, allegedly without prior expectation or systematic belief, the truth then emerges through unbiased scrutiny of the patterns that emerge.

Hypoxia Inducible Factor 1 (HIF-1)

This molecule is produced in response to hypoxic stress. It has a variety of cellular effects including activation of *p53*-dependent pathways and other effects on gene expression. It also increases production of **VEGF** (**p. 324**) and, therefore, promotes angiogenesis. It stimulates **erythropoietin** (**p. 119**) production. It is constitutively expressed and is formed of α- and β-chains. These chains will join when pO_2 is low. Only when the chains are joined in this way does the molecule have any biological effect. This mechanism ensures that the organism can respond rapidly to a change in pO_2, without needing to wait for new protein to be synthesised.

Hypoxia

Hypoxia defines a state in which a cell or tissue is relatively deficient in oxygen. The normal tissues of the body, with the exception of laryngeal cartilage, are normally well oxygenated. Tumours, on the other hand, often have areas in which cells are hypoxic: this can be demonstrated both histologically and by direct measurements of the partial pressure of oxygen (pO_2).

Hypoxia and hypoxic cells are important in oncology because:

- Hypoxic cells are relatively resistant to **radiation-induced cell killing** (**p. 256**).
- Hypoxic cells remaining after radiotherapy treatment might jeopardise cure.
- The presence of hypoxia may be an important differentiating factor between both tumour and normal cells.
- A variety of strategies is available for counteracting the adverse effects of tumour cell hypoxia.

- Bioreductive drugs may be specifically activated within areas of hypoxia: in theory, such treatment would be more damaging to tumour cells than to normal tissues.
- *p53* and hypoxia are interrelated: cells that lack normal *p53* do not undergo apoptosis when subjected to hypoxic conditions.

Hypoxia can be measured using direct measurements of pO_2 in tumours using microelectrodes (e.g. the Eppendorf apparatus as pioneered by Peter Vaupel). Oxygenation can also be estimated using techniques based on the quenching of induced fluorescence measured using implanted fibreoptic devices (Oxylite). The problem arises in agreeing on a definition of hypoxia. Various schemes have been proposed: one is simply to estimate the proportion of cells with pO_2 in the 'hypoxic' range (0–2.5 or 0–5 mmHg); an alternative is simply to quote a median pO_2, the median pO_2 in tumours is typically <30 mmHg, whereas normal tissues usually have a median $pO_2 > 45$ mmHg.

Radiobiological hypoxia (**p. 259**) is defined as cellular behaviour, in response to irradiation, that mimics that of cells hypoxic at the time of irradiation. Although lack of oxygen is assumed to be the cause of that behaviour described as radiobiological hypoxia, the assumption is hard to prove.

Much effort has been expended clinically in attempting to deal with the perceived problem of hypoxic cells within tumours limiting the **clinical effectiveness** (**p. 69**) of radiotherapy. Attempts have included irradiation in hyperbaric oxygen, hypoxic cell radiosensitisers, and methods aimed at improving blood flow and oxygen delivery to tumours, e.g. ARCON. Fractionation of radiotherapy, by allowing re-oxygenation of hypoxic tumour cells during a course of treatment, is in fact highly effective in dealing with the problem of hypoxic cells.

Idiotype

An idiotype acts as antigenic determinant. It dictates the specificity of a given antibody. The importance is that antibodies may develop to antibodies recognising a specific idiotype. These second antibodies will express an antigen that may resemble the original antigen recognised by the first antibody. If a third antibody is then formed to the second antibody, then this would effectively be an antibody to the original antigen. If the first antibody were exogenous, e.g. a murine antibody raised against a human tumour antigen, then the idiotypic network outlined above is a method whereby large quantities of human antibody to a tumour-associated antigen might be produced. This is potentially exploitable for both cancer treatment and cancer vaccines: indeed, some of the anecdotal successes of immunological approaches to treating cancer may have their origin in the through-the-looking-glass-world of anti-idiotype networks.

Ifosfamide

Ifosfamide is an alkylating agent, an oxazaphosphorine, related to **cyclophosphamide** (**p. 83**). It is a **prodrug** (**p. 246**) activated by the **cytochrome P450** (**p. 84**) system to the active metabolites, 4-OH-ifosfamide and aldoifosfamide. Further metabolism leads to the formation of acrolein and ifosforamide mustard. The active moieties form cross-links within the DNA molecules and thus interfere with cell division and proliferation. Ifosfamide is active against sarcomas, non-small cell lung cancer, lymphomas and germ cell tumours. Its main toxicity is haemorrhagic cystitis, a problem that can largely be prevented by the concomitant administration of the protective agent **Mesna** (**p. 195**). It reduces levels of pseudocholinesterase, an enzyme involved in cocaine metabolism. Cocaine users given ifosfamide will be at increased risk of cocaine toxicity. An alarming, but usually temporary, CNS syndrome is sometimes encountered in patients treated with ifosfamide, particularly those with impaired renal function and low serum albumin. The syndrome usually manifests as drowsiness, coma, psychosis, cranial nerve palsies, abject depression and ataxia.

Interleukin 2 (IL-2)

IL-2, also known as T-cell **growth factor** (**p. 144**), is a lymphokine produced by activated T-cells and some cell lines derived from lymphomas. It can stimulate the activity of natural killer (NK) cells and other immunocompetent cells. It has a molecular weight of 15 000. It has been used in the immunotherapy of cancer, particularly renal cancer and melanoma. Toxicity is considerable. It has been combined with re-infusion of tumour-infiltrating lymphocytes (TIL), the rationale being that the TIL are cells that recognise **tumour-associated antigens** (**p. 315**) and have the capacity to kill the malignant cells by immunological mechanisms. The concomitant administration of IL-2 will stimulate expansion and activation of TIL and thereby potentiate the immunological attack on the tumour.

The gene for IL-2 is on chromosome 4q26–q27.

Immunoblotting

The process in a **Western blotting** (**p. 327**) procedure where antibody probes are used to identify the sequences present. Peroxidase is tagged to the probe, linked to an antibody to the antibody to the antigen of interest. Using the biotin-avidin/streptavidin technique increases the sensitivity of the procedure.

Imputation

A process used to circumvent the problem of missing data, particularly in longitudinal studies. The simplest method is simply to take the arithmetic average of the measurements made both immediately before and after the missing value. More sophisticated estimates involve modelling the data set as a whole and then estimating the missing values according to the overall shape of the curve. The assumption with most of these methods is that the data follow some consistent pattern. This is not necessarily the case and there can be no totally reliable method for compensating for missing data. It is better to expend energy on ensuring completeness of data than it is to exercise ingenuity in compensating for its absence.

Inception Cohort

This describes a group of subjects recruited to a cohort study before the condition (often a disease) of interest has developed or **exposure** (**p. 124**) to the agent under study has occurred.

Incidence

Incidence is the number of new cases of disease arising in a given period in a specified population. This information is collected routinely by cancer registries. It is distinct from **prevalence** (**p. 244**), which indicates the number of people with a particular disease during a particular period. Prevalence includes both the incident cases as well as survivors whose disease was diagnosed before the period of assessment.

Induced-Fit Theory

A theory of enzyme action that, in contrast to models about a rigid enzyme as being analogous to a key being fitted into a lock, emphasises the plasticity of the conformation of enzymes. The substrate, and only the substrate for that enzyme, will induce conformational changes in the enzyme that produce the correct alignment of catalytic groups for the enzyme to exert its catalytic effect. The importance of this theory, which is supported by the facts, is that it demonstrates that very small conformational changes can produce major intracellular effects.

Inglefinger Rule

Named after the famous editor of the *New England Journal of Medicine* who first proposed it. The rule states that a journal will not accept for publication work that has been published elsewhere. The question that is begged is what is the definition of published: presented at a meeting and appearing as an abstract; featured on the front page of *The Times*; appearing in a press release from a university or funding body; put on a web site and sent over the Internet? Originally the rule was devised to protect the commercial interests of the journal. It was latterly adapted by Arnold Relman to ensure that information that reached the public had first been subject to **peer review** (**p. 229**). The assumption is that peer review is an effective mechanism for protecting the public from biased or fraudulent claims. The rule is no longer strictly applied, even by the *NEJM*: the text of 'unpublished' papers on radiochemotherapy for cancer of the cervix was made available, via the Journal's website, several months before the papers appeared in print. The main purpose of the rule today is to remind investigators that they have responsibilities to colleagues and to science as a whole: that there is more to the acquisition of knowledge than personal publicity and lucrative stock options.

Inhomogeneity/ Inhomogeneous

Terms used in physics to describe a medium that is not homogeneous. The tissues of the chest, soft tissue, air, bone, etc., are inhomogeneous. The bone and air constitute volumes of inhomogeneity within the soft tissue. This convention, soft tissue as the norm, is adopted since most measurements for radiation dosimetry are made in phantoms whose density is equivalent to that of soft tissue.

Insertional Mutagenesis

A potential hazard of **gene therapy** (**p. 137**): a genetic sequence inserted into the host genome might itself act as a mutagen and is, therefore, potentially oncogenic.

Integral Dose

A term defining the total energy imparted to a medium as a result of exposure to ionising radiation. The medium is usually taken, for clinical purposes, to be the patient as a whole. The integral dose becomes important clinically when large volumes are irradiated. Two different techniques may produce similar isodose distributions, but the integral dose with one, because of the use of multiple beams, might be higher than that with the other. It is usually preferable, all things being equal, to choose the technique with the lower integral dose. Integral dose is measured in Gy kg and, since $1\ \text{Gy} = 1\ \text{J kg}^{-1}$, the unit of SI measure is the Joule (J). Mayneord described the formula for integral dose, and its calculation requires knowledge of:

- Dose at the depth of maximum dose.
- **Focus skin distance** (**FSD**) (**p. 131**).
- Area of the radiation field.
- Total thickness of the patient.
- Half value depth.

Integrins α and β

These are **adhesion molecules** (**p. 8**). Increased expression can be associated with both **invasion** (**p. 167**) (in melanomas) and **metastasis** (**p. 198**) (rhabdomyosarcoma).

Intensity Modulated Treatment

This is a technique used in external beam radiotherapy to improve **dose distribution** (**p. 102**). The distribution of radiation intensity within each beam is allowed to vary. The classical method for doing this was to build a tissue compensator for each patient: the compensator was attached to the head of the treatment machine and was designed to produce a suitable pattern of differential attenuation. Compensators have been widely used to treat 'difficult' sites such as the supraglottic larynx and the upper third of the oesophagus. The problem is that it takes considerable time and effort physically to construct an appropriate compensator for each patient.

A more recent development is to use a **multileaf collimator** (**p. 210**) that can be programmed to move dynamically during each radiation exposure. As the beam, in imagination or reality, moves round the patient the intensity varies according to the thickness of the target as 'seen' by the beam: the intensity of the beam is at its greatest when, in the beam's eye view, the target is at its thickest. When the target is at its thinnest the beam is at its least intense. The simplest version of this approach uses each pair of collimators to define a window — the width of the gap and the speed of movement of each pair of leaves are selected so that, overall, the desired pattern of radiation intensity is produced during each exposure. Used in conjunction with a conformal approach to planning, intensity-modulated therapy offers a means of increasing the dose to tumour while minimising the dose to normal tissues. The technique is critically dependent on the validity of the algorithms used in the computer software that optimises the treatment. If there are greater **inhomogeneities** (**p. 161**) than those predicted by the algorithms then this could compromise tumour control.

Another IMRT technique involves the use of a small aperture beam that rotates around the patient. The concept is similar to that of tomographic imaging. The practical application is based on the multivane slit collimator. This involves vanes of lead being fired in and out of the beam path extremely rapidly as the beam scans the patient. The technique is known as MIMiC (multivane intensity modulating collimation).

The assumption implicit within the development of IMRT techniques is that they will allow the dose to the tumour to be increased while physically sparing normal tissues. This will be particularly true for diseases such as prostate cancer where the tumour target nestles within a concavity within the vulnerable normal tissue, the rectum. The further assumption is that the increase in dose to the tumour will improve the rate of uncomplicated cure.

Intensity

The intensity of a parallel beam of radiation is the amount of energy crossing a unit area, held perpendicular to the direction of the beam, in unit time.

Interaction

This term is used in a specialised sense in statistics: it describes the problem that arises when a treatment or intervention has different effects on different subgroups of subjects. The benefits from adjuvant chemotherapy will be greatest in those premenopausal patients with breast cancer. It may not be possible in a small comparative study looking at the effect of adjuvant chemotherapy including patients of all ages, because of the small numbers, to identify the interaction between menopausal status and benefit from chemotherapy. One of the advantages of **meta-analysis** (**p. 196**) is that through the power of large numbers, subgroups can be analysed and interactions can often be identified.

Interactions are also important in pharmacology and radiobiology. One treatment may influence the effects of another through biological or pharmacological interactions. Anti-emetics that work by antagonising $5HT_3$ receptors will, potentially, interfere with the mood-elevating effects of those anti-depressants which inhibit serotonin reuptake (SSRI) — and vice versa. Metoclopramide can act as a radiation sensitiser. Calcium channel blockers may abrogate some of the adverse effects of the MDR phenotype. Interactions can only be identified if all the medication that a patient is taking is known. With the increasing use of complementary therapies there is the potential for unexpected interactions to take place. What is the effect of aloe vera capsules on radiation-induced diarrhoea?

Interactions are in three main categories:

- No effect: $1 + 1 = 2$.
- Antagonism: one agent diminishes the effect of the other: $1 + 1 = 1.5$.
- Potentiation: one agent increases the effect of the other. There is a supra-additive effect: $1 + 1 = 3$.

The vocabulary in this area is complex (synergy, additivity, antagonism, etc.) and so are the methods for detecting whether interactions are occurring. Even in laboratory studies where elegant constructions such as isobolograms can be used, conclusions are by no means straightforward. In clinical practice, with a huge range of potentially interacting variables, some identifiable, some not even guessed at, it is often impossible to work out whether significant interactions are taking place.

Intercalation

Intercalation is a process whereby a molecule inserts itself into the space between the base pairs of the DNA double helix. This causes problems with replication, in particular **frameshift mutations** (**p. 132**). The main intercalating agents used clinically are the **anthracyclines** (**p. 16**). In experimental studies, the acridine dyes and ethidium bromide can be used as intercalating agents.

Intercept

The distance between the origin of a graph and the point at which a given line crosses the axis. In the equation of a straight line showing the relationship between y and x:

$$y = a + bx,$$

where a is the intercept on the y-axis and b is the slope of the line (figure 1).

Here, the intercept $(a) = 10$ and the slope $(b) = 15$: as x increases by 1 unit, y increases by 15 units.

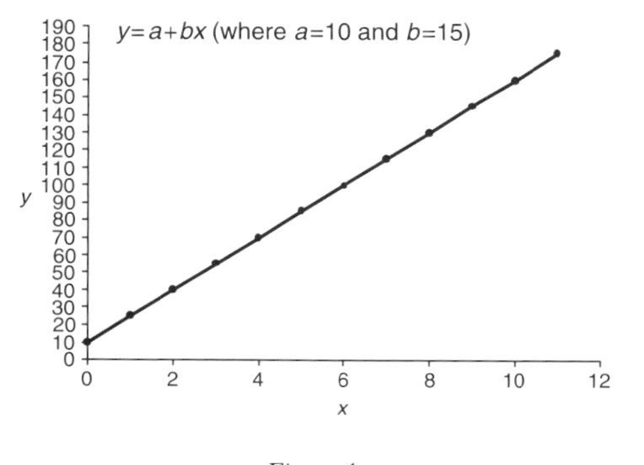

Figure 1.

Interferons

The interferons are a group of biologically active molecules originally isolated from patients with influenza. They were first introduced into medicine as potential anti-viral agents but are now more often used for their ability to modulate the effects of the immune system. There are three main classes: α, β and γ.

There are several types of α interferon designated α-2a, α-2b, α-n1 and α-n3. These have minor differences in amino acid sequence. The first two compounds are prepared in a highly purified form using recombinant DNA technology; α-n1 and α-n3 are obtained from stimulated white cells. α interferon has both direct antitumour activity as well as anti-viral effects. It has been used in the adjuvant therapy of malignant melanoma, in the treatment of AIDS-related Kaposi's sarcoma, hairy-cell leukaemia and non-Hodgkin's lymphoma. Other clinical uses include treatment of hepatitis B and C; treatment of venereal warts; and treatment of haemangiomas of infancy. It binds to cell surface receptors and has a wide variety of effects on cellular function. These include regulation of oncogene expression; activation of natural killer (NK) cells; prostaglandin synthesis; inhibition of cell division, including that of haemopoietic progenitors; and suppression of the **cytochrome P450** (**p. 84**) system. It may induce expression of tumour-associated antigens on malignant cells and thereby stimulate immune responses directed against the tumour.

β Interferon is used primarily in the treatment of autoimmune conditions. It antagonises the stimulatory effects of γ interferon on antigen expression. Its role in the prevention of relapses in multiple sclerosis is controversial: its main actions are to inhibit the production of γ interferon by lymphocytes and also to inhibit the effect of γ interferon on the increased expression of antigens by cells in the CNS. α and β interferons share a common receptor and are sometimes known as Type I interferons.

γ Interferon is known as Type II and has its own receptor. It has similar cellular effects to α interferon in that it has both antitumour and anti-infective properties. It is produced naturally by stimulated T-lymphocytes and enhances the function of macrophages, perhaps by increasing oxidative metabolism. It interacts with other components of the immune system such as IL-2, so that it also has regulatory effects on the immune system, hence its older name 'immune interferon'. The commercially available drug is manufactured using recombinant technology. It is virtually identical to the natural human form. It has been used, but without any real clinical benefit, in the treatment of renal cell carcinoma. Its main clinical use is in the management of certain immunodeficiency syndromes, such as chronic granulomatous disease, where there is evidence of abnormal phagocytosis.

When used clinically, the interferons will, in a dose-dependent fashion, produce flu-like symptoms: fever, chills, rigors, and pains in the muscles and joints. Myelosuppression may occur as well as alopecia and deranged liver function tests.

Interleukin 11 (IL-11)

IL-11 is a growth factor isolated initially from a culture of marrow stromal cells. It is now available in recombinant humanised form (rhIL-11). It causes proliferation of megakaryocyte precursors and might have a potential role in preventing treatment-related thrombocytopenia in patients treated with high-dose chemotherapy.

Interleukin 2 (IL-2)

IL-2, also known as T-cell **growth factor** **(p. 144)**, is a lymphokine produced by activated T-cells and some cell lines derived from lymphomas. It is constitutively expressed i.e. it is a gene expressed without the need for activation or induction.

It can stimulate the activity of natural killer (NK) and other immunocompetent cells. It has a molecular weight of 15 000. It has been used in the immunotherapy of cancer, particularly renal cancer and melanoma. Its toxicity is considerable. It has been combined with re-infusion of tumour infiltrating lymphocytes (TIL). The rationale being that TIL are cells that recognise **tumour-associated antigens** (**p. 315**) and have the capacity to kill the malignant cells by immunological mechanisms. The concomitant administration of IL-2 will stimulate expansion and activation of TIL and thereby potentiate the immunological attack upon the tumour.

The gene for IL-2 is on chromosome 4q26–q27.

Interleukins (IL)

A group of molecules, chemically unrelated, that function as signalling molecules in the coordination and propagation of immunological responses. The interleukins include:

Interleukin 1 (IL-1)

A protein with a variety of roles in mediating inflammatory and immune responses. The receptor to which it binds is a member of the immunoglobulin superfamily. It can be produced by a wide variety of cells: astrocytes, fibroblasts, macrophages, lymphocytes, epithelial cells, etc. The gene is on chromosome 2q14.

Interleukin 2 (IL-2, T-cell growth factor)

A glycosylated protein of only 133 amino acids. It promotes T-cell proliferation as well as proliferation of other cells such as macrophages and natural killer (NK) cells involved in effecting immune responses, and is itself produced by activated T-cells. The gene is on chromosome 4. IL-2 interacts with the IL-2 receptor (CD25, Tac antigen) on T-cells. It has been used therapeutically in the treatment of malignant melanoma and renal cancer, often in conjunction with administration of cultures obtained from tumour-infiltrating lymphocytes. The occasional spectacular success has been achieved but at considerable expense, both fiscal and toxic.

Interleukin 3 (IL-3)

A lymphokine that can induce the growth and proliferation of a variety of different cell types including both myeloid and erythroid precursors. It is produced by activated T-cells.

Interleukin 4 (IL-4)

IL-4 is a growth factor for lymphocytes and other cells, including mast cells. The gene is near chromosome 5q31. It is produced by activated T-cells.

Interleukin 5 (IL-5, eosinophil growth factor)

IL-5 is produced by helper T-cells and stimulates and supports basophils and eosinophils. The gene is very close to that for IL-4 at chromosome 5q31.

Interleukin 6 (IL-6)

IL-6 is an interesting molecule with therapeutic potential. It is 284 amino acids in length. It stimulates megakaryocyte precursors and, therefore, may be used in the **supportive care** (**p. 300**) of patients with thrombocytopenia. It also has some anti-viral effects. The gene is on chromosome 7q. For full stimulation of the receptor, a cofactor, gp130, is required.

Interleukin 7 (IL-7)

Stromal cells in the marrow and thymus produce IL-7. It stimulates both T- and B-cells and may be involved in the pathogenesis of acute lymphoblastic leukaemia. The gene is chromosome 8.

Interleukin 8 (IL-8)

IL-8 acts as an activator and chemotactic factor for neutrophils. It is produced by activated T-cells. It is a small molecule, only 72 amino acids in length, and the gene is on chromosome 4q12–21.

Interleukin 9 (IL-9)

IL-9 stimulates the growth of subpopulations of T-lymphocytes. It produced by activated T-cells and the gene is on chromosome 5q31–32.

Interleukin 10 (IL-10)

IL-10 has some similarities to a protein, BCRF1, associated with the Epstein–Barr virus. It is produced by helper T-cells and inhibits cytokine production by a separate subpopulation of helper T-cells.

Interleukin 11 (IL-11)

A growth factor derived from stromal cells that interacts with IL-3 and potentiates its ability to stimulate the growth of megakaryocytes.

Internet

A system of interconnected electronic communication based on personal computers (PC). By linking a PC into standard communication lines, telephone lines or other cabling, it is possible to link to other PCs and to the **World Wide Web** (**p. 327**). The usual way to this is by a modem or structured cabling, and linkage to an Internet service provider (ISP). The service provider effectively supplies an electronic postal service for outgoing and incoming e-mail, as well as providing access to the Web.

Interstitial Pneumonitis (IP)

IP is a clinical syndrome characterised by dyspnoea, cough, fine crackles throughout the lungs, fluffy pulmonary infiltrates on chest X-ray and hypoxaemia. It can occur 2–6 months after **total body irradiation** (**TBI**) (**p. 310**). It is often fatal. It is probably not just due to direct radiation damage to the lungs: infection (e.g. cytomegalovirus, CMV), drug toxicity (**cyclophosphamide** (**p. 83**), adriamycin), pre-existing lung disease (manifested as decreased lung volumes compared with predicted), **graft versus host disease** (**GVHD**) (**p. 143**) may also be important. The radiation damage is possibly mediated through damage to Type II pneumocytes (with consequent changes in surfactant), but vascular effects might also be important. The α/β ratio for IP has been estimated at between 3 and 6 Gy. Calculations suggest that sparing may occur down to fraction sizes of ~1 Gy. Repair half-time is also important: there is a slow repair component with a 4-h repair half-time. There is a vast amount of (non-randomised) clinical data suggesting that the incidence and severity of IP are reduced by fractionation. Obliterative bronchiolitis is a different syndrome. The clinical features may be similar but it is related to GVHD rather than to TBI.

Interval Cases

These are patients who present with cancer, undetected by screening, while participating in a **screening programme** (**p. 278**). They are apparently missed by the screening strategy and present, usually with symptoms, in the intervals between screening examinations. There are two main causes for interval cases: technical failures in the screening process or length time **bias** (**p. 28**). Technical failures in the screening process can occur as episodic, isolated, events or they can be systematic. Systematic failures can arise if the design of the whole screening strategy is faulty or because of persistent malfunction of a component of the programme. **Audit** (**p. 24**), particularly of interval cases, is a crucial feature of any screening programme.

Intrabody

A construct whereby the gene coding for the antigen-binding site of an antibody is introduced into a mammalian cell resulting in stable intracellular expression of the antibody. This technique can be used for **gene therapy** (**p. 137**). The antigen-binding site will express **epitopes** (**p. 118**) that will resemble the antigen against which the antibody was originally raised. If antibodies to tumour-specific antigens are used, then the use of intrabodies could be used as a form of tumour vaccination. The concept is similar to that of an anti-idiotype network.

Intra-operative Radiotherapy (IORT)

IORT is, as its name suggests, the administration of radiotherapy during a surgical operation. The rationale for this procedure is based on the belief that after surgery there may be viable tumour cells remaining in the tumour bed or at the excision margins. The persistence of these cells constitutes an obvious limitation to cure. If this is the case, then local radiotherapy given during the surgical procedure could be used to destroy these cells and thereby improve local control.

The logistics and practicalities of IORT are not simple. One option is to have a specially equipped operating theatre with its own linear accelerator. This is expensive both in capital cost and in terms of downtime for the linear accelerator. Another approach is to suspend the operation at the appropriate moment and then to transport the anaesthetised patient, with the wound open and the tumour bed accessible, to the radiotherapy department for treatment.

The usual treatment technique is to use electrons of appropriate energy and to place the applicator within the patient so that it is in contact with the tumour bed. Treatment is given as a single fraction, fractionated treatments being, for obvious reasons, impractical. The success of the approach depends on the trade-off between the physical advantages of placing the beam directly onto the tumour and the radiobiological disadvantages associated with treatment given in single fractions.

A variation on this approach, though not strictly intra-operative, is to place afterloading tubes into the tumour or tumour bed at the time of surgery. These can then be loaded, using ^{192}Ir wire or other suitable radioisotopes, during the postoperative period. This approach combines the ability to implant the tumour bed under direct vision with the physical and radiobiological advantages of brachytherapy.

Invasion

The ability to invade is characteristic of malignant transformation. It implies that malignant cells can transgress normal physiological and anatomical boundaries and, by so doing, disrupt the structure and function of adjacent tissues. Invasion is a process that involves reciprocal communication between malignant cells and their environment. Once a tumour has the potential to develop its own vasculature it can invade the surrounding tissue.
A sequence of essential events is involved:

- Tumour cells separate from each other.
- Extra cellular matrix is dissolved (proteolysis) and host cells are killed (cytolysis).
- Tumour cells move into the space created.
- Tumour angiogenesis enables the, now larger, tumour to remain viable and the sequence can repeat it self.

A wide variety of receptors and interactions is involved in each of these essential steps.

Cell separation

Cadherins (**p. 39**), e.g. cadherin E, are calcium-dependent molecules. They straddle cell membranes and bind to elements within the cytoskeleton called catenins. The cadherin–catenin system holds cells together and it is loss of expression and/or function that enables cells to separate from each other. N- and V-CAM are cell adhesion molecules also concerned in cell–cell adherence.

The integrins are a family of cell surface adhesion molecules that could both facilitate, by allowing cells to creep over the extracellular matrix, or prevent, by binding cells tightly to the basement membrane, invasion.

Dissolution of extracellular matrix

Enzymes, secreted by the tumour cells as inactive pro-enzymes, are activated and, by destroying the acellular matrix in the vicinity of the growing tumour, create a space into which the tumour can expand.

Although several enzyme systems are potentially involved in this process, the two most important are the matrix metalloproteinases and the plasminogen activation system:

- Metalloproteinases
  - — interstitial collagenases
  - — type IV collagenase
  - — stromolysin.
- Plasminogen activators (serine proteinases)
  - — urokinase type plasminogen activator (uPA)
  - — tissue type plasminogen activator (tPA).

Motility

To occupy the space created, the cancer cells need to move. Many of the factors that induce proliferation (mitogens) can also confer motility (motogens). Scatter factor is a 728 amino acid protein that can induce motility and is, in its sequence, identical to hepatocyte growth factor (HGF). Another motility factor (autocrine motility factor AMF) interacts with a receptor, gp78, to increase cellular motility. Just to illustrate how complex are the ramifications of intracellular communication and control — gp78 has **homology** (**p. 154**) with *p53*.

The result of the activation of mitogens is that the cancer cell can move, in an amoeboid fashion, using pseudopodia, out into the adjacent tissues.

Inverse Technique

This is a technique of radiotherapy planning that reverses the normal sequence of procedures. The method is derived from imaging. In the inverse technique the physicist starts with the isodose required and then works back towards the beam arrangement that would produce that pattern. The technique has the advantage that, for complex techniques, it might be quicker that traditional forward-planning measures. Its disadvantages are that there might be a lack of equivalence between the chosen isodose and the desired clinical outcome; the beam arrangement defined by the technique may not technically be possible or may involve excessive irradiation of vulnerable normal tissues. In practical terms, the method works well for small tumours — particularly when the problem can be defined solely in terms of a **dose distribution** (**p. 102**). It is less useful when larger volumes are to be treated or when integral dose is an important consideration.

Invisible College

A term that originates in the seventeenth century, when science was the preserve of gentlemen-amateurs working from home. An informal network of scientists evolved: men, and they were men, who were in regular contact by correspondence, but who rarely met. This invisible college finally achieved tangible, and visible, form with the foundation of the Royal Society.

Invisible colleges still exist: people know each other, either directly or through acquaintances, and reputations precede first-hand knowledge. This loose pattern of affiliation can extend into subjects such as peer review and the disadvantage is that the old boy (and they usually are boys) network may stifle progress in areas not considered relevant or fruitful by whatever invisible college is invited to comment on new proposals. The secret of such colleges is, of course, their invisibility.

Irinotecan

Irinotecan is an inhibitor of **topoisomerase I** (**p. 310**). It is a camptothecin derivative and, *in vivo*, is converted to a highly active metabolite 7-ethyl-10-hydroxycamptothecin (SN-38). SN-38, and to a lesser extent irinotecan itself, interfere with **topoisomerase I (p. 310)** by preventing the religation of the single-strand **DNA break (p. 101)**. The initial single-strand break is converted to an irreparable double-strand break by the formation of a complex involving DNA, topoisomerase I and SN-38 (or irinotecan). The inhibition of topoisomerase is most effective in the S-phase of the **cell cycle (p. 52)**.

Irinotecan has been widely used in the treatment of colorectal cancer, mainly because it is virtually the first drug active against that disease that does not rely on the inhibition of thymidylate synthetase for its action. Its main side-effect is diarrhoea. It can also produce a parasympathetic syndrome (flushing, visual disturbance, abdominal pain) that usually responds to the administration of atropine.

Irradiated Volume

This term has, in ICRU 50 terminology, a quite specific definition: it is the volume of tissue that receives a dose of radiation ≥20% of the specified target dose. As such, it is the largest defined volume in the ICRU nomenclature.

Isodose

A line, analogous to the isobars on a weather chart or the contour lines on a map, joining points of equal dose. Isodose charts can be produced for any beam produced by a radiotherapy treatment machine. The characteristics will depend on may factors including type of radiation, beam energy, field size, presence of **wedge (p. 327)** within the beam, etc. The isodose charts for overlapping beams can be added (50% from beam A + 40% from beam B will generate a point lying on a summated 90% isodose) and this will produce a dose-distribution for that particular arrangement of beams. Formerly, this was all done by hand and considerable experience was required to plan radiation treatment accurately and rapidly. Today, computerised planning systems are available and a more **heuristic (p. 150)** approach is possible.

Iterative

This simply means repeated over and over. It is a term often used in modelling to describe a simulation in which a process is carried out over and over again, as in a Monte Carlo simulation. Another approach is to find an value for an unmeasurable variable that best fits within an equation by repeatedly, iteratively, solving the equation substituting a series of values for the unknown quantity until the solution for the equation approximates most closely to the observed data.

Kaplan–Meier Method (syn. Product Limit Method)

This is a non-parametric method for constructing a **lifetable (p. 179)**. It maximises the use of the available data by recalculating the survival probability each time an event occurs. This is in contrast to the **actuarial method (p. 5)**, in which events are assumed to occur at the mid-time of specified time intervals. When events can be precisely dated, the Kaplan–Meier method is better than the actuarial method. Both methods adjust the survival estimates for missing, **censored data (p. 60)**. The Kaplan–Meier method produces a curve with steps — each step corresponds to an event. Censored observations are usually indicated on the curve. **Confidence intervals (CI) (p. 70)** can be calculated by a variety of methods: Greenwood's, Peto's, Rothman's transformation.

Kappa (κ)

This is a statistical technique used to assess the degree of agreement between two sets of observations, usually made by separate observers, upon the same data. It is, in the jargon, a measure of interrater agreement. It is computed from the observed and expected values in a frequency table, and is corrected for agreements that might simply have occurred by chance. A deficiency of κ is that all disagreements are treated equally. If one observer described a cervical smear as unequivocally malignant (C5) and another defined it as benign (C1) then this, potentially catastrophic, disagreement would be counted no differently than if one observer had graded a smear as C1 and the other had graded it as C2. Weighted κ is a means of incorporating the degree of disagreement into such a comparative analysis. The C1/C2 discrepancy is counted as less important than the C1/C5 discrepancy when a weighted κ is used. For a practical illustration of the use of κ, see **Measurement Error (p. 190)**.

Kerma

The kinetic energy released per unit mass (kerma) is a measure of radiation quantity ($J\ kg^{-1}$). It measures the first stage (interaction with matter to produce charged particles) of the two stage process whereby indirectly ionising radiation interacts with matter. The second stage (deposition of energy within matter by the charged particles) is measured as **absorbed dose (p. 1)**. Kerma and absorbed dose are often, but not necessarily, equivalent. If the secondary particles are energetic, and travel beyond the small volume used for measurement, then the kerma will be greater than the dose.

For a megavoltage beam, the absorbed dose does not reach its maximum until electronic equilibrium has been reached. Electronic equilibrium defines a state in which the number of electrons entering a volume of tissue is exactly equal to the number leaving. As a beam enters a medium, the electrons

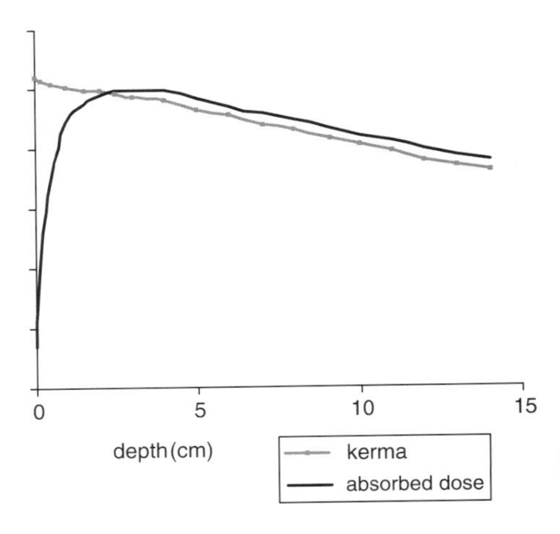

Figure 1 — Depth dose and Kerma data for a 10 MV X-ray beam. In the initial build-up region kerma is greater than absorbed dose. Thereafter, kerma is, in general, a little less than the absorbed dose.

have considerable forward velocity in the direction of the beam and so more leave a volume than enter. The absorbed dose, therefore, is less at the surface than is at as a depth of ~1–2 cm. This is the build-up region and accounts for the skin-sparing effect of megavoltage treatment. Kerma is maximal at the surface and then decreases linearly with depth: the absorbed dose is lower at the surface, builds to a maximum at a depth determined by the beam energy and then, finally, decrease in a linear fashion (figure 1).

Ki67

A cell-surface antigen that can be detected by appropriate monoclonal antibodies, and which is associated with cellular proliferation. The Ki67 index is a means of assessing the proliferative state of a tumour or tissue and may correlate with prognosis or response to treatment. One advantage of the antigen is that it is detected in paraffin-embedded sections and, therefore, can be used on archival material from tumours where the clinical outcome or response to treatment are known.

Kinase

An abbreviation for phosphokinase, a family of enzymes that catalyses transfer of a phosphate from ATP to a second substrate. The kinase is named according to the second substrate upon which it acts: tyrosine kinase transfers a phosphate from ATP to tyrosine residues; a protein kinase transfers phosphate from ATP to a protein.

Kinetochore

A specialised region of the chromatid during mitosis. Two kinetochores form at the **centromere** (**p. 61**) and are subsequently responsible for the accurate segregation of the chromosomes. They are slowly assembled during the **cell cycle** (**p. 52**), but are not fully functional until **metaphase** (**p. 198**).

KLAKLAKKLAKLAK

A peptide that can trigger **apoptosis** (**p. 19**) and which may have a potential role in cancer treatment. It is included in this glossary mainly because it is fun to say. Whether it is superior to 'giddy up' as a means of persuading a horse to gallop is, as yet, unproven.

Knockout Mouse

These animals have been developed to obtain animal models of autosomal-recessive disorders. This is in contrast to transgenic animals developed to provide animal models of autosomal-dominant diseases. In the creation of a knockout mouse, pluripotential embryonic stem cells (from the inner cell mass) are manipulated *in vitro* to delete the gene(s) of interest. The altered cells are microinjected into the blastocyst of a mouse embryo of ~3 days gestation. This is then implanted into a foster mother. The animal born subsequently contains progeny of the abnormal (manipulated) cells in all tissues, including germ cells. This chimeric daughter, effectively heterozygous for the gene deletion, can then be bred to produce mice homozygous for the deletion, i.e. they have no copies of the knocked out gene — they are the knockout mice. The knockout mice of relevance to oncology includes a ***p53*** (**p. 224**) deletion (a model for **Li–Fraumeni syndrome**, **p. 179**), a NF-1 deletion (a model for **neurofibromatosis**, **p. 213**) and a RB deletion (a model for **retinoblastoma**, **p. 271**).

Knowledge

Knowledge is data tempered by intellect and experience. It can be classified as follows:

- Things known to be true (and which are true).
- Things known to be true (and which are, in fact, untrue).

- Things we think we do not know (but, in fact, are known).
- Things we do not know (and which we do not know we do not know).
- Things we do not know (and know that we do not know).

Knudson Hypothesis

This concerns the development of tumours related to abnormal function of tumour suppressor genes. If only one allele is mutated, then the remaining normal gene should function as suppressor. If, however, an individual is born with one mutant suppressor gene and then acquires a mutation in the remaining gene, then that individual will have lost the ability effectively to suppress tumours. This is the two-hit hypothesis. The first (hit one) mutation is inherited, the second (hit two) is acquired in later life. The first hit is a **germline mutation** (**p. 141**); the second a somatic mutation. This is consistent with the clinical pattern of tumour development in patients with retinoblastoma.

Kruskal–Wallis Test

This is a generalised, multiple-sample version of the non-parametric rank sum test (Wilcoxon, Mann–Whitney). It is a non-parametric one-way analysis of variance. It tests whether multiple samples are drawn from the same population. For example, we might have data on age at diagnosis for four different centres contributing patients to a clinical trial. If the Kruskal–Wallis test gave a significant result, then this would imply unequal distribution of age across the four centres and would call into question the validity of the study — unless, of course, the trial design included **stratification** (**p. 298**) by centre.

L'Abbe Plot

A graphical method for demonstrating **heterogeneity (p. 150)** in a **meta-analysis (p. 195)**: the event rate in the control arm is plotted against the event rate in the experimental arm. If the individual trials cluster reasonably closely together, then they are reasonably homogeneous. Wide dispersion indicates heterogeneity.

Figure 1 shows the L'Abbe plot for a series of lung cancer trials included in a meta-analysis: there is one clear outlier suggesting that significant heterogeneity exists and this might weaken the conclusions of the analysis. In this example, the outlier was a post-surgical adjuvant study in operable disease. The other studies were all non-surgical studies in locally advanced disease.

Labelling Index (LI)

A term used in cell kinetics to describe the proliferative state of a population of cells. It estimates the percentage of cells in S-phase. This provides a method of providing some basic information about key parameters in the cell cycle. A pulse of tritiated thymidine is administered and shortly thereafter the cells or tissue are analysed autoradiographically.

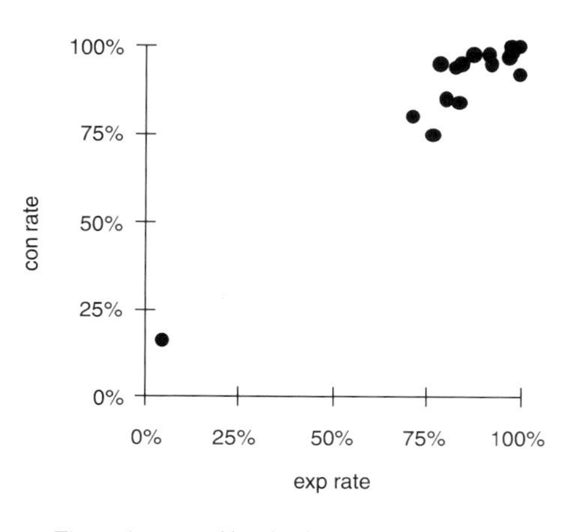

Figure 1 — L'Abbe plot for non-small cell lung.

LI can be related to the duration of S-phase (T_s) and the overall **cell cycle (p. 52)** time (T_c) by the equation:

$$\mathrm{LI} = (T_s/T) \cdot \lambda,$$

where λ is a constant in the range 0.5–1.0 used to define the age distribution of the cell population. A value of 0.69, corresponding to an exponential distribution over time, is customarily used.

Laddering

A classical hallmark of **apoptosis (p. 19)**. Gel electrophoresis of DNA from apoptotic cells classically shows laddering: a series of bands of DNA. The bands each correspond to fragments of equal molecular weight. These bands are each multiples of a unit weight corresponding to the distance between nucleosomes. This implies that the fragmentation of DNA in apoptosis is an ordered process and is in contrast to necrosis in which there is no pattern to the DNA fragmentation (and no laddering on a gel).

Lambda (λ) Phage

A bacteriophage derived from *E. coli* K-12 widely used as a cloning vector.

Langerhans' Cells

These cells form part of the antigen-presenting system and are a subtype of **dendritic cells (p. 93)**. They are positive for CD4 and CD1a; and are within the skin and mucosal surfaces. They can express CCR5 and may be crucially involved in the initial stages of infection with **HIV-1 (p. 151)**.

Linear Analogue Self-Assessment (LASA)

LASA scales are a form of **visual analogue scale (p. 325)** used for self-administered questionnaires. The subject is asked to indicate their response on a line, usually 10 cm long. The ends of the line are clearly labelled, 'anchored' in the jargon, with two contrasting statements such as 'a great deal'/'not at all' or 'strongly agree'/'strongly disagree'. The distance (cm) along the line to the subjects' marks is measured (cm) and the result is divided by 10 to give a score for that particular item. Calculated in this way, scores will all lie in the range 0–1.0.

Late Effects

A late effect is defined as an effect of treatment, not apparent during or immediately after treatment, which first appears ~3 months after treatment has finished, or according to the RTOG, as 'those toxicities which appear, or are persistent, 90 days or more from the start of therapy'.

The late effects of cancer treatment are less predictable than the acute effects. The lens of the eye is one of the most sensitive normal tissues to the late effects of radiation: doses as low as 8 Gy can cause cataract. Other tissues that are relatively sensitive to the late effects of radiation include the liver, kidney, lung and spinal cord. The sensitivity of growing tissues and organs to radiation poses particular problems in paediatric oncology where disturbances of growth and development are important late consequences of treatment. In general, the late effects of radiation on the skin and subcutaneous tissues are manifest as atrophy, shrinkage, fibrosis, pigmentation and telangiectasia. The shrinkage and fibrosis can affect adjacent tissues, e.g. the damage to the brachial plexus after radiotherapy to the supraclavicular fossa is largely due to subcutaneous fibrosis and entrapment of nerves rather than direct damage to the nerves themselves. For radiation therapy, there is a fundamental radiobiological difference between late effects and acute effects: the α/β ratio for late effects is typically ~3 Gy; for acute effects it is higher, typically ~10 Gy.

Table 1 illustrates some of the dose–response relationships for late effects in the radiation treatment of head and neck cancer.

Similar arguments apply to chemotherapy. The dose-limiting toxicity of chemotherapy is usually haematological and patients can be supported through even quite prolonged periods of cytopenia. The late effects of chemotherapy can cause problems. **Anthracyclines** (**p. 16**), such as **doxorubicin** (**p. 107**) and **daunorubicin** (**p. 88**), can cause damage to the myocardium. This damage is manifest clinically, long after the

Table 1

Tissue	BED ($\alpha/\beta = 3$ Gy)	Total dose at 2 Gy per fraction	Total dose at 2.75 Gy per fraction	Sequel
Lens	11	6	6	cataract
Retina	88	52	46	loss of vision
Optic nerve	99	59	52	loss of vision
Anterior eye	117	70	61	dry eye
VIII nerve	108	65	57	hearing loss
Cranial nerves	108	65	57	cranial nerve palsy
Temporal lobe	92	55	48	epilepsy dementia
Brain stem	88	52	46	dysarthria, mystagmus, ataxia, disturbed consciousness
Spinal cord	75	45	39	myelopathy
Temporal bone	107	64	56	ostenecrosis

end of treatment, as intractable cardiac failure. Although damage is unusual with total cumulative doses of **doxorubicin** (**p. 107**) of <500 mg m^{-2}, idiosyncratic damage can occur at lower doses. Conversely, some patients can tolerate much higher cumulative doses without suffering any adverse late effects on the heart.

Late effects are important because:

- There are ~1.4 million people in the UK who are long-term survivors following treatment for cancer.
- Late effects can cause premature death in patients cured of their malignancy.
- Late effects can cause unpleasant symptoms and thereby impair the quality of life following treatment for cancer.
- Unidentified, undiagnosed late effects can be an important, and potentially treatable, cause of distress in patients treated successfully for cancer.
- Management of late effects can be both expensive and time consuming.
- Late effects often limit the amount of treatment given and so late effects can be an important barrier to cure.
- Dissociation between acute and late effects of radiation can lead to therapeutic disasters when treatment regimens are changed.
- We may be under-dosing the majority of patients — basing our doses on the adverse experiences of an untypical minority.
- For patients to be adequately informed about the risks and consequences of accepting, or rejecting, treatment, they require accurate and comprehensible information about long-term effects.
- Counselling about fertility and conception must be based on sound knowledge of the risks to both the individual and the gene pool.
- To screen for problems, e.g. second malignancies, it is essential to know where and when they might occur.

Problems with assessing and reporting late effects:

- Until recently, there have been few appropriate scales available. Those that have been available have been applied only sporadically and inconsistently.
- Just because something cannot be easily measured, it does not mean that it is irrelevant or unimportant.
- Statistical techniques for analysing complication rates are, by no means, straightforward: crude calculations will underestimate the incidence of complications. Actuarial methods may overestimate their occurrence. How to deal with transient complications, e.g. necroses that heal, Lhermitte's sign, etc.
- Late effects may reflect bygone practices:
  - — 250 Kv XRT
  - — alternate fields alternate days
  - — large doses per fraction
  - — impossible to reconstruct original treatment.
- Dogs that do not bark: cases select themselves, but what of those many patients, treated similarly, who do not develop problems? How do we select controls?

Stochastic effects

These are effects that occur randomly. There is no dose-threshold below which they do not occur, but, as dose increases, then so the probability of the event occurring also increases. Increasing dose has no influence on the severity of the effect, only on its probability. Second malignancy after treatment is a classic example of a stochastic effect.

The derivation is from the Greek meaning to aim at a mark or to guess.

Deterministic (non-stochastic) effects

Effects that, for any given individual, will with increasing dose increase in severity once a threshold has been exceeded. Anthracycline-induced cardiotoxicity would be a reasonable example.

Figure 1 shows the problem of the last surviving patient suffering a complication: here are some data on complication rate calculated actuarially (the crude complication rate is 6/21, i.e. 29%).

The complication-free survival would be calculated actuarially as 50% (not 71%).

In contrast, figure 2 gives data for a data set in which the last surviving patient suffers a

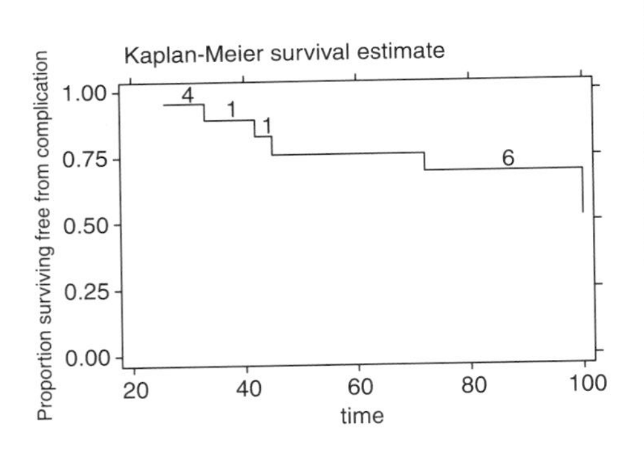

Figure 1.

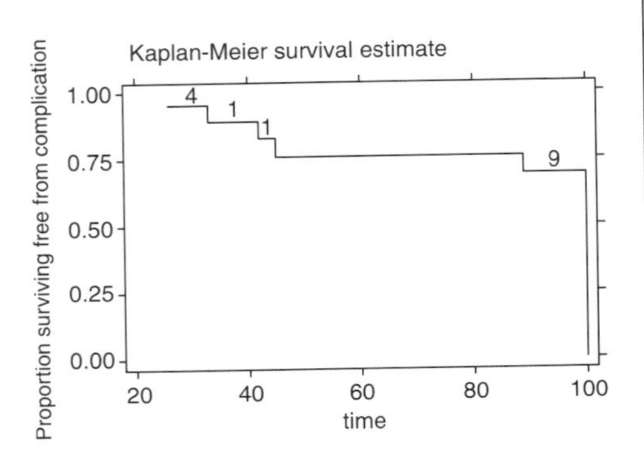

Figure 2.

complication — the crude complication rate is still 6/21 (29%).

The complication-free survival would be calculated actuarially as 0%.

Latent Membrane Protein (LMP)

LMP1 is a transforming protein associated with Epstein–Barr virus (EBV) infection: nasopharyngeal carcinoma; EBV-associated Hodgkin's disease; EBV-induced lymphoproliferative disease. It activates cells in a fashion similar to that produced by interleukin 4 (IL-4) acting on the CD40 receptor. The intracytoplasmic messenger for LMP1 is interactin, and the IL-4/CD40 interaction is with TRAF-3.

Law of Large Numbers (syn. Bernoulli's Theorem)

This law is simple in concept but it is difficult to express economically. It can be stated as: if the probability of an event occurring on a single trial is p, and if several trials are made independently under identical conditions, then the most probable observed proportion of the event occurring, expressed as a proportion of the total number of trials, is also p. This may seem obvious, but it is nevertheless crucial to much of statistical inference: particularly since, implicit within this formulation, is the idea that the more 'trials' there are, the more accurate will be the estimate of the true p. This leads directly on to considerations of sample size in clinical trials and it also provides a foundation for statistical **significance testing** (**p. 286**).

Lead Time Bias

Lead time bias is based on the fact that the date of diagnosis is a somewhat arbitrarily defined event in the natural

history of a tumour. In the absence of any effect from treatment, if the date of diagnosis is brought forward, then the time between date of diagnosis and death will be longer, and the survival rate will be improved.
The converse also applies.

Leaf Sweep

A technical term describing the dynamic movements of a **multileaf collimator (p. 210)** during intensity-modulated radiation therapy. The forward and the trailing leaves of the collimator can be programmed to move at different speeds and this will alter the intensity of the radiation fluence across the resulting beam. The easiest analogy is to consider the sun as viewed through a gap in ever-changing clouds.

Lectins

Lectins are **adhesion molecules (p. 8)**, usually of plant origin, that interact with carbohydrate molecules, e.g. on the surfaces of both white blood or endothelial cells. Lectins have found wide application as stimulators of proliferation (mitogens) in the culture of mammalian cells. They are also useful in blood grouping and in the immunocytochemistry of carbohydrates. Examples include concanavalin A, Pokeweed mitogen and phytohaemagglutinin (PHA).

Lectins are widely expressed by tumour cells that bind specific sugar molecules. Glycoproteins and glycolipids may, therefore, bind specifically to lectins expressed on the surface of tumour cells. This phenomenon is exploited in the **avidin–biotin (p. 25)** system for targeting tumours.

Length Time Bias

Length time bias arises because some tumours will grow so rapidly that they will present in the intervals between the periodic assessments of a **screening (p. 278)** programme. These rapidly growing tumours will cause death sooner than the more slowly growing tumours detected by screening. The symptomatic 'interval' tumours will, through this effect, be associated with worse survival than the screen-detected tumours.

Illustration of lead-time and length time bias (Figure 1)

Tumour A is steadily growing, its progress being uninfluenced by any treatment.
The arrows indicate the timing of tests in a screening programme. The horizontal lines indicate three thresholds: detectability by screening, clinical detectability and death due to tumour progression. A indicates the time at which the tumour would be diagnosed in a screening programme; B indicates the time at which the tumour would be diagnosed clinically, i.e. in the absence of any screening programme. If the date of diagnosis is used as the start time for measuring survival, then it is clear that, in the absence of any effect from treatment, the screening programme will, artefactually, add to the survival time. The amount of 'increased' survival = $y - x$ years, here just over 2 years. This artefactual inflation of survival time is referred to as **lead time bias (p. 176)**.

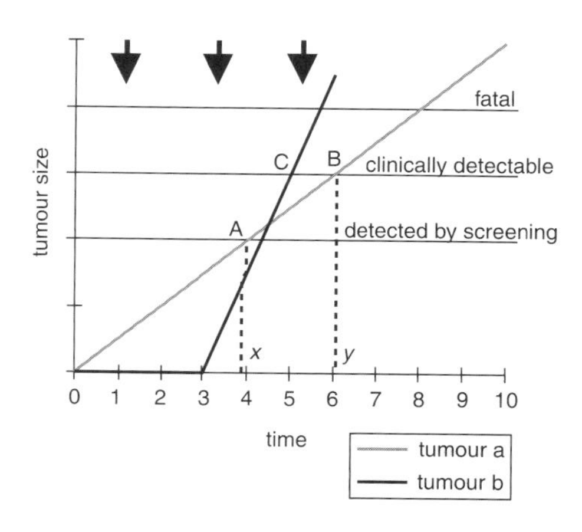

Figure 1.

Tumour B is rapidly growing. Again its progress is uninfluenced by treatment. It grows so rapidly that, in the interval between two screening tests, it can cross both the threshold for detectability by screening and that of clinical detectability. It will continue to progress rapidly after diagnosis and the measured survival time will be short. This phenomenon, whereby those tumours that are 'missed' by the screening programme are associated with decreased survival, is called length time bias.

Leptin

Leptin is produced in, and secreted by, adipocytes as a result of the expression of the *ob* gene. Leptin receptors are in the hypothalamus and defects in the leptin signalling network are associated with excessive eating and obesity.

Lethal Dose (LD_{10}, LD_{50}, $LD_{10/30}$, etc.)

LD is simply that dose that kills. The subscript indicates the proportion of experimental animals killed by the specified dose: the statement that the LD_{10} for drug A is *y* mg means that 10% of experimental animals will die when a group of animals is given *y* mg of drug A. The term after the '/' in the subscript indicates the time at which survival is evaluated. $LD_{10/5} = y$ mg indicates that 10% of animals will die within 5 days if a group of animals is dosed at *y* mg. The lethal syndromes following total body irradiation can be defined according to the time at which lethality is assessed. $LD_{10/5}$ will reflect the acute syndrome related to CNS toxicity; $LD_{10/10}$ will reflect the acute gastrointestinal mortality rate; $LD_{10/30}$ will reflect the mortality rate from myelosuppression. It has been suggested that the starting dose for drug studies in humans should be 33% of the mouse LD_{10} for that agent.

Leuprolide

Leuprolide is a synthetic analogue of gonadotrophin-releasing hormone (GnRH). After an initial increase in activity of follicle-stimulating hormone (FSH) and leutinising hormone (LH), which may cause **tumour flare** (**p. 316**), sustained administration of leuprolide causes down-regulation, inhibition, of secretion of LH and FSH. Its action is similar to that of **goserelin** (**p. 143**) and, like **goserelin**, it can be used in the treatment of hormone-sensitive cancers such as those of the prostate or breast.

LHRH (Leutinising-Releasing Hormone; syn. Gonadotrophin-Releasing Hormone)

LHRH is produced by the hypothalamus and stimulates the anterior pituitary to produce gonadotrophins: follicle-stimulating hormone (FSH) and leutinising hormone (LH). It is a decapeptide and several synthetic analogues are available: buserelin, goserelin, leuprolide. The continued administration of LHRH, as distinct from the physiological secretion, which is pulsatile, results in inhibition of FSH and LH release (down-regulation). This can be exploited therapeutically in the management of hormone-dependent tumours, e.g. breast and prostate cancers. LHRH analogue treatment is part of total androgen blockade (TAB) used to treat prostate cancer.

Down-regulation does not occur immediately and, indeed, there is an initial surge of gonadotrophin secretion at the beginning of treatment with LHRH analogues. This can produce an increase of tumour growth, tumour flare, often manifested as hypercalcaemia. This can be prevented by adding an anti-androgen for the first month of treatment with an LHRH analogue.

Li–Fraumeni Syndrome

A dominantly inherited syndrome in which there is mutation of ***p53*** (**p. 224**), so that tumour suppression is defective. As a result, a wide variety of tumours may arise: breast cancer, soft tissue sarcomas, osteosarcomas, brain tumours, adrenocortical carcinomas, leukaemia. There is, at least theoretically, an increased risk of radiation-induced malignancy following pervious treatment. *p53* is on chromosome 17p13.1. Lymphocytes from patients with Li–Fraumeni have decreased **responsiveness** (**p. 270**) to phytohaemaglutinin (PHA) and there are elevated levels of IgM in the serum.

There are clinical criteria for the diagnosis of Li–Fraumeni syndrome:

- A patient with sarcoma, and being <45 years of age, has a first-degree relative with cancer, of any type, presenting at <45 years of age, plus a third family member with sarcoma (at any age) or any other form of cancer at 45 years of age.

The risk of second primary tumour after the first malignancy in a patient with Li–Fraumeni syndrome is 50% at 20 years. Given the aggressive nature of many of the first tumours, the risk of second and third tumours is likely to be greater than the estimates: those who die from their first tumour are no longer at risk for a second tumour.

Life Expectancy

An unpleasant corruption of the phrase expectation of life': it refers to the average number of years that an individual of a given age might be expected to survive should current **mortality rate** (**p. 207**) patterns continue to apply. Life expectancy at birth is a special case; data on life expectancy are available from the Government Actuary's Department and is published each year in *Annual Review of Statistics* (London: HMSO) (figure 1).

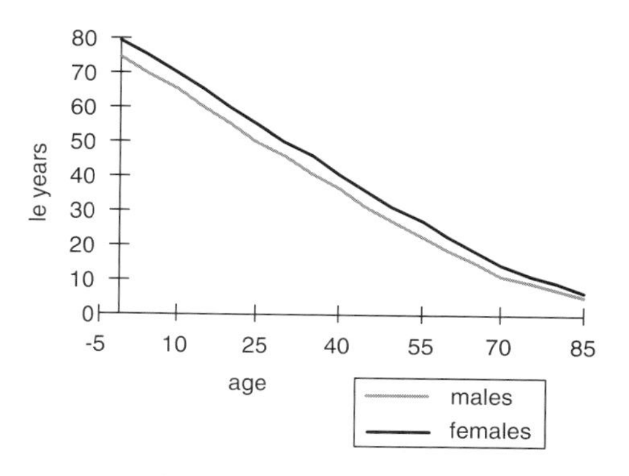

Figure 1 — Life expectancy.

The flattening of the life expectancy curve at >70 of age means that a 55-year-old man can expect to live until 77, whereas a 75-year-old male can expect to live until 84. This has implications for decision-making about treatment options for older patients: it is wrong to assume that simply because someone has reached 70 years of age that they are due to die shortly: the life expectancy for a 70-year-old woman is 14.6 years.

Lifetable

The term 'lifetable' is used to describe methods for summarising data on survival. The **current lifetable** summarises the mortality rate over a brief period (typically 1–3 years): it is usually used to estimate the age-specific mortality rate of a population. When derived from the general population, such tables are invaluable for constructing age-corrected survival curves. The **cohort** (**or generation**) **lifetable** describes the survival experience of a defined cohort: either a group of patients or a cohort of the general population born within a specified time interval. Although originally developed for the analysis of survival, lifetable methods can be used for other **endpoints** (**p. 114**) such as recurrence of cancer or the occurrence of complications. Edmund Halley, of comet fame, originally described the lifetable method in 1693. It has been subsequently developed and extended. The main methods used in oncology are the **Kaplan–Meier** (**p. 170**) and actuarial methods. The advantage of lifetable methods in general is that incomplete observations can be used to generate estimates of the complete survival experience.

Ligand

An entity, such as an ion or molecule, that binds to a receptor and which causes the receptor to exert its effect. Epidermal **growth factor** (**p. 144**), for example, is the ligand for the epidermal growth factor receptor; acetyl choline is the ligand for a ligand-gated ion channel, the nicotinic receptor, that allows Na^+ and K^+ to enter the target cells, thereby depolarising them.

Ligase/Ligation

Ligation, catalysed by DNA ligases, describes the formation of a 5′-3′-phosphodiester bond between two pieces of DNA.

Likelihood Ratio

A term is used in diagnostic testing and calculated as:

$$\text{Sensitivity}/(1-\text{specificity}),$$

where sensitivity is the probability that the test is abnormal in patients with disease and (1 − specificity) is a measure of the probability that a healthy person has an abnormal test result. The higher the likelihood ratio, the more likely that the test will accurately predict the presence of disease.

Likert Scale

This is used as a means of assessing and quantifying **responses to questions** (**p. 270**). Subjects are asked to respond to a statement by circling one of a series of possible responses labelled, usually, as follows: strongly disagree, disagree, undecided, agree, strongly agree. A Likert scale is, in one sense, similar to a **visual analogue scale** (**p. 325**) but with a restricted number of points. The problem with this analogy is that whereas with a VAS scale it might be reasonable to assume that a mark at 2 cm indicates a problem that differs by a factor of two from a problem marked at 4 cm, one can make no such assumption about a Likert scale. Category 4 (disagree) does not necessarily imply a twofold difference compared with Category 2 (agree).

Linear–Quadratic (LQ) Formula

The relationship between increasing radiation dose and decreased cell survival, in a reproductive sense, can be usefully described by an equation that has both linear (cell killing is proportional to dose) and quadratic (cell killing is proportional to $[\text{dose}]^2$) components:

$$f(d) = e^{-\alpha d - \beta d^2},$$

where $f(d)$ is the fraction of cells surviving a dose of radiation (d) and α and β are constants. α is in units Gy^{-1} and β is in units Gy^{-2}. It is possible to define a dose at which the cell killing from the $(\text{dose})^2$ component and the dose component are equal.

This dose is the α/β ratio, its units are Gy, since $Gy^{-1}/Gy^{-2} = Gy$.

By mathematical manipulation it is possible to define a biologically effective dose (E/α) or BED as:

$$\frac{E}{\alpha} = (nd) \times \left[1 + \left(\frac{d}{\alpha/\beta}\right)\right],$$

where n is the number of fractions and d the dose per fraction: nd, therefore, is the same a total dose.

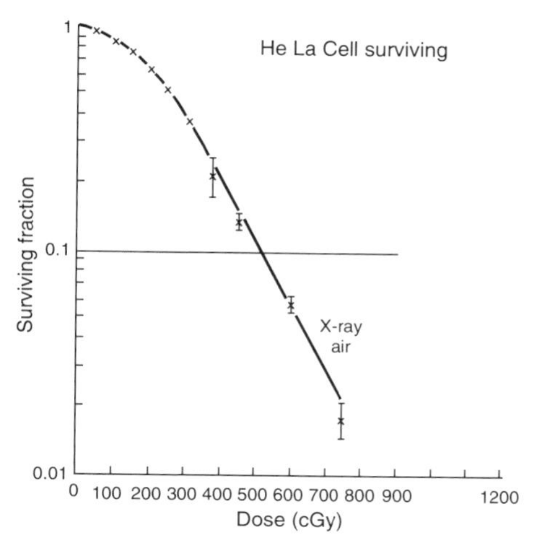

Figure 1 — First mammalian radiation cell-survival curve: HeLa cells irradiated in air with graded radiation doses from 0.75 to 7.5 Gy. The LQ formula provides a reasonable description of that part of the curve corresponding to doses in the range 0–500 rad (0–5 Gy).

Experimental observations show that the cell survival curve on a log–linear plot has an initial part, the shoulder, that is non-linear followed, at higher doses, by a linear component.

The LQ model has interesting resonances with biological **target theory** (**p. 304**) in that the linear component could represent a single hit on a single vulnerable target whereas the quadratic component could represent inactivation of adjacent targets that require multiple hits before they are irreparably damaged (figure 1).

The LQ relationship defines a curve that is continuously bending, even at high fractional doses. The LQ model, therefore, fits the data more accurately for the low fractional doses used in clinical radiotherapy but may be misleading when applied to the higher fractional doses used in experimental radiobiology (figure 2).

α/β ratios are, in general, higher for those cells which govern the early effects of radiation, α/β ratios are lower for those cells

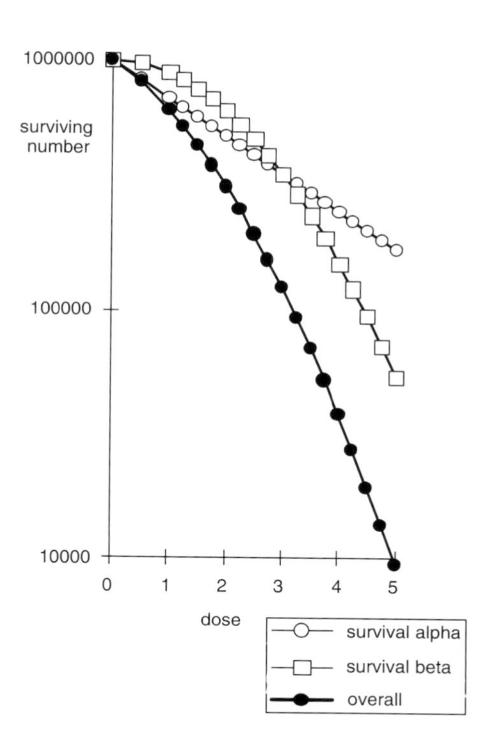

Figure 2 — Radiation cell-survival curve calculated according to the LQ equation with α and β components shown individually as well as the composite curve.

that are important in determining the late effects. The lower the ratio, the curvier the curve. Tumours have ratios similar to those of acutely responding tissues. When a radiation dose is divided up into a number of fractions there will be selective sparing of late damage, as opposed to acute reactions (or tumour cell kill). Fractionation, therefore, offers a method for improving the **therapeutic ratio** (**p. 307**).

Linkage

When two non-allelic genes are consistently inherited together, this indicates that they are close together on the chromosome. This is linkage and it should be suspected when genes are inherited together more than 50% of the time: this being the baseline rate for non-linked genes.

Linkage Disequilibrium

Linkage disequilibrium occurs when genetic markers or genes are inherited together more frequently than the chromosomal distance between them would predict. It can arise because there is locally reduced recombination or because of a **founder effect** (**p. 131**) in which insufficient time has elapsed for the expected equilibrium pattern to establish itself.

Linking Number

The linking number (Lk) is an integer used to define how one strand of DNA winds round another:

$$\text{Lk} = \text{number of double helical turns} + \text{writhing number}.$$

Type I topoisomerases will alter the linking number in steps of one, in contrast to Type II topoisomerases which change it in steps of two.

Liposome

A liposome is a phospholipid envelope that can be filled with a biologically active agent. The phospholipids used include phosphatidylcholine and distearyl-phosphatidylcholine. The hydrophobic tails of the phospholipid cluster together in an aqueous environment. This brings the hydrophilic heads together and a spherical structure is formed with a hydrophilic surface and a hydrophobic core. Physically, liposomes are small entities that can readily move from the intravascular compartment and through the extracellular fluid spaces. The surface characteristics of the liposome can be further modified according to need. Treatment with polyethylene glycol (PEG, **PEGylation**, **p. 230**) produces a liposome that is less likely to be recognised and trapped by the reticuloendothelial system. Size does matter: the biodistribution of the liposome will depend on its size. Several liposomally encapsulated formulations of drug are now available. DaunoXome is a liposomal preparation of **daunorubicin** (**p. 88**), Caelyx is a PEGylated liposome containing **doxorubicin** (**p. 107**). Both formulations are clinically useful in the treatment of Kaposi's sarcoma and there is increasing evidence of activity against other tumours: breast cancer, non-Hodgkin's lymphoma, angiosarcomas. The toxicity of the naked drug is considerably mitigated by liposomal encapsulation. Alopecia and nausea are uncommon and it is possible to give doses of anthracycline at >500 mg m^{-2} without encountering cardiotoxicity.

Literature

The biomedical literature is enormous, with ~16 000 journals currently being published. Five years' output corresponds to a tower of paper higher than Mount Everest (29 028 feet): few people can claim complete familiarity with the biomedical literature.

Literature Searching

A comprehensive unbiased examination of the available literature is an essential precondition for a **systematic review** (**p. 302**). The validity of a review's conclusions might depend critically on the original literature search. Literature searches based simply on retrievals from standard services such as Medline are notoriously incomplete; Medline searching identifies only ~50% of relevant trials. Ancillary strategies need to be employed to ensure comprehensive coverage: hand searching relevant journals; contacting key authors directly about unpublished data or data published by recondite journals; interrogating registers of randomised controlled trials.

Locus

The term 'locus' refers to the position of a specific sequence of DNA, e.g. a gene, on the chromosome.

LOD Score

A statistical method used in genetic mapping to assign a RFLP to a **linkage (p. 181)** group. The LOD score is $\log_{10}$ of the ratio: probability that data observed are due to loci being linked/probability that data could have arisen from unlinked loci. Linkage is defined as present if LOD >3.0. The odds on linkage being present and LOD score follow a simple relationship (figure 1). The relationship is simply the relationship between the probability of an event and the $\log_{10}$ of the odds.

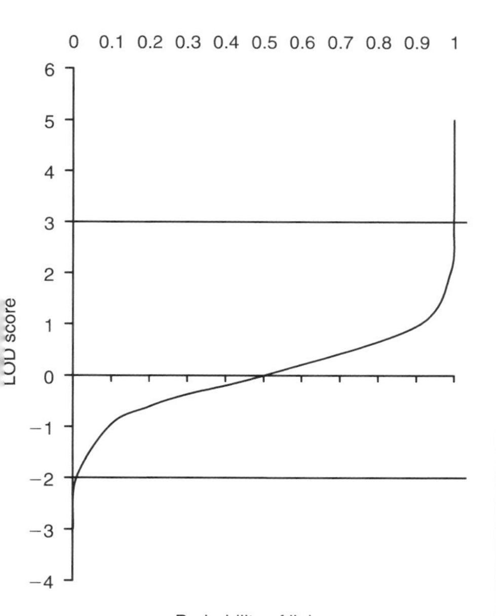

Figure 1 — The relationship between the probability of linkage existing and the LOD score. By convention, a LOD score of ≥3.0 (equivalent to a probability of linkage of 0.999) is taken to indicate linkage and, conversely, a score of ≤−2 (linkage probability of 0.09) is taken to reject the presence of linkage.

Logistic Regression (syn. Multiple Linear Logistic Regression)

A statistical modelling technique in which the purpose of the model is to use a series of explanatory variables to predict how the value of a dependent variable will be transformed. Logistic regression is used for categorical variables; linear regression is used for continuous variables. In logistic regression, the outcome is binary: a condition is present or absent. The calculations are complex and are usually based on a maximum likelihood estimate of the regression coefficients. Logistic regression can assess factors that predict prognosis, e.g. the chance of being alive 2 years after treatment (in which the binary outcome is alive/dead) or it can be used in diagnosis (in which case the binary outcome is disease present/absent).

Log-Rank Test (syn. Mantel–Cox Test)

The test compares survival curves in two or more groups of patients. The essential principle is each time there is a death in either group, to construct a 2 × 2 table and then to estimate observed versus expected deaths. By repeatedly performing this procedure until all deaths have been analysed, it is possible to compare the two groups in the total numbers of deaths observed divided by the number that would have been expected were there no difference in survival between the two groups.
The $(\text{observed} - \text{expected})^2/(\text{expected})$ value is summed for each group and adding the two summed values gives the log-rank statistic, which is analogous to a χ^2 with the **degrees of freedom (p. 91)** (d.f.) = 1 − number of groups compared. The log-rank test gives an indication of whether there is a significant difference in survival

between two groups. It gives no direct indication of the magnitude of that difference: that information is provided by the **hazard ratio (p. 148)**.

Long-Term Culture Initiating Cells (LTC-IC)

Cells with the capacity to form haemopoietic colonies in long-term culture and which can differentiate into various haemopoietic lineages.

Lonidamine

An indazole carboxylic acid derivative that decreases energy metabolism in tumour cells. It has no direct antiproliferative activity but could, in theory, potentiate the effects of other anti-cancer treatments, such as radiotherapy. Experimental evidence shows that it can impair repair of potentially lethal damage after radiation: this mechanism of action implies that its sensitising effect will be more pronounced the greater the number of treatment fractions. Given alone, it has some therapeutic effect against breast cancer and prostate cancer. A randomised study in head and neck cancer has shown a significant increase in disease free survival for lonidamine + radiotherapy compared with radiotherapy alone. The drug was originally developed as a male contraceptive: the aim being to depress the energy metabolism of sperm cells.

Lotka's Law

Only a small proportion of research projects produce major benefit: these rare studies are, however, crucial both for the advance of knowledge in general and for maintaining the viability of the individual unit or institution in particular.

Lrp (Lung Resistance Protein)

A protein that is associated with vesicles involved in transport between the nucleus and cytoplasm. It has been implicated in resistance to **cytotoxic drugs (p. 84)**.

Macrophage Migration Inhibitory Factor (MIF)

A ubiquitous cytokine intimately involved in inflammatory responses and which plays a key role in the pathogenesis of septic shock. Many cell types in response to stress release it, but the major sources of production are the monocytes and macrophages themselves. It activates T-cells and induces the expression of mediators of the inflammatory response. Glucocorticoids will induce the production of MIF, which will, in turn, reduce the anti-inflammatory actions of the steroid. In animal models, antibodies to MIF can protect against the lethal effects of septicaemia, e.g. in artificially induced peritonitis. Plasma levels of MIF are elevated in patients with septic shock — it remains to be seen whether this observation is of any diagnostic or prognostic utility.

Madness Among Oncologists

A recent survey has shown that ~25% of oncologists are 'mad', with symptoms of anxiety or depression. This rate of mental illness is no greater than that observed in medical students. No comparable data are available for surgeons or surgical trainees. The message is simple: medicine is in general a stressful career and oncology is no exception. Colleagues need to watch out for one another. If signs of stress are evident, then general support and encouragement supplemented, if necessary, by more specialised help should be available.

Male Pronucleus

The nucleus provided by the sperm, which is identifiable within the fertilised egg before it fuses with the nucleus of the egg. Transgenic material can be introduced into the male pronucleus by micro-injection. About 2–5% of the embryos so treated will incorporate the gene that has been artificially introduced into their germline DNA. This technique can be exploited in the production of transgenic animals.

Malignant Transformation

The lack of control of cellular proliferation is the hallmark of the malignant process. Contrary to common belief, the rate of cellular proliferation in a malignant tumour is not necessarily more rapid than that in normal tissues. The crucial difference is that the proliferation of normal tissues occurs in response to physiological need and, when that need is met, ceases. The proliferation of a population of malignant cells is inexorable and remorseless. It is subject to no physiological constraints. The main factors limiting growth and spread are the availability of living space and nutrients.

Rudolf Virchow first pointed out that cancer was a disease of cells, *omnis cellula e cellula*, each cell from a cell. He also used a metaphor drawn from politics to define the malignant process: proliferation in normal tissues is a well-ordered process, akin to a democracy; proliferation in a malignant tumour is a chaotic process, akin to anarchy. The political metaphor can be further extended: anarchists live outwith the law and are thereby denied its protection. The consequence for malignant cells is that, compared with normal cells, their ability to respond appropriately to environmental changes and insults might be impaired. This differential capacity in response to cytotoxic insult can be exploited therapeutically.

Since control involves communication, both within and between cells, it is apparent that approaches aimed at re-establishing normal control mechanisms in cells that have undergone malignant transformation will depend crucially on an understanding of how

information is exchanged among cells and between intracellular compartments.

It is now clear that there is no one defining event that inevitably leads to malignant transformation. Such transformation is a destination that can be reached by a variety of possible routes, many of which involve multiple steps. The corollary of this is that there is rarely one genetic abnormality pathognomonic for a particular type of cancer. Genetic mutation is fundamental to carcinogenesis: mutation can involve gain, in the sense of acquiring a means of escaping from normal control, or loss, in the sense of a normal control mechanism being deleted. The recent appreciation of the genetic basis of malignant transformation gives an underpinning for older ideas based on other perspectives. Several stages were recognised in experimental carcinogenesis: initiation, promotion and progression. In classical histopathology a similar progression could be described: hyperplasia and/or metaplasia, dysplasia, *in situ* carcinoma, and invasive carcinoma. A unified view, based on genetics and molecular biology, is now possible. It has been most elegantly described for colorectal neoplasia — based largely on the work of Vogelstein.

Figure 1 gives a simplified account of this sequence. An initial series of mutations takes place as a result of which the lining of the crypt cell changes from normal to hyperplastic: at least three genes (*APC*, *MLH*-1, *MSH*-2) are involved. These genes are concerned with DNA repair and cell–cell adhesion. A sequence of changes, involving *MCC*, *k-ras* and *N-CAM*, is associated with the change from hyperplasia to premalignant polyp. The change from premalignant polyp to carcinoma is associated with loss of function of normal (wild-type) *p53*. *p53* is a tumour suppressor gene that, in response to the detection of **DNA damage** (**p. 98**), triggers either repair of the damage or the death of the cell. The hypothesis is that cells with mutant ***p53*** (**p. 224**) do not die appropriately and

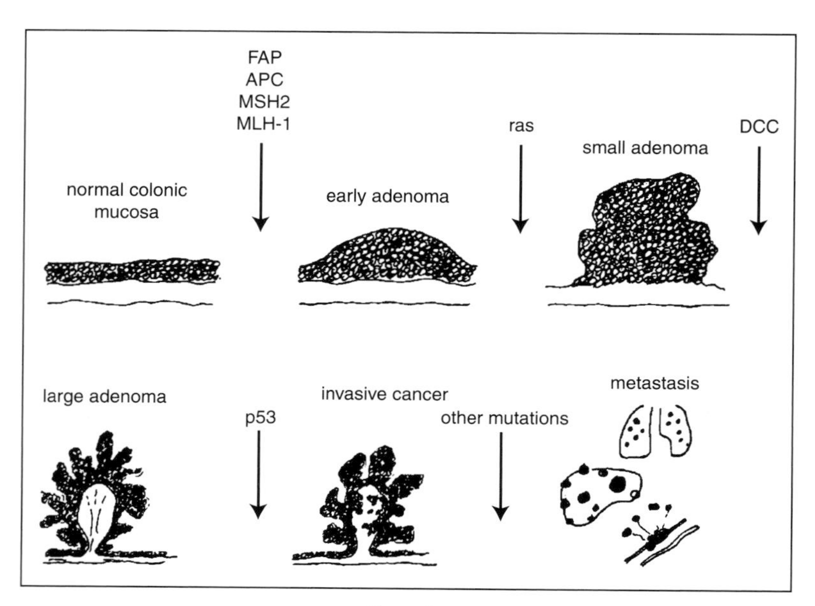

Figure 1.

thereby a further series of genetic abnormalities, associated with malignant transformation, might be propagated.

Mann–Whitney Statistic (syn. Wilcoxon Rank-Sum Test)

Unlike the **Wilcoxon Matched-Pair (p. 328) Signed Ranks test**, the Mann–Whitney can be used for unmatched, unpaired, data. It is a non-parametric test based on ranking the pooled data and then assessing whether it is possible that the observed rank order could have arisen by chance or whether there is a systematic difference between the two groups: with higher ranks predominating in one group and lower ranks in the other.

Mantel–Haenszel Procedure

This is a variation of the **log-rank test (p. 183)** and should be used when the number of events in the study is small. It incorporates a calculation of the variance as each event occurs. This is the hypergeometric variance. The actual differences between log-rank calculations and the Mantel–Haenszel calculations are often very slight.

Map/Mapping

Always remember that the map is not the territory.

A restriction map is deduced from the lengths of the DNA fragments produced when a series of **restriction enzymes (p. 271)** acts on the sequence to be studied. Pairs of enzymes are used: each alone (single digest) and then together (double digest). The observed fragment sizes are used to deduce possible permutations for the order in which the restriction sites might occur. From repeated analyses, the most likely permutation emerges. It is a necessary first step in the analysis of DNA sequences.

Gene map/mapping gives an indication of where genes are found on the chromosomes — it indicates their relative, but not their absolute, positions. A gene map can be obtained from a linkage map in which the frequency of recombination events for individual genes can be assessed and used to calculate distances between them (**centimorgans, cM**) (**p. 61**). If genes are not physically close to each other, are unlinked, then they have a 50% chance of being assorted separately at meiosis. If they are physically close to each other this probability falls: 1 cM is equivalent to a recombination frequency of ~1%. The **LOD** (**p. 183**) **score** is statistical measure of the degree of linkage between two loci. Mapping functions can interconvert distances (cM) (i.e. genetic distance) and distance in kilobases (base pairs, physical distance). Three main eponymous functions are used: Haldane's, Ott's, Kosambi's. As a rough guide, 1 cM = 1000 kb. Gene mapping in humans initially depended largely on studies of family pedigrees.

The matrimonial habits and impeccable record-keeping of the Mormons in Utah have been an invaluable resource, e.g. in studies of the linkage between haemachromatosis and HLA antigens. The CEPH (Centre d'Etude Polymorphisme Humaine) is a further resource for analysis of pedigrees in large human families. The X chromosome is the most completely mapped chromosome, mainly because of the ease with which sex-linked conditions can be traced through the generations. The development of RFLPs opened up gene mapping: by providing a regular series of markers along the chromosomes (analogous to the milestones along a road) RFLPs add a new precision and flexibility to the technique.

The Human Genome Project is a coordinated attempt to map the whole of the human genome: initially as a genetic map and finally as a complete physical map.

A **physical map** shows the sequence of base pairs along a length of DNA. The tools used in constructing a physical map are largely based on recombinant technology: mouse–human hybrids, direct *in-situ* hybridisation. Automated sorting of chromosomes after fluorescent tagging of specific DNA sequences is an emerging method for physical mapping.

Marginal Cost and Marginal Cost-effectiveness

The marginal cost-effectiveness is a method for comparing the **cost-effectiveness** (**p. 76**) of two competing strategies. Imagine that treatment A costs £15 000 and, compared with no treatment at all, provides on average an extra 1.75 years of life. This corresponds to a cost-effectiveness of £15 000/1.75 = £8571 year^{-1} gained. Imagine that treatment B costs £40 000 and, again compared with no treatment at all, provides an extra 3.25 years of life. The cost-effectiveness is £40 000/3.25 = £12 307 year^{-1} gained. The marginal cost compared with treatment A is £40 000 − 15 000 = £25 000, and the marginal effectiveness is 3.25 − 1.75 = 1.5 years. The marginal cost effectiveness is £25 000/1.5 = £16 666 per year gained when treatment B is used.

As cancer treatments become more expensive, without becoming appreciably more effective, the question of marginal cost-effectiveness becomes increasingly important. At what point do we stop paying for minor gains in survival: when marginal cost-effectiveness is £50 000 … £100 000 … £1 million year^{-1}?

Markov Model (Markov Process, Markov Chain)

The Markov process is a form of **decision analysis** (**p. 88**). It is widely used for modelling time-dependent processes. These could be biological processes, such as the growth of a tumour, or imposed procedures, such as follow-up policies. The great virtue of the Markov approach is that it enables the biological and the artificial to be combined: the natural history of a tumour can be incorporated into an assessment of varying strategies for staging and follow-up. Economic analysis can be a simple extension of the Markov technique.

The approach is to define a series of states in which an individual or entity can be at any particular time. The states should be mutually exclusive and should embrace all the possibilities. Three months after treatment for cancer, for example, the following states could be defined: alive with disease; alive, no evidence of disease; dead from disease; dead from another cause. The Markov approach divides time into equal periods (cycle length) and models how, over time, individuals move between states. These movements are governed by probabilities, known as transition probabilities, and by the fact that transitions to certain states, death being an obvious example, are one way. These states: death, lost to follow-up, etc., are known as absorbing states. Once a model has been designed, it is most simply run as a cohort study: 20 cycles each with a 6-month duration would simulate a follow-up experience of 10 years. The beauty of the system is that transition probabilities can be made time-dependent: higher probability of relapse early after treatment compared with later cycles.

Marrow Stem Cells

These cells, found in normal bone marrow and, after stimulation with chemotherapy and colony-stimulating factors, in peripheral blood, can reconstitute normal haemopoiesis in a recipient treated with **lethal doses** (**p. 178**) of drugs and/or radiation. The phenomenon is exploited in

high-dose therapy with **stem cell (p. 298)** rescue. The stem cells are obtained either from the bone marrow, by aspiration, or from the peripheral blood, using cell separation. **Marrow stem cells (p. 188)**, when defined in terms of their ability to reconstitute haemopoiesis in the **SCID (severe combined immunodeficient) mouse (p. 278)**, are **CD34$^+$ (p. 52)** and CD38$^-$. Similar cells, with stem-like properties, can be identified in patients with AML, between 0.2 and 100 cells/10^6 of leukaemic blasts are primitive cells (SCID leukaemia-initiating cells or SL-IC).

Maspin

This protein, which is a member of the serpin family, is encoded by a gene that functions as a tumour suppressor. It may function as an inhibitor of angiogenesis and prevents the migration of endothelial cells towards **growth factors (p. 144)** β-FGF and **VEGF (p. 324)**; it blocks mitosis in endothelial cells. Increased expression of maspin limits the growth and metastatic potential of breast cancer. Conversely, **down-regulation (p. 107)** of maspin is associated with tumour formation, cancers of the breast and prostate.

Matrix Metalloproteinases (MMP) and their Inhibitors

MMP are enzymes essential for tumour invasion and **metastasis (p. 198)**. They can dissolve the basement membrane and extracellular matrix and thereby enable tumour cells to gain access to the extracellular space and the circulation. They also have other roles in tumour progression: increasing the growth of the primary tumour; permitting circulating tumour cells to survive in the circulation; facilitating the growth of secondary tumours. Over 20 different MMP (1–24) have been described. They have a variety of enzymatic activities, collagenases, elastases and gelatinases, and include the stromelysins and matrilysin. They have a structure with three domains: one for secretion; one for maintaining latency; and an active site with a zinc-binding motif. They are secreted in an inactive membrane-bound form.

Scarcely surprisingly, given their potent biological effects, there are several physiological inhibitors of MMP. There are four main tissue inhibitors of MMP: TIMP-1–4, and the ubiquitous protease inhibitor α-2-macroglobulin will also inhibit MMP.

Since MMP are essential for metastasis and **invasion (p. 167)**, their inhibition ought to be therapeutically useful. Batimastat was the first MMP inhibitor to be used experimentally, but it was too toxic for clinical use. A similar compound, Marimastat, has been tested in clinical trials. It is an orally active low molecular weight (331.4 Da) peptide that inhibits MMP by chelating the zinc atom at the active site of the enzyme. Other synthetic inhibitors of MMP include AG3340 and BAY 12-9566. In general, the clinical trials of MMP inhibitors have been disappointing. This may reflect, in part, the fact that they might have their greatest therapeutic potential in the early phases of tumour development. Assessing the survival of patients with advanced disease may not be the most appropriate environment for testing the efficacy of compounds with such a subtle mode of action.

Max

A **transcription factor (p. 313)** that binds with the product of the ***MYC* oncogene (p. 211)**, and it is this binding that permits the *MYC–MAX* complex to bind to DNA.

Maximum Likelihood Estimation (MLE)

A technique used to estimate those values for a series of variables which, first, best fit the

observed data and, second, are consistent with a particular conceptual model.

The sequence of events for MLE is as follows. A data set is assessed quantitatively. The measurements are used to generate a series of estimates for a set of parameters that could best 'explain' the actual values observed. A 'likelihood value' can be assigned to each such estimate. 'Likelihood' in this sense indicates the probability that the values chosen correspond to the values observed. The likelihoods for the various estimates are compared and the estimate with the maximum overall likelihood is chosen. The estimates are generated heuristically, by trial and error, and iteratively, by repeating the process over and over, making what seem to be sensible changes until an overall maximum likelihood is reached.

MLE has a variety of uses. It can be used in: radiobiology, for finding α and β that, within the linear–quadratic (LQ) model, best fit a set of clinical or experimental data; molecular biology, for defining evolutionary trees from sequence analysis; and pharmacology, for estimating constants in multicompartment models.

Maximum Tolerated Dose (MTD)

MTD is the highest dose that can safely be tolerated. This term has been used loosely: sometimes described as the dose immediately below that which is usually lethal; sometimes described as the dose at which any detectable toxicity appears. This is misleading and there should be a clear distinction between MTD and the **toxic dose low** (**TDL**) (**p. 312**).

MTD is established for any potential new therapeutic agent initially through animal studies. Ideally several species should be used and the original recommendation was that the starting dose for studies on humans be set at 33% of the MTD in the most sensitive species. The concept of MTD is relatively straightforward when the limiting toxicity is haematological, since life-threatening levels of leucopenia and thrombocytopenia are easily defined. The problem arises when other toxic effects, such as CNS damage or gastrointestinal toxicity, are dose-limiting.

Rules based on LD_{10} and/or TDL have been adopted more recently. Starting dose should be:

- 33% of TDL in the most vulnerable large animal tested.
- 33% of the LD_{10} in the mouse.

MTD for humans can then be established in a **Phase I study** (**p. 232**), starting at a dose as defined above and then escalating dose according to a predetermined scheme such as a modified **Fibonacci sequence** (**p. 127**) (n, $2n$, $3.3n$, $5n$, $7n$, $9n$, $12n$, $16n$, ...), where n is the starting dose.

MDM2

This gene was originally discovered in a spontaneously transformed murine cell line. It is overexpressed in many sarcomas. The *MDM2* protein binds to, and inactivates, normal ***p53*** (**p. 224**). A foetal mouse which lacks *MDM2*, fails to develop, and there is no inhibition of *p53*-induced cycle arrest and **apoptosis** (**p. 19**). The animal can be rescued by mutant, ineffective *p53*.

When DNA is damaged *p53* is phosphorylated at serine residue 15, the phosphorylation interferes with *MDM2*/*p53* binding and so the *p53* is freed from its suppressor and can cause cell cycle arrest and apoptosis after DNA is damaged.

Measurement Error

'If you want to know the length of this board, measure it once and quit. If you measure it again,

you may get confused.' (anonymous high-school science teacher quoted by Feinstein)

The techniques available for making measurements in medicine are not infallible. It is unlikely that one observer, using the same ruler to measure the size of a breast tumour, will, even if the measurements are made within an hour of each other, record exactly the same value on each occasion. The tumour has a 'true' size, but the measurement process has errors associated with it and so measurements will only approximate to the 'true' size of the tumour.

It is useful to estimate the size of the error associated with a series of measurements, e.g. Dr X measuring the size of the primary tumours in a series of patients with breast cancer. This is particularly true if we are relying on reproducible measurements of tumour size to make decisions about treatment: mastectomy if tumour >30 mm, otherwise wide local excision and radiotherapy. If we have a series of measurements from each subject taken within a short interval, then we can calculate the within subject standard deviation (SD) (s_{ws}). The difference between any individual measurement and the 'true' value should, 95% of the time, be $<1.96 \times s_{ws}$. For example, if we measure a patient's tumour as 21 mm in diameter and, from making repeated measurements on a group of patients, it is known that $s_{ws} = 1.5$ mm, then there is a 95% probability that the 'true' size of the tumour is between $21 - (1.96 \times 1.5)$ and $21 + (1.96 \times 5)$ mm, i.e. between 18.1 and 23.9 mm. If s_{ws} were, as is more likely, 5 mm, then the 95% CI for the measurement of 21 mm is from 11.2 to 30.8 mm. It is obvious that a formal consideration of measurement error is exceptionally useful for formulating rational criteria for distinguishing between subjects.

Repeatability is variation on the same theme: repeatability $= (2)^{1/2} \times 1.96 \times s_{ws}$. The difference between measurements taken from the same subject will, 95% of the time, be less than the value calculated for the repeatability. For the 21-mm tumour, with $s_{ws} = 5$ mm, the repeatability is a disappointing 13.9 mm.

The message from this particular example is that if we wish to have a good estimate of the 'true' value, then we need to use either the mean of several observations or a better ruler (or, perhaps, train Dr X to use it properly).

Measurement error can produce placebo response rates. Moertel and Hanley showed in 1965 that, using a standard criterion of response (a >50% reduction in the product of two perpendicular diameters), measurement error would produce a 'response rate' of ~8%.

Figure 1 shows a plot of the relationship between measurements made by Dr A and Nurse X on a series of primary breast tumours. The **correlation** **(p. 75)** between the two observers is excellent, Pearson's correlation, $r = 0.9915$ ($p < 0.00001$). However, this is not the way to compare the observations made by the two observers. Correlation can be excellent, but this does not mean that the observers agree: it may

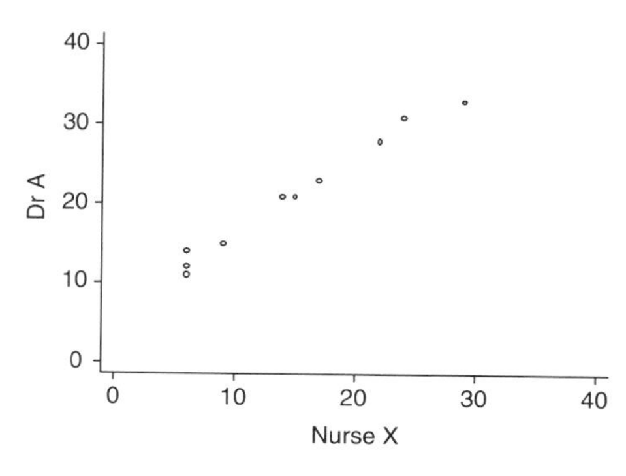

Figure 1.

simply mean that they disagree in a consistent fashion. Table 1 gives the data together with, since this is a set of hypothetical data, the 'true' measurements of tumour diameter.

There are important systematic differences between Dr A, who persistently overestimates the size of the tumour, and Nurse X who tends slightly to underestimate tumour size. How might we use statistical methods to encourage the emergence of these important differences?

Figure 2 shows the result of plotting the difference between each pair of observations (Dr A's measurement − Nurse X's measurement) against the average of each pair of measurements ({Dr A's measurement + Nurse X's measurement}/2). It is now evident that Dr A's estimates

Table 1

'True' diameter (mm)	Dr A (mm)	Nurse X (mm)
10	15	19
26	31	24
30	33	29
17	14	16
15	21	14
16	21	15
22	28	22
18	23	17
17	21	15
16	11	16
18	12	16

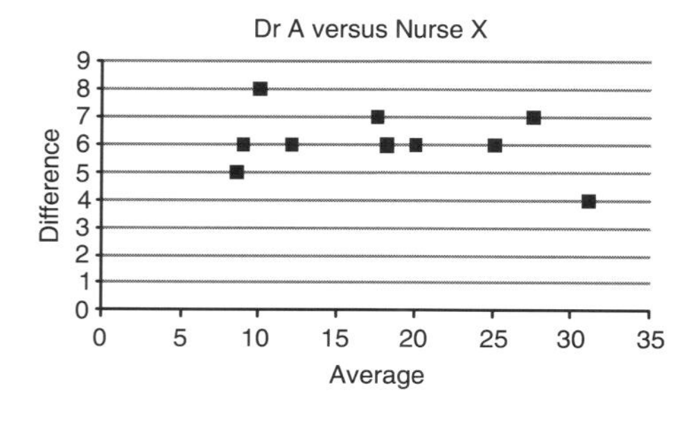

Figure 2.

consistently exceed those of Nurse A: this is suspicious and suggests that either Dr A or Nurse B might be making biased estimates. We could argue that, since the information on the 'true' size of the tumours is available, we should use these values rather than the average of the measurements. This is a contentious area. Although it might seem logical to use the best possible standard against which to assess deviations, the approach can, in practice, produce peculiar results and inconsistencies.

If we consider how the information on tumour size will be used to define the T-stage, a possible reason is found why Dr A might be biased: if she consistently overestimates tumour size, then she will have more patients with **T2** (**p. 303**) tumours and fewer patients with T1 tumours. **Stage shift** (**p. 294**) means that both the patients whom she stages as having **T1** (**p. 303**) tumours and the patients she stages as having T2 tumours will do better than if the T-staging had been left to Nurse X. Table 2 shows the T-stage data based on the estimates of size made by the two observers.

The test that one can use to compare the performance of two observers in classifying categorical data, such as T-stage, is κ (**kappa**) (**p. 170**), a measure of interrater agreement. Using the above data, $\kappa = 0.35$, indicating that the agreement between Dr A and Nurse X is not particularly impressive. The interpretation of κ is not straightforward. When $\kappa = 0$, it shows that the agreement between observers

Table 2

Dr A: T-stage	Nurse X: T-stage	Dr A: T-stage	Nurse X: T-stage
1	1	2	1
2	2	2	2
2	2	2	1
1	1	2	1
2	1	1	1
		1	1

Table 3

κ	Strength of agreement
<0.2	poor
0.21–0.40	fair
0.41–0.60	moderate
0.61–0.80	good
0.81–1.00	very good

is neither less, nor more, than could be expected by chance. When $\kappa = 1$, then agreement is total. A negative κ indicates that there is more disagreement than could be explained by chance. Landis and Koch have proposed the following interpretations for intermediate κ (table 3).

κ is yet another derivative of the contingency table: the frequency of chance agreement is calculated; this is subtracted from 1.0; the frequency of chance agreement is subtracted from the observed agreement. κ is calculated as (observed − chance)/(1.0 − chance).

The interpretation of the observations made by Dr A and Nurse X is now very different: from almost perfect correlation we have moved to agreement that is only fair. This highlights the difference between correlation and agreement: in clinical classification, tests of correlation are of only limited value. **Correlation (p. 75)**, and a high correlation coefficient, only implies that two variables are, in some way, related. Correlation should not be confused with agreement, nor should it be confused with causation.

Media

A term used to describe the newspapers and television. Its strict definition should include any means for transmitting information and would thus include the Internet, mailshots, etc. Advances in cancer research and treatment are announced at press conferences, often tied to a presentation at a scientific meeting or publication in a journal. Through press and television coverage information may reach the public before it is fully accessible to the scientific community. This can have unfortunate consequences and is one of the main arguments in favour of the **Inglefinger Rule (p. 160)**. Many years ago Karl Kraus pointed out that journalists were people who had nothing to say but knew how to say it. Times have changed but little; those who would solicit media coverage for their work would best sup with a long spoon.

Medroxyprogesterone

Medroxyprogesterone is a synthetic progestagen. It is used in the treatment of breast cancer and may also be useful in the management of menopausal symptoms and in treating the wasting and anorexia sometimes associated with cancer or **HIV (p. 151)** infection. For antique and obscure reasons, the drug has become an established therapy for the management of metastatic renal cancer. Recent data suggest that biological therapies, interferon-α or **interleukin 2 (p. 164)**, separately or in combination, are more effective.

Medroxyprogesterone is a respiratory stimulant and has been used in treating the Pickwickian syndrome (which did not actually affect Mr Pickwick himself, but rather the fat boy). The depot form (depot Provera) has been used as a long-acting contraceptive: particularly in the developing world and for those with learning difficulties.

Megestrol

Megestrol is a synthetic progestagen used in the treatment of breast cancer and, occasionally, in the treatment of menopausal symptoms or anorexia. It has several actions

including lowering serum levels of both leutinising hormone (LH) and oestrogen. It may also antagonise cachexin and other possible mediators of wasting syndromes associated with HIV infection or malignancy. It causes weight gain, acneiform skin eruptions and fluid retention.

Meiosis

The specialised type of nuclear division that occurs in germinal tissues, such as the ovary or testis, when **diploid cells** (**p. 98**) divide to form **haploid** (**p. 148**) **cells**. A diploid cell is one that contains two homologous sets of chromosomes (2n); a haploid cell contains only one, unduplicated, set of chromosomes (n).

The cell enters meiosis with two sets of paired chromosomes (4n). During the first phase, division I or reduction division, the chromosomes are reduced in number. Homologous chromosomes form pairs and the **chromatids** (**p. 63**) become apparent. A tetrad of chromatids is formed, held together by chiasmata. The chiasmata are structures essential for the next phase of the process: recombination, crossing over. Genetic material is exchanged between homologous chromosomes: the genetic cards are shuffled, apparently at random. The diversity so produced is, perhaps, one of the reasons for the evolutionary success of sexually reproducing organisms. Which is simply to say that there might be more purpose to sex than simply enjoying ourselves.

During division II, the chromosomes separate, as in mitosis, to form, from each diploid pair, two haploid cells. By the end of Division II, there are four haploid cells each with n chromosomes. Meiosis starts with one cell with 4n chromosomes and ends with four cells, each with n.

Multiple Endocrine Neoplasia (MEN) Syndromes

These syndromes are characterised by the development of multiple tumours, usually of the endocrine glands. They are inherited as autosomal dominants. Mutations of the *Ret* gene, a gene concerned with embryological development on chromosome 10, are often associated with the MEN type 2 syndromes. MEN type 1 is due to abnormalities of the *MEN*1 gene, located at chromosome 11q13, close to the *BCL* gene. The gene product, menin, is thought to function as a tumour suppressor gene:

- MEN type 1 (Wermer's syndrome): tumours of the endocrine pancreas, parathyroid and pituitary.
- MEN type 2A (Sipple's syndrome): phaeochromocytoma and medullary carcinoma of the thyroid.
- MEN type 2B: mucosal neuromas plus phaeochromocytoma and medullary carcinoma of the thyroid.

Quite apart from their exceptional interest biologically, these syndromes have important practical consequences. Patients with endocrine tumours should be screened for other endocrine tumours and, if they turn out to have an MEN syndrome, then their families should be tested for inherited mutations. Any members with the genetic abnormality should be followed up and screened regularly for occult tumours: serum calcitonin for medullary carcinoma of the thyroid; urinary catecholamines for phaechromocytoma.

Menin

The product of the *MEN*1 gene. It may function as a tumour suppressor gene. *MEN*1 is abnormal in multiple endocrine neoplasia (MEN) type 1.

Merkel Cell Carcinoma

A skin tumour of neuroectodermal origin — it may present as a nodule deep to the skin and thus be mistaken for a metastasis from an occult small cell cancer of the lung. The primary management of these tumours is usually surgical, but postoperative radiotherapy has a significant role to play in preventing local recurrence.

MeSH Terms

These are the medical subject index headings used by the Medline database run by the National Library of Medicine. MeSH headings can refine and facilitate **literature searches** (**p. 182**): they are not totally reliable as any search can only be as good as the individual who originally assigned the index term to the article. Another aspect of the problem of classification was neatly illustrated by the improbably large number of references in *Index Medicus* with B Chir as an author.

Mesna

Mesna, sodium 2-mercaptoethanesulphonate, is an agent used to protect against the toxic effects of the oxazaphosphorine drugs, **cyclophosphamide** (**p. 83**) and **ifosfamide** (**p. 159**). These drugs produce toxic metabolites, acrolein, chloroacetaldehyde and 4-OH-ifosfamide, which, when excreted in the urine, damage the urothelium and cause haemorrhagic cystitis. Mesna forms thio-ether bonds with acrolein and the resulting complex is excreted and causes no damage to the lining of the urinary tract. Because its action is local and specific, mesna has no protective effect on the bone marrow, or on any of the other toxic effects of therapy with oxazaphosphorines.

Meta-Analysis — Methods for Combining Data

The simplest approach to combining data from several clinical trials would simply be to tot up the basic data from all the trials and calculate **confidence intervals** (**CI**) (**p. 70**) and/or perform **significance tests** (**p. 286**) on the resulting grand totals.

The problem with this simple method is that a large trial, but one in which there are relatively few events in either arm, will dominate the analysis and in so doing bias the results. This is because in a statistical sense, events and information are equivalent, silence (lack of events) is uninformative: sample size calculations include a consideration of number of events.

The principle of combining data from clinical trials for a **meta-analysis** (**p. 195**) is that the results of individual trials should be combined in such a way that the weighting accorded to each trial should in some way depend on the number of events observed in that trial.

In practice, there are two main methods for weighting and combining results. The **Mantel–Haenszel method** (**p. 187**) assumes that the studies are homogeneous, that is the magnitude and direction of the treatment effect is the same for all trials. The **DerSimonian and Laird method** (**p. 93**) allows for **heterogeneity** (**p. 150**) among trial results.

- Fixed effects method: the assumption is that all trials assess the same degree of effect and it is only because of the play of chance that the observed results differ from each other (Mantel–Haenszel).

- Random effects method: the assumption is that all trials are slightly different from each other. The observed results differ not only just because of the play of chance, but also because there are genuine

differences between studies (DerSimonian and Laird).

In any **meta-analysis (p. 195)** the observed effect has several components:

$$\text{Observed effect} = \text{average true effect} + \text{random error} + \text{systematic error.}$$

The fixed effect methods (e.g. Peto) assume the null hypothesis to be true for each trial and also assume that the true effect is equal for each trial. The random effects methods (e.g. **DerSimonian and Laird**) (**p. 93**) accept that true effects may differ, that studies may be heterogeneous. The random effects methods are, therefore, more conservative.

The nomenclature used, particularly in the breast overview, for reporting the results of meta-analyses is sometimes confusing. The 'percent reduction in odds of event' is not immediately comprehensible to clinicians. The sensible duellist chooses his weapon carefully and the 'percent reduction in odds of event' makes differences between treatments seem more dramatic: if percent reduction in odds of recurrence is 27%; the odds ratio = 0.73. The actual rate difference (by DerSimonian and Laird) is only 6%. A 6% reduction is less striking than a 27% reduction; the difference lies in presentation, not in substance. Remember that the **odds ratio** (**p. 217**) is really being used as an approximation to the risk ratio: the assumption here is that the number of events is relatively small compared with the total number of patients:

$ad/bc = a(c+d)/c(a+b)$, when c is small relative to d and a is small relative to b.

A simple way round the problem of nomenclature in meta-analyses is to use the **number needed to treat** (NNT) (**p. 216**). This is simply the reciprocal of the absolute rate difference (ARD), e.g.:

if ARD = 6% reduction in death, then one would need to treat 1/0.06, i.e. 17, patients to prevent one death.

Meta-analyses are, potentially, highly influential and should be subject to the same rigour, in terms of design and execution, as any individual **randomised controlled trial** (**p. 261**).

Meta-Analysis and Overviews

These terms are often used interchangeably: in North America the term meta-analysis is favoured, whereas in the UK, the term overview is often preferred. Whatever the term chosen, the concept and rationale are relatively straightforward (practice is, however, another matter).

Individual clinical trials are often too small to have adequate statistical power: does a negative trial result represent truth or simply reflect a small sample size? Even for trials that are, by traditional statistical criteria, positive, how confident can we be that the absolute magnitude of the true difference is the same as that shown by the trial? For small trials the answer to both these questions is 'not very sure at all'. Clinical trials often investigate similar questions in similar populations of patients. Protocols and selection criteria will certainly differ between trials but the basic question, e.g. concerning the role of adjuvant chemotherapy in node-positive breast cancer, will often be the same. By combining the results of these similar trials into, as it were, one large trial, the power of the analysis will be increased. Even a series of entirely negative studies might, when combined, yield a positive overall conclusion: an example, which Marx might have enjoyed, of the transubstantiation of quantity into quality.

Provided the analysis has been rigorously performed the results from an overview will give a good indication of the value, or lack thereof, of a particular intervention. There are, however, considerable practical difficulties in conducting an overview. The primary aim of the procedures used to perform an overview is to reduce bias. The quality of an overview is assessable, just as the quality of an individual clinical trial is assessable. An overview should have a specific protocol set out in advance — **data-dredging** (**p. 87**) once the results from a few interesting studies have been combined is reprehensible, tempting and must be specifically prevented. The best prevention is to have a protocol for conducting the overview — as with all protocols the protocol should precede the analysis. A protocol written after the analysis is counterfeit currency — not as rare as one would like to think in science.

The details given in the protocol for a **meta-analysis** (**p. 195**) should include:

- Aim of the overview with appropriately specified **endpoints** (**p. 114**).
- Method for identifying trials.
- Account of homogeneity/**heterogeneity** (**p. 150**) of trial results.
- Criteria for including or excluding trials.
- Endpoints used and how data are to be extracted.
- How **censored data** (**p. 60**) are going to be handled.
- Type of analysis: per-patient or literature-based.
- Statistical method used to combine results from individual trials.
- Definition of difference to be regarded as statistically significant.
- Procedures for **sensitivity analysis** (**p. 283**).
- How the problem of publication bias is going to be addressed.

Table 1

Heading	Subheading	Contains
Introduction		explanation of the clinical problem, the biological basis of the intervention and the rationale for performing the review
Methods	searching	detailed account of sources of information
	selection	inclusion and exclusion criteria
	assessment of **validity** (**p. 322**)	quality criteria applied
	data abstraction	processes and procedures used (single or duplicate; masked or unblinded)
	characteristics of studies	describes study design, patients included, interventions, outcomes, estimate of homogeneity
	quantitative data synthesis	principal measures of effect (ARD, OR) method for combining results, estimate of homogeneity
Results	flow chart	identified studies ⇒ studies included in final analysis
	study characteristics	descriptive data for each trial
	quantitative data synthesis	report on data extraction agreement/disagreement, presentation of summary measures of extracted data
Discussion		summary of key findings and implications for clinical practice and future studies. Discussion of potential biases. Suggestions for agenda for future research

Ideally a **meta-analysis** (**p. 195**) should not rely on data from the literature but should be based on complete follow-up data on every patient entered into each clinical trial. (This is an enormous, but achievable, task.) Stewart and Parmar have clearly shown that literature-based analyses may overestimate the benefits of treatment. This overestimation is not simply due to publication bias, other forms of bias, such as data-extraction bias and language bias are important.

The QUORUM statement has been published in *The Lancet* with the intention of improving the quality of reporting the results of meta-analyses of randomised controlled trials contains a checklist and flow chart for evaluating and reporting a meta-analysis (table 1).

Metaphase

The third phase of **mitosis** (**p. 201**), it takes ~30 min. The fully condensed chromosomes migrate to the equator of the mitotic spindle. It ends when the sister **chromatids** (**p. 63**) separate and start to move towards their respective mitotic poles. Spreads of cells in metaphase are particularly useful for studying chromosomes and their aberrations.

Metastasis

Metastasis is the process whereby a tumour spreads to sites remote from its original formation. For metastasis to occur, cells from the primary tumour:

- Must gain access to a route for dissemination:
  - — enter blood vessel: haematogenous spread
  - — enter lymph vessel: lymphatic spread
  - — enter subarachnoid space: CSF spread
  - — enter body cavity (e.g. peritoneum): transcoelomic spread

The mechanisms are essentially the same as those used for invasion.

- Cells must be distributed:
  - — evade immune attack
  - — remain viable in a hostile environment
  - — find a suitable resting place

There is good evidence of a seed and soil relationship: certain tumour types will metastasise preferentially to certain sites. Metastatic sites do not simply reflect blood flow, there is preferential homing. Brain metastases are common in patients with small cell lung cancer but rare in patients with colorectal cancer. The critical initial interaction is between the cancer cell and the endothelial cells of the target organ.

- Cells must establish themselves at the new site:
  - — endothelial cells retract exposing basement membrane
  - — cancer cell binds to basement membrane
  - — cancer cell dissolves basement membrane
  - — cancer cell moves out into tissues
  - — cancer cell proliferates
  - — whole cycle begins again.

This whole complex process, from tumour angiogenesis to successful implantation and round again, is under genetic influence. Genes that have been implicated at various stages in the process include: *ras*, *p53*, *mos*, *raf*, *fms*, *src*, *fes*. **Matrix metalloproteinases** (**p. 189**) are important at many stages in the formation and establishment of metastases.

Methotrexate (Aminopterin)

Methotrexate is an anti-metabolite derived from folic acid. It is a competitive inhibitor of dihydrofolate reductase, the enzyme that converts folic acid to reduced cofactors involved in the transfer of single carbon units. The transfer of single carbon units is essential for the synthesis of precursors of several nucleotides, particularly thymidylic acid. Methotrexate is S-phase-specific and treated cells will accumulate at the G_1/S boundary. At low serum concentrations (<100 μmol) methotrexate enters cells using an active transport mechanism specific for reduced folates. At high serum levels, methotrexate will enter cells by passive diffusion. This is the rationale for high-dose therapy (with **folinic acid** (**p. 131**) rescue). Once it has entered the cell, methotrexate undergoes polyglutamation, a process that occurs more efficiently in many tumour cells. The polyglutamated methotrexate remains in the cell for longer than the native drug, effectively a depot effect, and this explains some of the selectivity of methotrexate for tumour cells.

Methotrexate is effective orally as well as intravenously. It has a broad spectrum of antitumour activity with clinical usefulness in leukaemia, head and neck cancer, and cancer of the cervix. It is also useful as an immunosuppressive agent and has been used to treat arthritis and Crohn's disease. It is also used, because of its anti-proliferative effects on the basal cells of the skin, to treat psoriasis. It can be given intrathecally to treat neoplastic infiltration of the meninges.

Folinic acid can counteract the toxic effects of methotrexate and can treat inadvertent overdosage or can be used electively as part of a high-dose regimen with folinic acid rescue. The folinic acid can restore intracellular levels of reduced folates.

The excretion of methotrexate is predominantly renal and unexpected accumulation of the drug may occur in patients with renal impairment. Methotrexate is weakly acidic and excretion is more effective if the urine is maintained at an alkaline pH.

Resistance to methotrexate can arise via a variety of mechanisms. These include increased intracellular levels of dihydrofolate reductase, decreased active transport and decreased **polyglutamation** (**p. 238**).

Metoclopramide

A drug is widely used in oncology as an anti-emetic. It has both a central action, through its antagonistic effect on dopamine-mediated pathways, and a peripheral effect. It improves gastric emptying. In high doses it can produce Parkinsonian side-effects including oculogyric crisis. It may also act as a radiosensitiser, which, when given the ubiquity of its use in oncology, has interesting implications.

Microsatellite Instability

Microsatellites are sequences of repetitive DNA. The units of repeat are short, only one to three bases in length. For any given individual the microsatellites are of constant length, even though they might be widely dispersed throughout the genome. Polymerase activity may be imperfect and during replication microsatellites may be lengthened or shortened. This alteration is usually detected and corrected by the **base–base mismatch DNA repair system** (**p. 26**). In **hereditary non-polyposis colorectal cancer** (**p. 150**) (HNPCC) the system is defective and so tumour cells may contain microsatellites whose lengths differ from what, for that individual, is the norm.

This variation is known as microsatellite instability. Microsatellite instability can be regarded as a manifestation of defective DNA repair. Tumours exhibiting microsatellite instability are sometimes known as RER tumours (replication error). The presence and extent of microsatellite instability may be useful in assessing prognosis for those tumours, such as colorectal tumours, for which DNA mismatch repair is known to be important.

MIMiC (Multivane Intensity-Modulating Collimation)

This is a form of radiotherapy. The beam rotates round the patient and treats a slice, the thickness of which is governed by the field length and within which the distribution of radiation intensity at depth is governed by the movements of the leaves of a multivaned collimator as it rotates around the patient. A series of slices is treated so as to include the whole of the target volume — this is a form of intensity-modulated therapy. The practical application is based on the multivane slit collimator. This involves vanes of lead being fired in and out of the beam path extremely rapidly as the beam scans the patient.

Mimicking Cancer

The diagnosis of malignant disease is based almost entirely on the pathological demonstration of the presence of malignant cells: clinical suspicion is not enough. There are several non-malignant conditions that can mimic cancer:

- Tuberculosis.
- Syphilis.
- Systemic lupus erythematosis (SLE).
- Diverticular disease.
- Retroperitoneal fibrosis.

As a general rule, a patient should not be convicted of having cancer without there being a tissue diagnosis. In exceptional circumstances, it may be necessary to treat without proof, but this is unusual and may, in part, explain some of the miracle recoveries from 'cancer' occasionally described.

Min ± Mouse

A mouse model that mimics **familial adenomatous polyposis coli** (**p. 125**). The mouse has a deletion of one functional copy of the *APC* tumour-suppressor gene. The mouse, therefore, is predisposed to cancer of the colon.

Minimal Toxic Dose

A concept used in Phase I studies of new treatments for cancer. It is the dose level at which at least one of three patients treated with that dose shows reversible toxicity.

Mitochondrial Membrane

Changes in the permeability of the mitochondrial membrane might be an important trigger to **cell death** (**p. 55**) by **apoptosis** (**p. 19**). Some anticancer agents may act, in part, via this route: **lonidamine** (**p. 184**), photosensitisers, **KLAKKLAK** (**p. 171**).

Mitomycin C (MMC)

Mitomycin C is an antitumour antibiotic derived from *Streptomyces caespitosus*. It is inactive until metabolised in the liver. Its activated form functions as an alkylating agent that affects DNA, but not RNA. It is a bioreductive drug also capable of forming **free radicals** (**p. 132**) and, for this

reason, it may be synergistic with radiation. It may produce prolonged myelosuppression and regimens usually involve dosing every 6, as opposed to the more customary 3, weeks. It may cause haemolytic-uraemic syndrome and evidence of intravascular haemolysis may be an important warning sign of impending catastrophe. It has a broad range of activity: head and neck cancer, anal cancer, lung cancer, gastrointestinal cancer.

Mitosis

An essential prerequisite for cellular proliferation is to provide a complete set of genetic instructions for each daughter cell. A cell has to double its DNA content. Morphologically, this is manifest as the chromosomal changes during mitosis: **prophase (p. 249)**, metaphase, **anaphase (p. 14)**, **telophase (p. 306)** and cytokinesis:

- Prophase (time 0–17 min): chromosome condensation, separation of **centromeres (p. 61)**.
- **Prometaphase (p. 248)** (takes ~5 min following completion of prophase): breakdown of nuclear envelope, **kinetochores (p. 171)** form, jiggling.
- Metaphase (takes ~30 min) plate forms.
- Anaphase (lasts ~5 min after completion of metaphase): migration of chromosomes to the poles.
- Telophase (for 20 min after anaphase): new nuclei form.
- Cytokinesis: new cells separate.

The morphological changes that accompany mitosis are the most dramatic manifestation of the changes occurring during cell division they must, however, be underpinned by a series of kinetic and biochemical events: events that are usefully summarised using the concept of the **cell cycle (p. 52)**.

Mitotic Index (MI)

A term used in cell kinetics to indicate the proliferative state of a population of cells. The percentage of cells in mitosis:

$$MI = (T_m / T_c) \cdot \lambda,$$

where λ is a constant in the range 0.5–1.0 used to define the age distribution of the cell population, T_m is the duration of mitosis and T_c is the duration of the **cell cycle (p. 52)**.

Mitotic Spindle

The mitotic spindle is a temporary structure formed during mitosis. Its presence enables the chromosomes to migrate to the opposite poles of the cell so that two daughter cells, each with a full set of chromosomes, can be produced.

Mitoxantrone

Mitoxantrone is a cytotoxic drug structurally similar to the **anthracyclines (p. 16)**. Chemically it is an anthracendione. However, its spectrum of clinical activity is distinct from that of **doxorubicin (p. 107)**. It is used in the treatment of breast cancer, lymphomas/leukaemias, sarcomas and ovarian cancer. It has very little activity against germ cell tumours or non-small cell lung cancers. In combination with prednisolone, it is a useful treatment for men with hormone-refractory metastatic cancer of the prostate.

It acts as an intercalating agent, preferentially localising between guanine and cytosine. It also inhibits **topoisomerase II (p. 310)** as well as affecting the cytoskeleton. It is metabolised in the liver and biliary excretion is an important mechanism for its elimination. The dose should be reduced in patients with biliary obstruction or deranged liver function tests. It causes myelosuppression and is cardiotoxic. The incidence of cardiotoxicity

rises sharply above a total cumulative dose of 150 mg m^{-2}. It is a pleasant shade of blue; this may colour the urine and the nails.

MLL Gene

Abnormalities affecting *MLL* are particularly associated with acute monoblastic leukaemia (AML-M5) and acute myelomonocytic leukaemia (AMML-M4). *MLL* is identical to a gene formerly described as *ALL*1. It is at chromosome 11q23, and 11q23 translocations are often associated with leukaemia. A zinc finger **domain** (**p. 102**) at 11q23 is typically translocated to chromosomes 4 or 9 or 19. The resulting fusion product has similarities to a DNA methyltransferase. Up to 70% of the leukaemias associated with treatment using inhibitors of **topoisomerase II** (**p. 310**) are associated with translocations involving 11q23. The malignant cells in this particular type of leukaemia have a translocation t(11;16)(q23;p13.3) that produces a fusion gene *MLL-CBP*.

Model

The essential principle of modelling is to infer a rule, or set of rules, of putative general applicability from a series of observations. The model so derived will have some explanatory power; it may also, but this is not a prerequisite, provide clues about the underlying biological mechanisms involved. A model, after all, is just a way to make things easier for ourselves. Nature has no need to oblige us by making her methods compatible with our models. A model is a form of map and, as Bateson pointed out, the map is not the territory. The advantage of modelling is that in enables complex information to be presented in a relatively simple form: it is a form of shorthand. Models can have predictive value and the conclusions from modelling should wherever possible be checked against reality. There is a certain seductive charm to modelling — graphs without data points are always clear and easy to understand. The relevance of such graphs to reality is, of course, the key issue.

Any argument by analogy, e.g. leucine zippers, DNA as a spring with action at a distance, is to some extent an act of modelling.

Models of relevance to oncology include:

- **Neural networks** (**p. 212**) to model intracellular communication.
- α/β ratios to model the radiation cell survival curve.
- Models used for computerised planning of radiation therapy.
- Models used to indicate the structure of a protein from knowledge of its amino acid sequence.
- Cox models and prognostic indices.
- Pharmacokinetic models.
- Epidemiological models, e.g. for spread of disease caused by a transmissible agent.
- Monte Carlo simulations in radiation biology, pharmacology and departmental management.

By definition, a model cannot go much beyond that which is already known — new knowledge can easily make a model redundant. The most fruitful models are those in which there is some consonance between the mathematics and the underlying biological reality. The **linear–quadratic** (**LQ**) **model** (**p. 180**) is more productive scientifically than a **Cox model** (**p. 79**) that attributes an exponent of 0.435 to a categorical variable. The Cox model can give useful prognostic information but does not tell much more than that. The LQ model not only provides a useful approximation of the radiation cell survival curve, but also suggests possible mechanisms, in terms of **target theory** (**p. 304**), for

radiation-induced cell killing. Models used in conjunction with **sensitivity analysis** (**p. 283**) can usefully show where the important gaps in knowledge are. Models can sometimes mislead and this can have adverse consequences for patients, e.g. the use of the NSD model in radiation biology that ignored the importance of fraction size in influencing the late effects of radiation.

Modelling is often an important intermediate step in scientific creativity: Kary Mullis was once a computer programmer and the PCR reaction that he developed is, in an important sense, modelled on the iterative recursive loops used in many computer programs.

Molecular Biology

Molecular biology, despite its 60-year history, remains an extremely difficult discipline to pin down. In a sense it can become all consuming: in that anything that concerns biology and molecules might be called molecular biology. In essence, molecular biology implies a reductionist approach — that the structure, function and possibly the behaviour of organisms, no matter how complex, can be explained in terms of their component macromolecules. The term was first coined ~1939 in a report to the Rockefeller Institute and no doubt it reflected contemporary developments.

Molecular biology owes its origins to a confluence of several distinct scientific streams: **physics**, particularly X-ray crystallography: **chemistry**, particularly that of DNA; and **genetics**, particularly in its mechanistic aspects. Particular individuals, institutions and publications also played their role in bringing these different disciplines to bear on the problems of life, its definition, origins and propagation. The **phage group** (**p. 233**) were particularly interested in the relationship between genetics and the chemistry of nucleic acids. (Delbruck, Luria, Hershey).

The other discipline involved in the formation of molecular biology was X-ray crystallography (Bragg, Astbury, Bernal, Kendrew, Perutz, Pauling, Wilkins, Franklin). A particularly conspicuous trend was the tendency of those who had trained as physicists to move into the investigation of biological problems. It was no accident that the first major advance in molecular biology, the elucidation of the structure of DNA, was achieved by Watson (a biologist turning his hand to crystallography) and Crick (a physicist investigating a biological problem). Watson was Luria's pupil, Luria's friendship with Kendrew placed Watson in the Cavendish Laboratory at Cambridge, sharing a room with Crick. There is some irony in the fact that, just as uncertainties (Heisenberg, Bohr, Einstein) were starting to undermine the physicists' **deterministic** (**p. 95**) view of things, some physicists migrated into biology and imposed an overly deterministic view. The consequences of this, perhaps misleading, viewpoint still permeate much of molecular biology, particularly that which has been applied to medicine. A simple **reductionist** (**p. 263**) view is dominant even though experience suggests that life is rarely that simple.

A consequence has been a view of life that is occasionally overly deterministic and reductionist. However, **determinism** (**p. 94**) in biology may be starting to crumble, in spite of some of the bizarre claims emanating from the human genome project. The so-called central dogma of molecular biology (gene → RNA → protein, and never in the reverse order) has already been overthrown by the enzyme **reverse transcriptase** (**p. 272**). Post-transcriptional modification of nucleotide sequences means that it is not always be possible to deduce the amino acid sequence of a protein, and hence its structure and likely function, simply from a knowledge of the nucleotide sequence of a gene.

There are at least two main discernible themes in molecular biology: the relationship of structure to function — how this particular physical structure might have this potential biological function; and the question of communication and control. This last point finds its most direct expression in the question: if all cells contain the complete genetic instructions to make a human being, why are some cells neurones, others intestinal epithelium, others form red blood cells, etc.? As issues of control, of signalling, promoting, inhibition, of negative and positive feedback loops, of cell–cell communication, of action-at-a-distance, become increasingly important and complex, then some of the certainty previously expressed by molecular biologists is bound to dissipate. Life is complex, it is not just a matter of, if A then B then C then D, following on in an orderly and predictable sequence. A may sometimes lead on to D via B and C, but on other occasions may lead to $(P \times R \times T)^2$, by routes that are completely undetermined. Since cancer is, in large measure, a disease of disordered control at the cellular level, it is easy to see how a knowledge, based firmly on molecular biology, of the mechanisms whereby cells communicate and are controlled, might add considerably to the understanding and management of cancer.

Molecular Biology and Radiation Therapy

Clinical radiotherapy was effectively treating cancer long before the four, or five, R's of radiobiology were described or defined. Radiotherapy has been a clinically driven speciality with the basic science either being grafted on later or developing synchronously. This historical pattern is now likely to change. The recent advances in basic science, particularly molecular biology and genetics, are likely to generate clinically testable hypotheses and clinical radiotherapy will become science-driven as opposed to science-supported.

The older model of radiation action simply considered **DNA damage (p. 98)**, and particularly double-strand breaks, to be the important mechanism by which ionising radiation exerted a biological effect. The biological effects of radiation are much more complex and this very complexity is an asset since it implies that there are many possible ways in which we might make radiotherapy more effective.

Radiation can act on receptors (?location) to produce signal transduction:

- Protein kinase C.
- Tyrosine kinase.
- MAP kinase.
- *raf*-1.

This may, in turn, cause activation of genes, presumably mediated by RRE (radiation responsive elements) in promoter sequences.

Below is a list of genes and gene products implicated in the response to irradiation:

- *A-T* gene product activates c-*Abl*.
- *RB* gene product.
- ***p53*** **(p. 224)**.
- *GADD*45.
- *AP*1 genes: *fos*, *jun*.
- Zinc finger genes: *EGR*-1.
- *Rel* family of **transcription factors (p. 313)**: NFκB.
- Genes concerned in **apoptosis (p. 19)**: *bcl*2, c-*myc*, ***p53*** **(p. 224)**.
- 'Radiation response elements', e.g. *CArG*.
- Secondary responses: TNF; bFGF; TGF-β; IL-1β; *raf*-1 gene product.
- *cdc*25.

- *wee*1.
- *mik*1.
- *nim*1/*cdr*.
- *rad* 9.
- *WAF*-1/*CIP*-1 system.
- *XRCC*-1.

The effects of *p53* on the response to irradiation are by no means simple: experimental data have suggested that loss of wild-type *p53* is associated with increased radioresistance, increased **radiosensitivity** (**p. 260**) or that it has no effect on sensitivity to radiation. The interaction between *p53* and *mdm* may be pivotal.

Many of the events triggered by irradiation are concerned with allowing the cell time to activate and implement mechanisms for repairing damage: delay at the G_1/S border is critical in this respect.

Radiation can, through the binding of the **ATM** (**p. 23**) (mutated in **ataxia-telangiectasia**) (**p. 23**) gene product to the SH3 **domain** (**p. 102**) of c-*Abl*, activate c-*Abl*. The activated c-*Abl* affects RNA polymerase II, SAPK, *Crk* and, possibly, *p53* and *RB* protein.

BRCA2 binds to *Rad*51: the *Rad*51 knockout mouse is hypersensitive to radiation, so is the *BRCA2* knockout mouse. *Rad*51 is probably concerned with repair after **DNA damage** (**p. 98**). There is, here, a tantalising link between a gene associated with susceptibility to cancer (*BRCA2*) and a gene associated with increased susceptibility to the biological effects of radiation (*Rad*51).

A gene switched on by radiation, and which is a promoter, can be placed upstream of a gene which codes for a cytotoxic protein, e.g. *TNF*. This genetic construct can then be placed into a suitable vector and administered either locally or systemically. Local irradiation to the tumour will switch on local production of the cytotoxic protein and there will be no systemic cytotoxic effect. The interaction between radiation and radiation responsive elements in effect is used as a switch to turn on the local production of a cytotoxic agent. So far this approach has only worked in experimental animals but the early results suggest that the theory may be borne out in practice.

How does a cell know that it has been irradiated?

- Nuclear signal from damaged DNA mediated via *GADD*45.
- Cytoplasmic signal via reactive oxygen intermediates and NFκB activation.
- Membrane signals: lipid peroxidation, ion fluxes, etc.

This is a highly simplified account of how therapeutic irradiation might affect cells: the important message is that the interaction is not only more complex than has been supposed, but also that it is more complex than can be supposed.

Molecular Biology in Summary

What is it?

The practise of biochemistry without a licence (Chargaff). A science that recognises that the essential properties of living beings can be interpreted in terms of their macromolecules (Monod).

What are its concerns?

Structure and its relation to function; information and communication; complexity disguised as chaos; flexibility and the demands of a changing environment.

Where did molecular biology come from?

It grew out of the fusion of physics, biochemistry and genetics in the latter 1940s.

Most spectacularly, with the discovery of the double helix, it illustrated what could happen when a physicist who had been working on sonographic detection of submarines was put to work with a youthful bird-watching biologist with an interest in the problems of genetics.

What is the relevance of molecular biology to cancer?

Cancer represents the failure of those normal mechanisms that control cellular proliferation. It is not just one disease. It may originate as a result of a single cell undergoing a series of transformations. It is a genetic ferment: cancers can change as they progress. **Invasion (p. 198)** and **metastasis (p. 167)** are hallmarks of the process. Molecular biologists bring the following abilities to bear on the complex series of problems involved in the origins and perpetuation of malignant transformation. They can:

- Locate and identify genes.
- Clone genes.
- Identify mutations.
- Identify gene products.
  - — **growth factors (p. 144)**
  - — receptor proteins
- Tumour suppressor genes, etc.
- Construct 'new' genes.
- Insert genes into cells.
- Identify sequence homologies.
- Create strange and wonderful animals.
- Bring an evolutionary perspective with the recognition of:
  - — duplication and redundancy
  - — multistep, multifactorial processes
  - — utility rather than elegance in some of the biological solutions
  - — the fact that something which is advantageous in one context may be disadvantageous in another (and vice versa)
  - — the conservation of that which is crucial and fundamental.

Why is an evolutionary background useful?

Because a tumour is an evolutionary environment, mutations are frequent. There is an important relationship between the generational age of a tumour and cell loss. This provides abundant time for spontaneous mutations to occur, some of which might confer resistance to therapy. Evolutionary processes such as natural selection can be used to interpret the interplay between tumour and treatment: with treatment as a selection pressure and the survival of the fittest tumour cells as a factor limiting cure.

Given the nature of cancer, and given the tools provided by molecular biology, then what programmes would we propose to improve the management of cancer?

Gene therapy (p. 137) is the most obvious candidate, but there may be more immediate returns, in terms of improved management of cancer, on investment in molecular biology. These would include: identifying new targets for cancer treatment such as **angiogenesis (p. 14)**; the mechanisms controlling **invasion (p. 167)** and **metastasis (p. 198)**; tumour suppressor genes (restore function or inhibit the inhibitor of the inhibitor); **oncogenes (p. 218)**; *mdr* genes; **growth factors (p. 144)**; receptor modification; **apoptosis (p. 19)**; surface markers; enhanced immunogenicity; **cell cycle (p. 52)** control mechanisms. Improved ability to predict prognosis and response to treatment would be an important, and immediately applicable, contribution of molecular biology to **cancer**

management (**p. 40**). It would enable **Pareto** (**p. 228**) improvements to be achieved: we could invest the savings made by not using futile or unnecessary treatments in the treatment of those who have a genuine, and predictable, chance of achieving benefit.

Molecular Epidemiology

Molecular epidemiology defines a new discipline. The synthesis of descriptive epidemiology, genetics and molecular biology, sometimes termed 'molecular epidemiology', has greatly improved the knowledge of how and why cancer occurs. The insights achieved through this melding of disciplines will, within the foreseeable future, have practical consequences in terms of how cancers are to be prevented. For example, if we could identify individuals unduly susceptible to environmental carcinogens, we could provide them with specific advice about how they might lower their personal risk of cancer.

Molecular Medicine

A term used to describe the practical application of advances in molecular biology to medical care; a portmanteau term since it could be all things to all people, a rubric so ill-defined as to subsume anything and everything. It is a popular term with those trying to sell a product: research proposals, academic departments, books, conferences, etc. The most obvious question today would be: when is medicine not molecular?

Molecular Phylogeny

The use of nucleotide and protein sequence data to investigate interrelationships and patterns of descent among living species. The demonstration that the gene sequences of humans, chimpanzees and gorillas are >98.5% identical suggests that the three lineages diverged comparatively recently, <7 million years ago, a conclusion that suggests that man's true place in nature is less elevated than we had previously assumed. Studies in molecular evolution have confirmed Darwin's extraordinary prescience and emphasise the usefulness of Darwinian evolution as an overarching theoretical structure.

Knowledge of evolutionary tree structure does not come directly from inspection of DNA sequences but requires modelling to construct, from the sequence data for the individual species, a tree consistent with the basic data. The simple, **heuristic** (**p. 150**), method of **maximum likelihood estimation** (**MLE**) (**p. 189**) is inadequate for the task and other methods are being developed.

Monomer

A **domain** (**p. 102**) or complex of domains that forms part of the **tertiary structure** (**p. 306**) of a protein. A monomer is a subunit of a complex protein.

Mortality Rate

The number of cancer deaths occurring in a given period in a specified population, e.g. the mortality rate for prostate cancer could be expressed as deaths per 1000 men year^{-1}.

Motif

A motif is an arrangement of α-helix and β-pleated sheets that forms part of the domain of a protein. There are only a few ways in which sheets and helices can combine in a stable structure and, therefore, many completely different proteins will contain similar motifs. Examples of motifs include: the zinc finger motifs found in proteins, such as transcription factors, which bind to DNA; the leucine zipper motifs, also found in proteins, which influence

the transcription of DNA; and the hairpin β-motif; the β–α–β-motif. If domains were houses, then motifs would be structures such as windows or drainpipes that are similar from house to house — even though the houses themselves might be completely different in overall architecture.

MPF (M-Phase Promoting Factor, Maturation Promoting Factor, Mitosis [Meiosis] Promoting Factor)

MPF is a molecule with functions that are as diverse as they are crucial. It causes sudden destruction of cyclin at anaphase/metaphase transition, it stimulates its own activation, an amplificatory burst triggers **mitosis** (**p. 201**), it has protein **kinase** (**p. 171**) activity, it condenses chromosomes and changes microtubule function and it can change **centrosome** (**p. 61**) structure and function. A complex between a mitotic cyclin and a protein kinase (Cdc2) forms. Its ability to trigger M-phase was first demonstrated in the ***Xenopus*** **oocyte system** (**p. 329**) when injection of cytoplasm from an M-phase unfertilised egg into a G_2 oocyte induced mitosis and maturation in the recipient oocyte.

A similar system with cdk2 (p33), cdk4 and **cyclins** (**p. 82**) A, D and E is a component of the G1 checkpoint and is responsible for the move from G_1- to S-phase.

Magnetic Resonance Imaging (MRI)

A technique, based on the principle that nuclei, such as the hydrogen nucleus, with an odd number of protons and neutrons will, when placed in a strong magnetic field, behave like magnets and will line up along the direction of the applied field. The nuclei will also wobble, 'precess' in the jargon, along an axis in the same longitudinal direction as the applied magnetic field. If a brief pulse of a radiofrequency field is then applied, then the nuclei will move first away from, and then back towards, the orientation imposed by the magnetic field. The pulse will also produce an initial period where the nuclei all wobble in phase followed by a return to less coordinated wobbling. The energy changes can be detected by a signal coil placed nearby and, if a series of such coils is placed in a circumferential array around the subject, then the pattern of changes in energy distribution can, as in CT scanning, be interpreted using a computer algorithm and used to create an image. Three main components are used to form the image: **T1** (**spin–lattice, longitudinal relaxation time**) (**p. 303**) is the time taken for the nuclei to resume their orientation in the magnetic field; **T2** (**spin–spin, transverse relaxation time**) (**p. 303**) is the time taken for the synchronised, in-phase, nuclear wobbling to re-establish the out of phase precessions that existed before the application of the RF field; the intensity of the signal, which depends on the concentration of rotating nuclei in the region. By altering the RF applied, the relative contributions of T1 and T2 can be altered: T1- and T2-weighted images. In a T1-weighted image, obtained using a sequence of RF pulses termed 'saturation recovery', fat appears white while water appears dark. The reverse is true for a 'spin-echo' sequence designed to produce a T2-weighted image: water, e.g. CSF, appears white while the grey matter of the CNS is grey.

Open coil designs can be used to permit MRI-guided surgery. Unlike images obtained using X-rays, there are no known health hazards associated with MRI and it is, therefore, a logical method for image-guided therapy or for screening healthy individuals for disease. MRI used to be called nuclear magnetic resonance (NMR), but 'nuclear' was considered too off-putting.

(MRP) Multidrug Resistance-Associated Proteins

Multidrug resistance-associated proteins are proteins associated with multidrug resistance distinct from the classical *mdr*-1 gene/P-glycoprotein system. They are 190 kD in weight and are associated with cell membranes. There are several such proteins: MRP1 was the first to be described, followed by MRP2 (cMOAT — the multispecific organic anion transporter). The spectrum of resistance differs between MDR and MRP. MRP usually confers cross-resistance to **anthracyclines** (**p. 16**), epipodophyllotoxins and **vinca alkaloids** (**p. 324**) but less resistance than with MDR to taxols and **mitoxantrone** (**p. 201**). MRP functions mainly as a transporter of glutathione conjugates and this property may be useful in distinguishing between drug resistance mediated by MDR and that mediated by MRP.

Muir Torre Syndrome

This syndrome is defined by the occurrence of sebaceous gland tumours, carcinomas of stomach and endometrium in association with **hereditary non-polyposis colorectal cancer** (**HNPCC**) (**p. 150**).

Multidisciplinary Care

This is certainly a cliché, and it is equally certainly crucial to the successful management of patients with cancer. The concept of multidisciplinary care explicitly recognises that no one is omniscient.

The management of cancer is a complex business and, to ensure the best care for an individual patient, a wide variety of skills is required at every stage of the process. No individual possesses the knowledge or expertise to provide total care and the concept of the multidisciplinary team, working together to provide the repertoire of skills needed for that particular patient, has evolved. Working in such a team has both advantages and disadvantages for the team members, but their views and comfort are not what is at issue: the issue is what is best for the patient. If at one hospital visit the patients can have all their needs assessed and dealt with, then this is clearly preferable to trailing around for several weeks attending different clinics and specialities. The components of the team will depend on the tumour site and the needs of the patients, but the essential principles that govern any successful multidisciplinary team are:

- Mutual respect for professional competence and skills.
- Good communication.
- Flexibility concerning roles within the team.
- Lack of an arbitrary hierarchy.

There should be, from the patient's view, no abrupt transitions in care. This is the concept of seamless care and it has been introduced to draw attention to some of the problems experienced by patients in the past. Patients should not have to cope with statements implying abandonment: 'There is nothing more I can do for you so I am going to ask Dr X to see you.' Such statements put both the patient and Dr X in an invidious position. **Seamless care** (**p. 280**) would imply that, before the time to change the emphasis in management had occurred, the patient would have met Dr X and been aware that she was an important member of the team. Seamless care is an attractive concept but it is often difficult to accomplish: not all members of a team can be available for every consultation. Nevertheless, regardless of the practical difficulties, it is target to be aimed at, even though it might not always be achieved.

Multileaf Collimator (MLC)

A collimator is an object placed within a beam of radiation to modify the shape or intensity of that beam. A MLC allows a beam to be shaped using finger-like processes, leaves, which can be introduced into or withdrawn from the radiation beam. This can be done statically, the beam is the same shape throughout a treatment, or dynamically, in which the beam changes shape during treatment. MLC is an essential prerequisite for **conformal therapy (p. 71)**. Virtually any beam shape is possible. Another advantage is that it eliminates the need for the creation of customised lead blocks; this is labour saving in a variety of ways.

Multiple Drug Resistance (MDR)

The spontaneous acquisition of a mutation conferring resistance to multiple chemotherapeutic agents may be an important limiting factor in cancer treatment.

Many tumours, e.g. head and neck cancers, exhibit a rapid response to chemotherapy. However, the response is short-lived and, in spite of further treatment, the tumour regrows. A spontaneously acquired mutation could increase a cell's ability to resist cytotoxic damage and a subclone of relatively resistant cells could, by the transmission of this mutation from generation to generation, arise within a tumour. The persistence and eventual expansion of such subclones after initially effective therapy could provide a major practical barrier to successful treatment.

A transmembrane protein, P-glycoprotein, has been identified in many tumour cell lines. The protein acts as a pump and removes cytotoxic drugs from cancer cells before they have had a chance to reach the nucleus and have any effect on cell division. The gene family that controls P-glycoprotein is known as *MDR* (multiple drug resistance). In humans, *MDR* are on the long arm of chromosome 7: there are so far two members *mdr*1 and *mdr*2. P-glycoprotein itself belongs to a superfamily of transport proteins, the ABC (ATP binding cassette) transporters. Through activation of a mutation conferring activation of the *MDR*/P-glycoprotein system, a tumour may be resistant to drugs to which it has never been exposed. Such a mutation could be acquired at any time during the long preclinical phase of tumour development. The system itself is evolutionarily conserved and presumably represents an important defence mechanism against environmental stress.

An important feature of this system is that it is not specific to particular drugs — a variety of different agents, with little in common chemically, can be removed by the same system.

Commonly used **cytotoxic drugs (p. 84)** for which resistance can be mediated by the *MDR*/P-glycoprotein system:

- Adriamycin.
- Mitomycin C.
- **Vincristine (p. 324)**.
- **Vinblastine (p. 324)**.
- Taxol.
- **Etoposide (p. 119)**.
- Actinomycin D.
- Mitoxantrone.

The activity of the *MDR*/P-glycoprotein system can be modified pharmacologically.

Several commonly used drugs can decrease the activity of the *MDR*/P-glycoprotein system and offer, in theory, the promise of

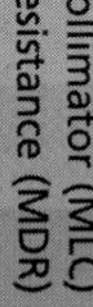

circumventing the problem of drug resistance:

- Calcium channel blockers: nifedipine, verapamil.
- Cyclosporin A.
- Phenothiazines.

There is at least one other separate system that can also confer multiple drug resistance: the **MRP system** (**p. 209**).

MYC Oncogene

This oncogene is highly conserved, in humans it is at chromosome 8q24.12–q24.13. This is close to the breakpoint involved in Burkitt's translocations. Lymphomas with translocations involving 8q24 express high levels of the *MYC* product. β-Catenin will activate *MYC*. The *MYC* product has a molecular weight of 65 000 and acts on the nucleus as a transcription factor. In the presence of the **Max** (**p. 189**) transcription factor it will bind to DNA and stimulate cellular proliferation and transformation. Among the genes activated by the *MYC*/Max complex are ***CDC25A*** (**p. 52**) and ornithine decarboxylase. *MYC* will activate **telomerase** (**p. 305**). It also inhibits the expression of a growth restraining gene, *GAS*1. The product of the wild-type *APC* gene will repress *MYC* and loss of this repression may be an important feature of malignant transformation in the FAP syndrome.

Named Patient Use

A procedure whereby patients can be treated with drugs not licensed for use. It is based on the principle that when an individual's life is seriously threatened by disease, then it is legitimate to take risks. The protection afforded to the general public by the licensing process is not a relevant issue when, if an individual were to be left untreated, their death would be inevitable. This is also known as the 'compassionate use' basis. The prescriber takes full responsibility for the consequences, adverse or otherwise, of the intervention.

Neoadjuvant

Neoadjuvant treatment is treatment given before any definitive local or locoregional treatment, e.g. treating large breast cancers with chemotherapy before surgery; treating germ cell tumours with chemotherapy before any surgery to the retroperitoneal nodes; and preoperative radiotherapy for rectal cancer. There are several mechanisms by which this approach might improve results:

- Sterilisation of malignant cells at the periphery of the tumour might make surgical margins less critical.
- Shrinking the tumour might improve resectability or permit a less radical operation, e.g. wide local excision rather than mastectomy.

One problem with neoadjuvant treatment is that, by delaying definitive treatment such as surgery, it might facilitate the emergence of metastases containing resistant clones — the so-called 'leaner and meaner' problem.

Neural Network (Artificial Neural Network, ANN)

A neural network represents an attempt to use computers to assist with decision-making and prediction. It is a mathematical model in which input can be related to output using a series of rules which are formulated by induction based on analysis of a set of samples for which both input and output are either specified or known. In practical terms, a neural network is a software program that can be 'trained' using real data. The trained program can then be used to predict outcome for any given set of data entered. A three-layer model uses a layer of input nodes, a layer of hidden nodes and an output layer. Each input node connects with each node in the hidden layer, and vice versa. Each hidden layer node connects to the output nodes. The connections each have a weight attached to them and this weight can be either positive or negative. The hidden layer effectively integrates the weighted information from the input nodes and derives a prediction that appears at the output node. The advantage of this overall approach is that it is relatively free of assumptions: in particular there is no assumed linear relationship between change in an input value and outcome. Equally there is no logical connection between input and output: arbitrary events can occur within the model and this is acceptable provided the model fulfils its basic function, which is to predict outcome from input. The approach is, therefore, both **heuristic** (**p. 150**) and pragmatic.

The approach has been used to predict survival for patients with head and neck cancer. The input data consisted of tumour size, nodal stage, tumour stage, tumour resectability and haemoglobin. The output was probability of being alive 2 years after treatment. The model was reasonably good at predicting survival: when the sensitivity rate was 80%, the specificity rate was 60%. The corresponding specificity rate, given a sensitivity rate of 80%, for a **logistic regression** (**p. 183**) model was only 40%.

Neurofibromatosis

his autosomal dominant condition, ormerly known as von Recklinghausen's isease, exists in two forms: NF1 and NF2. NF1 is diagnosed clinically on the basis of wo or more of the following criteria:

- Café au lait macules (children, more than five of at least 0.5 cm diameter; adults, more than six that are at least 1.5 cm in diameter).
- Neurofibromas: either one plexiform neurofibroma or two or more neurofibromas of other type.
- Multiple axillary or inguinal freckles.
- Abnormal long bones (thinning of the cortex) or dysplasia of the sphenoid wing.
- Bilateral gliomas of the optic nerve.
- Two or more hamartomas of the iris (Lisch nodules).
- First-degree relative fulfilling these criteria.

he genetic abnormality is on chromosome 7q11.2. The gene product is a protein, eurofibromin, which activates guanosine iphosphatase.

he malignant tumours associated with NF1 clude: malignant Schwannomas, strocytomas, neuroendocrine tumours, rimitive neuroectodermal tumours, leukaemia, **Vilms' tumour** (**p. 328**) and other sarcomas.

NF2 is caused by a mutation at chromosome 2q12.2. The clinical diagnosis is made if ne of two major criteria is present:

- Bilateral VIIIth nerve masses seen on **gadolinium** (**p. 134**)-enhanced MRI; or
- First-degree relative with NF2 plus one of the following:
  - — imaging evidence of an VIIIth nerve mass
  - — plexiform neurofibroma
  - — two or more neurofibromas
  - — two or more gliomas
  - — posterior subcapsular cataract at a young age
  - — two or more meningiomas
  - — imaging evidence of spinal cord or intracranial tumour.

The main malignant tumour associated with NF2 is glioma.

Neutron

An uncharged subatomic particle, which, together with protons, forms the nucleus of an atom. Neutrons contribute to the mass number (A) but have no effect on atomic number (Z). The mass of the neutron is 1.6747×10^{-27} kg (1.00866 amu). Its mass is about the same as that of a proton, though the neutron has slightly more mass.

High-energy neutrons can be used for particle therapy, lower-energy (0.025 eV, thermal) neutrons can be used for **Boron capture therapy** (**p. 35**). Therapy with high-energy neutrons was once considered the great hope of clinical radiotherapy: the main biological advantage was perceived to be the fact that neutron-induced cell killing was virtually independent of oxygen tension. In clinical practice, however, there was little evidence that for most tumour sites neutron beam therapy produced any improvement in **therapeutic ratio** (**p. 307**). In fact, for many tissues, especially subcutaneous tissues and the central nervous system, the adverse effects of neutron therapy were greater than initial radiobiological studies had predicted. This was in some part because of the fractionation trap: the **relative biological effectiveness** (**RBE**) (**p. 264**) for neutrons rises as fraction size decreases and this can cause unexpectedly severe damage to normal tissue, particularly those in the penumbra of the beam.

Only a few highly specialised centres are persevering with therapy using high-energy neutrons.

Nicotinamide

This relatively simple compound improves the oxygenation of tumours, mainly by increasing the uniformity of blood flow. It may specifically reduce areas of acute, as opposed to chronic, hypoperfusion. It may also inhibit repair of radiation-induced damage through inhibition of the enzyme **adenosine diphosphoribosyl transferase** (**ADPRT**) (**p. 9**).

Nicotinamide will decrease spontaneous arteriolar constriction, as well as phenylephrine-induced constriction, in human tumour vessels. It also inhibits contraction of smooth muscle in the gut wall, which may explain some of the gastrointestinal toxicity associated with nicotinamide.

The main clinical use of nicotinamide has been as a component of the **ARCON** (**p. 20**) schedule. Unfortunately a Phase I/II study suggests that nicotinamide, given at a dose sufficient to produce improvement in tumour oxygenation, may produce unacceptable vomiting and nausea.

Nilutamide

Nilutamide is an anti-androgen used in the treatment of prostate cancer. Its mode of action is similar to that of **flutamide** (**p. 131**). It binds tightly to androgen receptors and blocks the access of androgen to the receptors. It sometimes interacts with alcohol and causes hypotension and facial flushing. Its other main side-effects are gastrointestinal problems, particularly constipation.

Nitric Oxide (NO)

The gas was once thought to be, in biological terms, rather uninteresting. However, its biological versatility is now increasingly apparent. It acts as a signalling molecule for short periods (**half-life** (**p. 14**' 5–10 s). Deamination of arginine by NO synthetase produces NO, which can then diffuse rapidly through plasma membranes and act directly on the cytoplasm of neighbouring cells. These effects can be direct, such as relaxation of smooth muscle fibres, or indirect, mediated by guanylyl cyclase and its effects on cyclic GMP. Carbon monoxide is another gas that can function as a messenger.

NK1 Receptor

A central nervous system receptor involved in mediating vomiting. NK1 receptor antagonists are a potentially new class of anti-emetic compounds; the anti-emetic effects of dexamethasone may be mediated by NK1 inhibition.

Non-stochastic (Deterministic) Effect

When the severity of an effect is directly proportional to dose, then it might be assumed that a direct causal relationship exists between dose and effect. There may be a threshold dose below which the effect is not observed, but thereafter increasing the dose increases the damage in a fairly predictable fashion. A non-stochastic effect is one for which severity is directly dependent on the dose of the causative agent. An example would be the dose-dependent cardiotoxicity of anthracyclines: there is a rapid increase in the severity of cardiomyopathy with total doses of doxorubicin >500 mg m^{-2}.

Normal Tissues: Their Importance in Oncology

The tolerance of normal tissues to insult is the major dose-limiting factor in the treatment of cancer. This applies whether the treatment is with surgery, radiation or **cytotoxic drugs** (**p. 84**). The response of normal tissue to injury is influenced by the organisational pattern of a tissue, as well as by the intrinsic sensitivity of critical cells within the tissue. There are several different patterns of organisation in terms of cell renewal:

- Simple renewal: in which there is simple cell-for-cell replacement, as one cell is lost it is replaced by simple division of a neighbouring cell:
  - liver cells divide every 1–2 years: can divide in response to insult (unless pretreated, e.g. with non-lethal doses of irradiation)
  - endothelium: slow turnover except at sites of turbulence; important changes in tumour angiogenesis, neovascularisation, wounding
  - thyroid: slow turnover (9 weeks).
- Hierarchical systems: a relatively small number of **stem cells** (**p. 298**) are, ultimately, responsible for maintaining tissue integrity. These cells are often inconspicuous, might be difficult to define morphologically, and sometimes can only be identified in functional terms. The **Hayflick hypothesis** (**p. 148**) of tissue **senescence** (**p. 282**) assumes that the total number of stem cell divisions might be limited. The mechanism may involve shortening of the **telomere** (**p. 306**).

The stem cells give rise, often via a series of amplificatory divisions, to progeny that differentiate, first, into committed precursors and then into fully differentiated cells. These differentiated cells, unlike their precursors and in contrast to the mature cells in a simple renewal system, often have lost the ability to divide: examples include erythrocytes and spermatozoa:

- Bone marrow.
- Gut.
- Skin/mucosa.
- Testis.

The concepts of 'stemness' and clonogenicity are semantically difficult to disentangle. Full clonogenicity implies that each cell produced can give rise to two further cells by simple division. If every cell division produces two progeny each of which can give rise to two more progeny, then **self renewal probability** (**p. 282**) is defined as 1.0. If there is no cell loss then the clone expands according to a simple arithmetic progression.

Northern Blotting

A technique, similar in principle to **Southern blotting** (**p. 292**), used to identify RNA sequences. Using specific probes, tissue-specific expression of mRNA can be identified. Quantitative estimates of, for example, increased expression of c-*myc* RNA can be made. This can be used as a marker for c-*myc* expression and, as such, might have some prognostic value.

Nottingham Health Profile

An instrument used to assess general health status. Part 1 covers six main areas:

- Sleep.
- Physical mobility.
- Energy.
- Pain.
- Emotional reactions.
- Social isolation.

In Part 2 the respondent is asked to estimate how their ability to function in each of seven areas has been affected by their health. The areas are:

- Employment.
- Household activities.
- Social life.
- Home life.
- Sex life.
- Hobbies and interests.
- Holidays.

As with all such instruments, it easy to nit pick: the fact that sex life is listed separately from hobbies and interests might raise the odd eyebrow.

Nuclear Hormone Receptor Superfamily

A group of **transcription factors** (**p. 313**) activated by **ligands** (**p. 180**) that influence a wide variety of processes: metabolic regulation, cell differentiation, specification of cell migration patterns. The main categories are:

- Steroid receptors.
- Retinoid receptors.
- Thyroid hormone receptors.
- Orphan (ligand unknown) receptors.
- **Peroxisome proliferator-activated receptors** (**PPAR**) (**p. 242**).

Number Needed to Treat (NNT)

This measure of outcome for a comparative trial of treatment has two compelling advantages: it is both easy to calculate and to understand. NNT is simply the number of patients we need to treat with superior treatment to prevent one event. It is calculated simply as the reciprocal of the **absolute rate difference** (**p. 1**). Survival with adjuvant polychemotherapy for breast cancer is 46.6% at 10 years; the corresponding figure for controls is 39.8% (data from breast overview):

- Absolute rate difference = 46.6 − 39.8% = 6.8%.
- Number needed to treat = 1/0.068 = 15.

To prevent a death, we need to treat 15 women with adjuvant therapy. When trial results are expressed this simply it becomes much easier to make sensible decisions on the relative balance between the effectiveness of a treatment and its morbidity rate. Methods are now available for calculating **confidence intervals** (**CI**) (**p. 70**) for the estimated NNT.

Occam's Quote

This is the original version of the principle of **parsimony (p. 228)**, a principle both useful and misleading in scientific investigation. 'Pluralitas non est ponenda sine necessitate' (plurality should not be proposed unnecessarily). If there is a perfectly adequate solution or explanation, it is unnecessary to seek or manufacture additional ones.

Octreotides (Endocrine Cyanides)

This group of compounds inhibits peptide hormones such as ACTH, TSH, VIP and gastrin. Those available for therapeutic use include somatostatin, vapreotide and lanreotide. The octreotides' most obvious use is in the management of symptoms related to excess peptide hormone associated either with endocrine neoplasia or ectopic production. Uses include management of VIPomas, ectopic ACTH, gastrinoma (Zollinger–Ellison). The side-effects of octreotides are predominantly gastrointestinal, diarrhoea and steatorrhoea, but long-term use can cause gallstones. Octreotides, by inhibiting excess small bowel secretion, may be useful in the conservative management of small bowel obstruction in **terminally ill (p. 306)** patients.

Octreotide labelled with indium-111 can be used to image tumours of neuroendocrine origin, both primary and metastatic.

Odds Ratio

The popularity of the odds ratio in clinical statistics owes more to the ease with which it can be calculated than to the ease with which it can be understood. In its primordial form, it starts out as an approximation to relative risk ratio in a retrospective epidemiological study. Consider a **case-control study (p. 47)** of residential radon **exposure (p. 124)** and the incidence of lung cancer:

Exposed	Cancer	No cancer
Yes	a	c
No	b	d

Thus,

$$\text{Relative risk ratio} = \{a/(c+a)\}/\{b/(d+b)\} = a(d+b)/b(c+a).$$

The **relative risk (p. 264)** ratio in this context becomes somewhat arbitrary since it depends on how many patients with cancer (a + b) there are relative to the number of controls (d) one chooses to study. In this type of study, (a) is usually very small relative to (a + c), and (b) will be similarly small relative to (b + d). The relative risk ratio could be written as: ad/bc without any appreciable loss of precision. The expression is the odds ratio or, as it is sometimes called, for obvious reasons, the cross-product ratio. This means that a cross-sectional case control study can provide information every bit as useful as a more cumbersome and time-consuming cohort study. Another advantage of the odds ratio is that it is independent of the orientation of the table: if we had laid out our table:

Cancer	Exposed	Not exposed
Yes	a	c
No	b	d

the odds ratio is still ad/bc, whereas the relative risk ratio would depend on the orientation of the rows and columns. Put into words, the odds ratio can be defined as: 'the ratio between the odds of the event occurring in the exposed group and the odds of the event occurring in the non-exposed group'. Do not confuse odds and

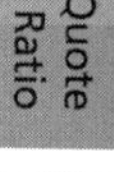

probabilities, the two are of course interrelated, but not identical.

$$\text{odds} = \frac{\text{probability}}{(1 - \text{probability})}.$$

Clinicians are not used to judging the effects of treatment in terms of the relative odds of success and failure. Another difficulty is that the odds ratio will only approximate the relative risk ratio, the real measure of interest, when the event rate is low: if (a) is small relative to (a + c), and (b) is small relative to (b + d). This is not always the case in clinical trials and, in such circumstances, it might be better to measure the relative risk ratio directly or to use some other measure of effect: **absolute rate difference** (**p. 1**); **number needed to treat** (**p. 216**).

The odds ratio has been pushed even further into the realms of incomprehension by using the expression 'relative reduction in odds of event'. This is simply 1 − odds ratio expressed as a percentage. Using data on mortality rate after adjuvant CMF chemotherapy for early breast cancer, an odds ratio = 0.74 is found. The percent reduction in odds of death is 26% (100 − 74%). This is a more impressive than the 8% absolute rate difference or the NNT (number needed to treat) of 13, calculated from the same data.

Oestrogens/Anti-oestrogens

The effects of oestrogen on breast cancer are paradoxical. Oestrogen therapy was the first effective, albeit palliative, therapy for metastatic breast cancer. Today, it is hardly used: its role usurped by anti-oestrogens and other forms of endocrine manipulation. Oestrogen, in the form of stilboestrol, is a cheap and effective treatment for disseminated prostate cancer. It is unpopular because of an association with excess cardiovascular mortality rate. This can be avoided by limiting the dose to 1 mg day^{-1} and administering aspirin prophylactically.

Anti-oestrogens include tamoxifen, droloxifene, raloxifene and toremifene. **Tamoxifen** (**p. 304**) is the most widely used anti-oestrogen and has a role both in the adjuvant treatment of beast cancer and in the hormonal manipulation of patients with disseminated disease. It is well tolerated: the main side-effects are weight gain and menopausal-type symptoms. Long-term use is associated with an increased risk of endometrial cancer. It might be protective against osteoporosis.

Tamoxifen is now used in the chemoprevention of breast cancer in women at high risk of developing the disease. Early results from the USA suggest a genuine benefit; longer follow-up will be required before any definitive conclusions can be drawn. Such studies also beg the question: what is the appropriate definition of high risk?

Off-Label Use

This is when a drug is used to treat a condition for which it is not licensed. The use of the anti-helminthic agent levamisole to treat colorectal cancer is an example of off-label use. This action is also known as **unlabelled use** (**p. 319**) (but this term is slightly inaccurate since the bottle still has a label on it). It is related to compassionate use, but in the case of off-label use the drug is at least licensed for something.

Oncogenes

Oncogenes are genes involved in the production of cancer. Oncogenes were first identified in viruses known to be associated with malignant disease. The classic observation on transmissible chicken sarcomas by Peyton Rous in 1911 eventually led to the identification of an oncogenic sequence, v-*src*, in the virus. Once a series

of viral oncogenes had been identified, it became apparent that homologous sequences could be identified within the normal mammalian genome. The prefix 'v-' identifies viral sequences; 'c-' identifies cellular sequences. Oncogenesis by viral oncogenes is, in fact, a relatively rare event. The viral oncogenes served to draw attention to a fundamental point: there are dormant nucleotide sequences within a human genome that, if activated, have the potential to cause cancer. These are the proto-oncogenes.

Proto-oncogenes

Proto-oncogenes lurk in the genome and can be activated into fully functional oncogenes — within the human genome are the seeds of our own destruction. Evolutionary analysis shows that cellular oncogenes are highly conserved. The most likely explanation for this is that, at a very basic level, they are concerned with the control of growth and differentiation. Once mechanisms evolved to carry out these functions, they were conserved.

Deranged proto-oncogenes, or cellular oncogenes, could cause cancer through a variety of mechanisms. There are two main mechanisms by which an oncogene could cause cancer: by stimulating growth or by failing to suppress abnormal growth.

Growth stimulation

- The oncogene could code for a protein that functioned as a growth factor and which directly stimulated the growth and division of the target cells. For example, the c-*sis* oncogene codes for a protein that has a sequence **homology** (**p. 154**) to platelet-derived growth factor (PDGF).
- The oncogene could act as an intracellular messenger and amplify events occurring at the cell membrane so that minor stimuli could produce overreaction at the nuclear level. The *ras* family of oncogenes, through its effects on adenyl cyclase and GTPase, might act in this way.
- The oncogene could increase levels of membrane receptor for **growth factors** (**p. 144**). The signal at the cell surface could be amplified simply by over-production of the receptor protein or by producing a structurally abnormal protein that locks the switch into the 'on' position. The *erb* and *fms* oncogenes act mainly through their effects on cell surface receptor proteins.
- The oncogene could act directly on the nucleus and trigger inappropriate proliferation. The precise mode of action of oncogenes, such as *myc* and *fos*, that act at the nuclear level is unknown. This is mainly because it is so difficult to investigate such events experimentally.

Failure to inhibit abnormal growth (tumour suppressor genes)

Mutations are acquired all the time. This is a consequence of normal cell turnover — that which is lost has to be replaced through proliferation of normal cells. Mistakes in DNA replication are probable each time a cell divides. Some of these mutations might give rise to malignancy and, during evolution, cells have evolved mechanisms to eliminate undesirable mutation. The genes involved in these protective mechanisms could be described as tumour suppressor genes — since they stifle the malignant process at the time of its inception. *p53* is a classic example of a gene that functions as a tumour suppressor. The **retinoblastoma** (**p. 271**) gene has a similar suppressive role — its function is to switch off cellular proliferation. The retinoblastoma gene product will not allow cells to proceed

through the **cell cycle** (**p. 52**) until they have correctly completed G1-phase. The gene that is abnormal in **familial polyposis coli** (**p. 125**) also functions as a tumour suppressor. Mutant ***p53*** (**p. 224**) or mutations in the retinoblastoma gene allow potentially malignant cells to multiply. The repair of damaged DNA is under genetic control. The genes *MSH2* and *MLH1*, abnormalities of which are part of the adenoma/carcinoma sequence involved in the formation of colorectal cancer, are concerned with DNA repair. If their activity is impaired, then **DNA damage** (**p. 98**), which should have been temporary, might become permanent.

Ontogeny

A term for the embryonic development of an individual. It is often contrasted with **phylogeny** (**p. 235**), the development of the species.

ONYX

ONYX-015 is a mutant adenovirus with activity against malignant disease. The virus has been designed specifically as an anti-cancer agent. The reasoning is as follows. Virus infection can activate ***p53*** (**p. 224**), which in turn might activate **apoptosis** (**p. 19**). Small DNA viruses, however, contain genes, ***E1A* and *E1B*** (**p. 111**), that can inhibit the activation of *p53* and any consequent apoptosis. In ONYX-015 a part of *E1B*, coding for a 55 kDa protein, is deleted. If ONYX-015 infects a cell with normal *p53* then the *p53* in the absence of normal viral *E1B* gene product can inhibit viral replication: virally induced cell killing is, therefore, minimal in cells with normal *p53*. The mechanism for this inhibition is, as yet, unknown.

Many cells in human tumours contain mutant *p53*, this mutant *p53* cannot inhibit the replication and cytopathic effects of ONYX-015. The net result is selective killing by the virus of tumour cells with their mutant *p53*, and selective sparing of normal cells with their wild-type *p53*. The experimental data, so far, bear out this hypothesis. ONYX-015 appears to be selectively toxic to human tumour cell lines, either grown in culture or as xenografts. It is encouraging that several of these lines were derived from tumours resistant to conventional chemotherapy.

Osteoprotegrin (OPG)

OPG is a member of the TNF receptor family. It potently inhibits osteoclasts and, therefore, has the potential to block skeletal destruction by cancer. Bone pain in metastatic disease is due to a tumour-induced imbalance between bone resorption and formation. RANK is a receptor that controls osteoclast activity. Osteoprotegrin ligand (OPGL) normally interacts with RANK, OPG binds and holds OPGL, which then cannot interact with RANK and osteoclast activity is thereby inhibited.

Ovary and Oocytes

The ovary has a unique system of organisation. By puberty there are only ~4000 ovarian germ cells per ovary compared with 6 million at the fifth month of intrauterine development. They rest, as primary oocytes, in the **prophase** (**p. 249**) of meiosis until shortly before ovulation. The final meiotic division eventually produces a mature ovum and three polar cells. These go on to participate in the formation of ripening ovarian follicles. The intermediate stages of follicular development are more radiosensitive than the mature or immature stages. Thus, after ovarian irradiation, there can be a temporary period of fertility followed by sterility, followed by a regain of fertility. This pattern reflects the differing radiosensitivity of differing stages of follicular development and, in contrast to the pattern observed in the testis, has nothing to do with recruitment of stem cells. The number of primary oocytes declines with age; hence, the increased sensitivity to irradiation of the ovary with age. The ovary may be surprisingly resistant to irradiation in younger women: pregnancy can occur after doses of 8 Gy. Radiotherapy was used in the past to treat infertility: 341/796 women treated with ~2 Gy in 3f fell pregnant; 27/61 women treated with ~2.3 Gy in 3f subsequently conceived. It must be remembered that these women had previously been defined as 'infertile'. Ovarian ablation is inevitable with fractionated doses >24 Gy and, in women >40 years of age, nearly all will experience ovarian failure with doses >6.5 Gy.

Overview

A term often used as an alternative to **meta-analysis** (**p. 195**). Strictly speaking, an overview implies any *tour d'horizon* and does not imply either a systematic approach or any statistical pooling of results, both of which are prerequisites for a valid meta-analysis.

Oxaliplatin

Oxaliplatin is a cytotoxic drug derived from ***cis*-platinum** (**p. 66**). It is the first platinum analogue to have shown significant activity against colorectal cancer. It inhibits DNA synthesis by forming **adducts** (**p. 7**) and this ability is mainly due to the presence of a bulky DACH (1,2-diamineocyclohexane) moiety. Its main toxicity is peripheral neuropathy. It causes less renal damage than *cis*-platinum and is less myelosuppressive than **carboplatin** (**p. 46**).

Oxidative Stress

Oxidative stress is an important environmental carcinogen. The mechanism involves the loss of an electron from an atom or molecule. Reactive oxygen species (ROS) are produced endogenously from mitochondrial electron transport and exogenously from ionising radiation.
A variety of insults can act via this route: ischaemia, infection, ionising radiation. Oxidative stress can be mediated by free radicals, activation of **cyclooxygenases** (**p. 83**) (**COX**), lipooxygenases, release of metal ions (copper, iron), release of haem and its breakdown products, and loss of protective systems, e.g. reduced glutathione.

Each day, each person's cells receive ~10 000 oxidative hits to their DNA from ROS. The main ROS are the hydroxyl radical, the superoxide radical and hydrogen peroxide. The DNA damage produced by the ROS is potentially carcinogenic and, unsurprisingly, intracellular enzyme systems have evolved as protective mechanisms to cope with the problem of oxidative stress. Abnormalities of the enzymes involved, such as catalase and glutathione peroxidase, could be implicated in increased risk of cancer and, in such cases, self-synthesised carcinogens might be as important as exogenous carcinogens.

Oxygen Enhancement Ratio (OER)

Hypoxic cells are more radioresistant than well-oxygenated cells. The extent to which this phenomenon occurs *in vitro* is defined as the oxygen enhancement ratio (OER). For any given level of effect, e.g. 1% survival:

$$OER = \frac{\textit{radiation dose under hypoxic conditions}}{\textit{radiation dose under well-oxygenated conditions}}.$$

This is, for X-irradiation, typically taken as ~3. This is an oversimplification since OER will depend on the level of effect chosen. For HeLa cells, OER = 2.3 for 10% survival and 2.0 for 50% survival. OER will not, however, be constant for all tissues and fraction sizes. Even *in vitro* there are problems: OER for G_2-phase cells has been estimated as ~2.3; for S-phase cells it is nearer to 2.9. In high-dose experiments an OER = 2.5–3.0 is assumed; for the doses used in clinical radiotherapy 2.0 is probably more realistic.

OER is an example of a dose-modifying factor — in this case the modifying factor is oxygen and radiation is the treatment with which it interacts.

OER and **relative biological effectiveness (RBE)** (**p. 264**) are both related to linear energy transfer (LET): OER falls as LET rises from 10 to 100 keV μm^{-1}, RBE rises to a maximum at the upper limit of this range and thereafter falls — the overkill effect. The overkill effect simply reflects the fact that you cannot kill a cell twice: energy deposited in an already dead cell is wasted and cannot contribute to biological effectiveness (figure 1).

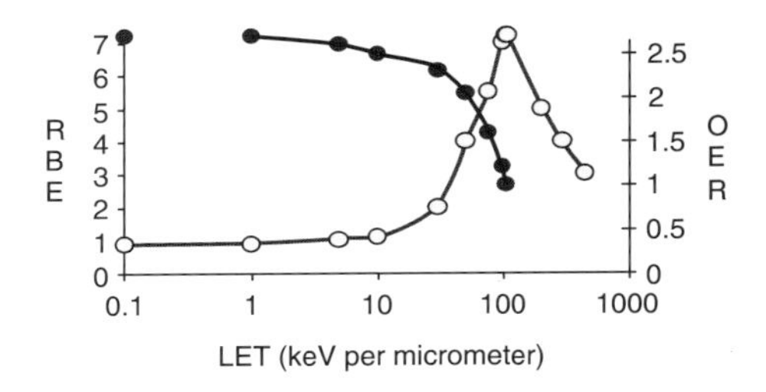

Figure 1.

p

Probability (p) is widely used as a measure of statistical significance. p has its uses (perhaps), but on the whole its use creates confusion rather than clarification. Its popularity stems, in large part, because it is often possible to perform a small-scale, not too demanding experiment and obtain, using a ***t*-test (p. 303)**, a χ^2-test or other parametric test, a statistically significant answer. The issue of clinical significance is ignored in such an approach.

What does p indicate? It simply indicates, according to some hypothetical distribution of the true differences, whether the difference observed between groups could have arisen by chance. A 'significant' $p = 0.05$ indicates that there is only a 1:20 (5%) probability that the difference observed has arisen by chance; $p = 0.001$ shows that the probability of the observed difference having arisen by chance is 1:1000 (0.1%). Note that p indicates nothing about the magnitude of the difference, but, as clinicians, it is the magnitude of the difference with which we are concerned. This is best exemplified in **meta-analyses (p. 195)** where, because of the large numbers involved, very small p values might accompany very small **absolute rate differences (p. 1)**.

In the discussion below, probability is used in the sense of likelihood, and chance indicates that events can occur at random without any particular cause or reason.

A friend has a coin that she tosses three times, each time it comes up heads. She asks you if you think the coin is double-sided. She is effectively asking you what the probability is that the three heads in a row could have arisen by chance: what is the significance of her result? A coin has no memory, it does not say to itself 'I've just done two heads in a row, now it's time for tails.' Each coin toss is independent of every other toss. Since there are only two sides to a coin, the probability of heads at each toss equals the probability of tails at each toss. Probabilities for each toss must add up to 1.0. The coin cannot land on its edge, it must either be tails or heads. For each toss, therefore,

$$\text{probability of heads} = \text{probability of tails} = 0.5.$$

The probability of three heads in a row is:

$$0.5 \times 0.5 \times 0.5 \times = 0.125.$$

There is a 12.5% chance that an ordinary coin tossed three times will come up heads each time. By convention, statisticians regard as significant an event that has a <5% chance of occurring: $p = 0.05$. Statistical significance is, therefore, achieved when $p < 0.05$.

For an alternative view of what constitutes significance, see **Bernoulli's Balls (p. 27)**. Returning to coin tossing: the chance of four heads in a row, with a normal, unbiased, coin is 0.0625; for five heads in a row $p = 0.031$. Thus, to a statistician, if your friend tossed heads five times in a row, either she, or her coin, would appear suspicious.

Here is another example. do ripe olives float in the Mediterranean or do they sink? Three olives are thrown into the sea and they all float. I have not, however, proved to a statistician's satisfaction that all olives float. The fourth olive sinks; not all olives float. This is an example of how statistics can apply to everyday life. Even four in a row would not be statistically significant. A Type I statistical error implies that significance has been claimed inappropriately: in the coin tossing, claiming that three heads in a row indicated some type of cheating would be a Type I error.

p has achieved a currency out of all proportion to its importance or usefulness. It has aspects of a pagan deity: scepticism and common sense are discarded in pursuit of a 'significant' p.

p offers many opportunities for abuse and misinterpretation, **confidence intervals** (**CI**) (**p. 70**) around an estimate are usually a better method of assessing whether an observation could have arisen through chance or whether the result has genuine meaning.

Abuse and misinterpretation of *p*

- Multiple peeking, multiple **endpoints** (**p. 114**): these are variations upon the same theme. If an infinite number of monkeys bashes away for long enough at an infinite number of typewriters, then, eventually, the works of Shakespeare might emerge. Or, if not Shakespeare, some other work of literature.
- Spurious excitement: a highly significant *p* does not mean that a result is important, only that it is unlikely to have arisen by chance. If as you arrive home three nights in a row you find a penny on your doorstep, you can be fairly certain that this has not occurred by chance. Your demented aunt is simply trying to be kind, but is the unexpected acquisition of three pennies really important?

p16

The *p16* gene functions as a tumour suppressor gene. It is on chromosome 9p21. Its gene product is a cyclin-dependent **kinase** (**p. 171**) inhibitor (CDK1). The nomenclature is a trifle confusing in that p16 is also known as *INK4A, MTS*-1 or *CDKN2A*. Its role as a tumour suppressor gene first came to light in a study of patients with familial melanoma. Now it is realised that *p16* abnormalities might be found in a wide range of tumour types.

p16 acts on the G_1-phase of the **cell cycle** (**p. 52**) and binds either to CDK or inhibits cyclin D. It acts upstream of the retinoblastoma gene product: since the binding of cyclin D to CDK forms a complex which phosphorylates and activates the pRB protein. When *p16* in non-functional cells accumulates at the G_1/S boundary, *p16*s function can be affected in several ways: deletion, point mutation and abnormal methylation. Its biological role may be concerned with cell **senescence** (**p. 282**). The p16 null mouse can develop normally but it has a high incidence of spontaneous tumours. The slow accumulation of *p16* over the generations may result in a gradual slowing down of cell division through the generations.

p53

p53 is a tumour suppressor gene, possibly the most important one. Over 50% of human tumours contain abnormal *p53*. The dire consequences of developing abnormal *p53* function are illustrated by the **Li–Fraumeni syndrome** (**p. 179**) in which multiple malignant tumours occur at an early age. The only reason that patients with Li–Fraumeni syndrome develop only a few tumours each is that they do not survive long enough for the catastrophic effects of abnormal *p53* to wreak their full havoc.

Mutant *p53* is like a self-fuelling fire. Once a mutation has occurred, and a malignant clone has arisen with defective *p53*, then the rapid expansion of the clone, and the many mitoses involved, will give opportunity for further mutations to occur. Because *p53* is abnormal, cells bearing these mutations will be allowed to proliferate in a way that, had wild-type *p53* been around, would not have been permitted. The net result is a rapid expansion of malignant cells containing a wide variety of mutations.

p53 can be regarded as the guardian of the genome, its product is the shape up or ship out molecule. *p53*-dependent pathways cause cells that have suffered some genotoxic insult either to come out of cycle, so that they have time to repair the damage (shape up), or to undergo immediate death by apoptosis (ship out). It not difficult to envisage the evolutionary importance of a molecule that prevents an organism from acquiring cell lineages containing a wide variety of potentially harmful mutations.

p53 protein contains three **domains (p. 102)**: a central DNA binding domain, a transactivation domain at the amino-terminal and an oligomerisation domain at the carboxy-terminal. The molecule has five evolutionarily conserved boxes (I–V). The majority of *p53* missense mutations associated with cancer occur in the central DNA binding domain, boxes II–V. A simple model would envisage that loss of DNA binding would lead to failure of transcription of *p53* target genes and that this would lead to failure of tumour suppression.

p53 gene expression is switched on by a variety of environmental insults — **hypoxia (p. 157)**, ionising radiation, UV radiation — as well as by **DNA damage (p. 98)**. p53 protein, in turn, affects a variety of genes: *mdm*-2 (expression of which has an inhibitory effect on *p53* expression); genes controlling **apoptosis** (*bax* **(p. 26)**, *fas*, *IGF-BP3*, *PAG 608*, death receptor *DR*5); genes controlling the cell cycle (*p21*, *GADD*45, *14-3-3σ*) and other genes whose effects are, as yet, unclear such as cyclin G. Activation of **caspases (p. 47)** by *p53* expression will occur along a variety of pathways: **TRAIL (p. 313)**/DR5, **Fas/fas ligand (p. 126)**, bax-induced mitochondrial leakage, which, in turn causes caspase activation by permitting **cytochrome c (p. 84)** to enter the cytosol.

The mechanisms and reasons whereby *p53* induces a cell to undergo apoptosis, as opposed to cell cycle arrest, are ill understood. Some effects are tissue- and insult-specific: ***cis*-platinum (p. 66)** is particularly likely to induce *p53*-mediated **apoptosis (p. 19)** in germ cell tumours expressing wild-type *p53*; lymphocytes often undergo *p53*-induced apoptosis, unless they are dependent on IL-3 when, in the presence of *p53* and IL-3, they will opt out of the cell cycle rather than die by apoptosis.

The restoration of normal wild-type *p53* function to tumour cells is a tempting target for **gene therapy (p. 137)**. One approach has been to use direct intratumoral injection of an adenovirus construct containing wild-type *p53*. This has had some limited success in the treatment of locally advanced non-small cell lung cancers. The **ONYX (p. 220)** system represents a somewhat more sophisticated, in a biological sense at least, approach. It exploits the ability of cells with normal *p53* to destroy viruses and the impaired viral cytopathic effect in cells that contain normal *p53*. Another approach is to investigate the possibility of using small peptides that mimic the effects of *p53*. This approach, in a sense at least, may mimic the final common pathway whereby radiation and **cytotoxic drugs (p. 84)** kill malignant cells. As with all cancer treatments the critical problem will always be that of achieving **selective toxicity (p. 282)**.

p73

This is a recently discovered protein with structural and functional resemblance to **p53 (p. 224)**. It was discovered serendipitously: in a search for mediators of insulin signalling it turned up as a false-positive cDNA clone. The *p73* gene is on chromosome 1; *p73* over-expression induces p21. Deletions involving the p73 region of chromosome 1 can be associated with cancers in humans: neuroblastoma, colorectal cancer, melanoma and breast cancer. The indications are that *p73* will also prove to be a tumour suppressor gene. It may also take over *p53* function if *p53* is missing, as in the knockout mouse. This may explain why the *p53* knockout mouse can develop normally, at least until its malignancies develop.

Paclitaxel

Paclitaxel is a **taxane** (**p. 305**). It is semisynthetic in origin. Originally, it was isolated from the bark of the Pacific yew tree; it is in fact produced by a fungus, *Taxomyces andreannae*, that grows beneath the bark. The Pacific yew is rare and there were major problems with the supply of taxanes until another method of production was devised. The needles of the European yew, *Taxus baccata*, can be used to provide a precursor that can then be used to synthesise paclitaxel. Paclitaxel blocks **mitosis** (**p. 201**) by stabilising the microtubules so that the spindle cannot be dismantled. Cells treated with paclitaxel are arrested at the G_2/M-phase boundary, which is when cells are at their most radiosensitive. Paclitaxel is, therefore, a potential radiosensitising agent. The morphological consequence of paclitaxel treatment is the presence of bundles of microtubules within cells, regardless of their mitotic state. Other effects of paclitaxel include the inhibition of angiogenesis and increased expression of TNF-α. It can promote **apoptosis** (**p. 19**) by a variety of mechanisms including: activation of cdc (cyclin dependent kinases); and the stress activated protein kinase pathway that involves the c-*jun* gene product.

Resistance to paclitaxel can arise both through the MDR mechanism and as a result of the tubulin subunits polymerising more slowly. The combination of paclitaxel and platinum has been particularly effective in the treatment of ovarian cancer. Other tumours in which paclitaxel seems particularly active include: breast cancer, head and neck cancer, Kaposi's sarcoma.

Paclitaxel is a toxic drug. It causes myelosuppression and an acute hypersensitivity reaction, with fluid retention, may occur. This latter reaction can be mitigated if dexamethasone is given before the paclitaxel. Other problems include: cardiac arrhythmias, fevers and myalgias.

Palladium (^{103}Pd)

This radioactive isotope is a pure γ-emitter that has been used in the brachytherapy of prostate cancer. The γ-emissions have energies in the range 0.020–0.0227 MeV; it has a **half-life** (**p. 147**) of 17 days and it is available as seeds and can be used for permanent interstitial implantation.

Palliative Treatment

There is no justification for a policy of routinely treating all patients simply because they have cancer. If a patient has no symptoms to relieve, then treatment may simply be meddlesome and might make the patient feel worse rather than better. In some circumstances, however, it may be sensible to give treatment to an asymptomatic patient to forestall symptoms that will almost inevitably develop. An example would be to give local radiotherapy to a patient with a lung cancer that is severely narrowing a bronchus and which, were it allowed to progress, would obstruct a major airway.

Ideally, palliative treatment should relieve symptoms without, itself, producing any side-effects. This ideal is often difficult to achieve. If there is an art to caring for patients with cancer, then this is where it lies — there can be no off-the-shelf approaches, each patient's problems and priorities are unique and require management tailored to that individual.

Paradigm

A paradigm is a template or example, usually used in a fairly broad sense, to describe a way of doing things: the **randomised controlled clinical trial** (**p. 261**) is currently the optimal paradigm for clinical research.

Paradigm Shift

When first coined by Thomas Kuhn, this was a novel and exciting description of how important science moved forwards. Overuse has turned the expression into a cliché: every dull discovery is now hailed as a 'paradigm shift'. The real paradigm shift would be, albeit reluctantly, to abandon the use of this once splendid expression.

Parallel Architecture

This describes a system in which the functional units are arranged side by side so that, if one unit fails, the units to either side can take over its function, the result being that the system as a whole is relatively unaffected. The kidney is the classic example of parallel organisation. Even though each nephron is fairly radiosensitive, most of the kidney has to be damaged before the organ fails. This is in complete contrast to a structure, such as the spinal cord, which has a **serial architecture** (**p. 284**) (figure 1).

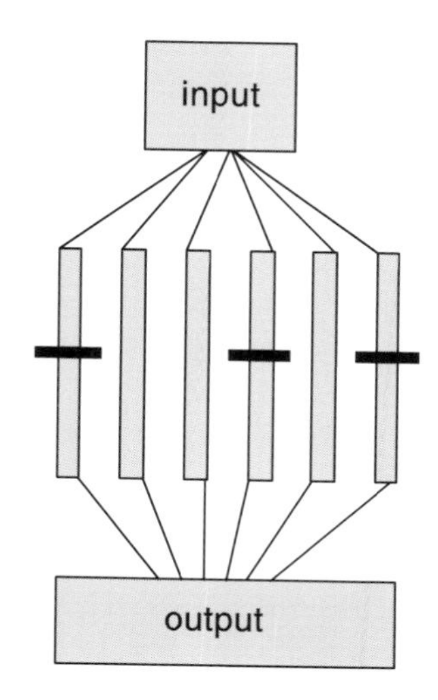

Figure 1 — Parallel architecture. Even though three of the subunits are damaged, the remaining three are still undamaged and the system as a whole remains functional.

Parametric and Non-Parametric Tests

These terms sometimes cause confusion. Parametric tests, and the ***t*-test** (**p. 303**) is a good example, rely on data sets conforming to known distributions: the statistical properties of the distributions are known and so the distributions, and by extension the data, can be compared. The distributions can be defined in terms of a minimum number of parameters, typically the mean and **standard deviation** (**SD**) (**p. 296**); hence, the term 'parametric'. Non-parametric (syn. distribution-free) methods make no assumptions about the shapes of the distributions of the two sets of data. They work on the principle of putting the pooled data into rank order and then assessing whether the two groups of data are randomly intermingled, or whether one group tends to be at the top of the order and the other towards the bottom. The **Wilcoxon-signed ranks test** (**p. 328**) would be an example of a non-parametric test of this type. Before computers were widely available, the non-parametric methods were tedious and time-consuming, and parametric methods were often preferred simply because of convenience. Now, in terms of effort, there is little to choose between the two types of method. Parametric tests tend to be more powerful statistically. Through the use of a known distribution, a few data points are converted into an infinite number (since a line is an infinite number of points). If there is a choice between performing either a parametric or a non-parametric test, the usual advice is to perform the former, since it is more likely to be significant. The proviso is that the data in the groups to be compared should follow approximately normal distributions. If we do not know how the data are distributed, then it is probably better to use a non-parametric method. If, even after transformation the data are skewed, or if the values to be compared are scores rather than measurements, then non-parametric methods must be used.

Pareto, Paretian Principle

Pareto was a nineteenth-century Italian economist who worked in Switzerland: he is credited with formulating the principles that underpin the economic aspects of welfare and are, therefore, directly applicable to concepts of value and justice in the economics of healthcare. The Paretian principles are:

- The existing distribution of resources is a reasonable starting point for exploration of alternatives, i.e. there should be no radical overthrow of the status quo simply for the sake of it.
- The system as a whole, when it reaches a point at which one individual cannot be helped unless this is at the expense of another individual, is at its maximum efficiency.

A Paretian improvement is one that follows from the second principle: it is an improvement that benefits an individual without depriving or harming any other individual. Given finite resources, true Paretian improvements are hard to identify: questions of opportunity cost arise. Pareto also had a useful comment to make about 'mistakes' (commenting on Kepler): 'Give me a fruitful error every time, full of seeds, bursting with its own corrections.'

Parsimony (syn. Occam's Razor)

The principle of parsimony is also sometimes known as **Occam's razor** (**p. 217**). In essence, it implies that if an obvious explanation is already present, then there is little point in expending energy looking for more recondite solutions. It is a version of lumping that might be useful but which might also mislead, e.g. polyuria and polydipsia in a patient with breast cancer is not necessarily caused by hypercalcaemia as a complication of the cancer — it might be due to diabetes mellitus arising as a coincident condition.

Partial Response

A term used in the evaluation of cancer treatment. The criteria that must be fulfilled for a response to be classified as a partial response are:

- There is a 50% reduction in the product of the diameters of at least one assessable index lesion. Assessment can be clinical or radiological.
- This reduction in size lasts at least 4 weeks.
- No new lesions develop.
- No other lesions grow in size during this period.

Partial responses (PR) are often added to **complete response** (**CR**) (**p. 70**) rates and the resulting total quoted as response rate (RR), i.e. RR = CR + PR.

Pascal's Wager

Blaise Pascal (1623–62) was a French mathematician, an invalid, a philosopher and part-owner of the first bus company in Paris In Pensee 680 (*Infinity Nothingness*) he writes apropos of wagering, on the toss of a coin, whether God exists:

> Let us weigh up the gain and the loss by calling heads that God exists. Let us assess the two cases: if you win, you win everything; if you lose, you lose nothing. Wager that he exists then, without hesitating!

This approach to decision-making has important parallels in oncology. If, without treatment, a patient is doomed to die and if there is an intervention available which has virtually no side-effects, then the gamble, heads treat/tails do not treat, is analogous to Pascal's wager. When there is little to be lost and much to be gained through intervention, then the logical action is to intervene. One application of Pascal's wager in oncology is the practice of treating women with metastatic adenocarcinoma of unknown primary with tamoxifen.

PCR-ISH (Polymerase Chain Reaction with *In Situ* Hybridisation)

PCR-ISH is a technique that adapts the principle of PCR so that it can be used on tissue sections, even paraffin-embedded ones, to detect the site of genes that are produced in low copy number. Genes that would not show up on conventional *in situ* hybridisation can be detected by PCR-ISH. The technique has demonstrated HHV8 sequences in endothelial and spindle cells from patients with Kaposi's sarcoma.

PDQ (Physicians Desk Quarterly)

An information service concerning cancer and its management provided by the National Cancer Institute. It can be obtained commercially on CD-ROM or access can be obtained via the World Wide Web. It is menu-driven and it is possible to obtain information for patients and relatives, staging information, treatment information, and information about active clinical trials for a wide variety of tumours.

Peer Review

Peer review is one of the main mechanisms through which the integrity of the scientific literature is maintained. It is also an important factor in the assessment of research proposals.

Papers intended for publication are sent out to two, or perhaps three, individuals considered expert in the issues addressed by the article. Their opinions about the scientific worth of the article and its suitability for publication are sent to the editor. Their comments are often sent anonymously to the authors. Few journals have the, ideal, policy of sending the referees articles that have been adapted so that their authorship is unidentified. It is, therefore, often possible for professional jealousies and prejudices to undermine a process that should be above such petty issues. Another drawback of the system is that outstanding work, but which does not conform to prevailing norms, may be rejected. At its very worst, peer review could lead to a process of self-sustaining mediocrity.
Peer review cannot always detect fraud: downright deceit is probably rare and attracts attention out of all proportion to the frequency of its occurrence. However, truth will, usually, out: sometimes because of peer review, sometimes in spite of it.

Peer review implies that judgement is made by someone of equal standing to those whose work is being judged: the question of how to define 'equal standing' is difficult. In what respect equal: intelligence, reputation, achievement? **Experts** (**p. 123**) may not make the most appropriate reviewers: they may well be perfectly capable of judging the technical quality of a piece of work but be completely unable to pronounce sensibly upon its relevance. We all have our hobby horses and will try to ride them whenever possible. Expert reviewers can define whether the job has been done correctly; they will be less secure in defining whether the job done is the correct one. The trick, however, is 'to do the right things right'.

Kostoff (www.dtic.mil/dtic/kostoff/index.html: Research Program Peer Review: Principles, Practices, Protocols) has identified several critical factors for peer review of research:

- Commitment by the review manager to quality.
- Review manager's motivation to conduct technically credible review.
- Competence and objectivity of reviewers.
- Normalisation and standardisation across disciplines.
- Selection of evaluation criteria:
  - — research merit
  - — research approach
  - — team quality
  - — relevance
  - — overall quality.
- Anonymity (reviewer/reviewed).
- Cost.
- Ethics.

The advantages of peer review have been summarised by Chubin:

- An effective resource allocation mechanism.
- An efficient resource allocater.
- A promoter of science accountability.
- A mechanism for policymakers to direct scientific effort.
- A rational process.
- A fair process.
- A valid and reliable measure of scientific performance.

Other points might be:

- The quality of the basic work is better: people work better when they know they are going to be scrutinised.
- The quality of the submitted report is improved: people know that they have to persuade a potentially sceptical expert.
- The quality of the published work is improved: constructive suggestions made by reviewers can improve the quality of the scientific literature.

The process of peer review is, with the development of evidence-based medicine, coming under increasing scrutiny. 'Sed quis custodiet ipsos Custodes?' (But who is watching the guards themselves?), as the Roman Juvenal wrote in the first century AD.

PEGylation

PEGylation is an ugly word used to describe the treatment of a **liposome** (**p. 182**) with polyethylene glycol. The treatment coats the surface of the liposome with methoxypolyethylene polymers and is believed to improve the targeting of the liposome by allowing the cell evade detection and engulfment by the reticuloendothelial system. The Gulf War (1992) appears to have influenced the vocabulary used. There is talk of 'Stealth' liposomes and a product called 'Evacet'.

Pentostatin

This drug is used for treating hairy cell leukaemia; formerly it was known as 2′-deoxycoformycin. It is an antitumour antibiotic derived from *Streptomyces antibioticus*. Structurally, it is similar to **fludarabine** (**p. 129**) and cladribine. It inhibits adenosine deaminase, presumably, since chemically it is very similar to adenosine, by acting as a false substrate. As a result of this inhibition, levels of dATP rise and this in turn inhibits ribonucleotide reductase. The consequent depletion of deoxynucleotides interferes with synthesis of DNA and RNA.

The main toxicity is: myelosuppression, skin rashes, headache, lethargy and fatigue.

Pentoxifylline

Pentoxifylline is an agent used in treating peripheral vascular disease. It is a xanthine derivative structurally similar to theophylline and caffeine. It increases the deformability of erythrocytes and also lowers plasma fibrinogen levels. It is one of the few therapeutic agents that has any activity in treating radiation-induced fibrosis. This may be due, at least in part, to its ability to suppress the transcription of TNF-α which can be induced by lipopolysaccharides.

Peptide Mimetics

A technique for developing new compounds with potential therapeutic activity. Random peptide sequences are generated, e.g. in phage. The sequences are screened to see if they relate to any known receptors. Those peptides that have an associated receptor are then assessed to see whether they have any therapeutic action. For example, if we find a peptide with affinity for the thrombopoietin receptor, we would test whether the peptide is useful for treating treatment-related thrombocytopenia.

Percent-Labelled Mitoses (PLM)

This is one of the most accurate and complete methods for obtaining detailed information about the cell cycle. It is cumbersome and time-consuming and cannot be used in clinical practice. It does have a useful role in validating more feasible methods for assessing **cell cycle** (**p. 52**) parameters. Cells are labelled in S-phase using a radiolabelled DNA precursor (e.g. tritiated thymidine) and serial samples are taken subsequently. By plotting, over time, the proportion of mitoses radiolabelled, it is possible to calculate the overall duration of the **cell cycle** (**p. 52**) as well as the absolute durations of each of its components: S-phase, G_2, **mitosis** (**p. 201**), G_1. Today, it has been largely superseded by techniques based on **FACS** (**fluorescence-activated cell sorting**) (**p. 129**).

Peripheral Blood Stem Cell Rescue (PBSCR)

This is a means of reconstituting haemopoiesis in patients treated with high-dose chemotherapy. It is based on the finding that after stress to the marrow, circulating stem cells are present in numbers sufficient to reconstitute haemopoiesis. The sequence is as follows: treat patient with **cytotoxic drugs** (**p. 84**); amplify the overshoot of circulating haemopoietic precursors by timing **G-CSF** (**p. 135**) administration appropriately; harvest PBSC during rebound.

The stem cells are harvested using a cell separator and can be identified through their expression of the **CD34** (**p. 52**) antigen. They can be re-infused later, after high-dose cytotoxic therapy.

Peripheral Primitive Neuro-ectodermal Tumour (PNET)

This is an example of a small round cell tumour. It is similar to Ewing's sarcoma of bone and soft tissues. The cells are small and uniform and usually stain positively for glycogen. There are minor immunohistochemical differences between PNET and Ewing's. PNET is usually positive for neurone-specific enolase (NSE), vimentin and HBA 71 (a product of the *MIC*2 gene). A characteristic chromosomal translocation, t(11;22)(q24;q12) is found in >90% of Ewing's sarcomas and PNET.

As a result of this translocation the *EWS* gene, which produces an RNA polymerase, is fused with the *FLI*1 gene, whose product belongs to the ETS family of **transcription factors (p. 313)**. The result is overexpression of chimerical proteins with transforming properties. The identification of this specific abnormality means that FISH and PCR can be used as diagnostic tests for PNET on very small samples, e.g. those obtained by **fine-needle aspiration (FNA) (p. 128)**.

Peutz–Jeghers Syndrome (PJS)

An inherited syndrome including polyposis of the bowel and circumoral freckling. The polyps are hamartomatous, most frequently affecting the jejunum, and have limited malignant potential. However, the overall risk of developing cancer for patients with PJS is 50%. The gene is on chromosome 19p13.3; the gene product is a serine threonine **kinase (p. 171)**.

Phage Group

A group of biochemists and geneticists active during the 1940s. They studied bacteriophage, viruses which parasitise bacteria. Their work provided the cornerstones for molecular biology. Members included: Andre Lwoff, Gunther Stent, Seymour Benzer, Salvador Luria, James Watson and Max Delbruck.

Pharmacodynamics

The study, over time, of the effects, both beneficial and adverse, that a drug has on an organism. Its ultimate goal is estimation of quantitative effects, in terms of both intensity and duration, of drug therapy. It indicates what the drug does to the organism; pharmacokinetics indicates what the organism does to the drug.

Pharmacokinetics

The study of the time-courses of the processes whereby an organism handles a drug. The acronym 'ADME' summarises what is involved: absorption, distribution, metabolism, excretion. Knowledge of pharmacokinetics allows the construction of a narrative picture over time of what the organism does to the drug. This in turn, gives us an indication of what the drug might do to the organism — pharmacodynamics.

Phase I Study

The purpose of a Phase I study is to determine the **MTD (maximally tolerated dose) (p. 190)** that can safely be given to patients using a given schedule and route. A Phase I study starts using a dose that, based on the MTD or toxic low dose (TDL), in animal experiments is assumed to be safe in humans. Patients are entered, three at each dose level, until toxicity, usually myelosuppression, is noted. The dose increments usually use a predetermined scheme, such as a modified **Fibonacci escalation (p. 127)**.

Phase II Trial

A Phase II trial is designed to establish the clinical spectrum of activity of a new treatment. Treatment is used in a group of patients with a variety of different tumours. The **endpoint (p. 114)** most commonly assessed is response rate (the sum of complete and partial responses divided by the number of patients studied). There is controversy about the optimal size of Phase II studies.

According to **Gehan's rule** (**p. 135**), there is no point in continuing a Phase II study if no responses have been observed after treatment of 14 consecutive patients. This assumes that we would reject as ineffective any drug that produced a response rate of <20% and that we also accept the risk that, 5% of the time, a false-negative conclusion will be drawn. This is regarded by some as too restricted a view.

Larger Phase II studies, sometimes with >100 patients, have been used in an attempt to define precisely the activity of a new drug. The legitimate criticism of approaches based on the use of **Gehan's number** (**p. 135**) is that they will not identify drugs active against specific forms of cancer. For example, if the new drug is only active against lymphoma, and only one of the 14 patients in the negative phase II study actually had lymphoma, then there is the risk of prematurely discarding a potentially useful drug. Another problem with Phase II studies is that they often recruit patients who have already been extensively treated. Patients enrolled in Phase II studies may have tumours that have acquired a **multidrug resistant** (**p. 210**) phenotype. Once again, this could lead to a potentially useful drug being inappropriately discarded, since a drug active against a newly diagnosed tumour may be completely inactive against a heavily pretreated tumour.

The usual design for a Phase II study is a prospective single-group study, although there is often some attempt, usually tacked on retrospectively, to make comparison with an *ad hoc* group of historical controls. The randomised controlled design is being increasingly used for Phase II studies. The main reason for this appears to be to speed up the process by which drugs are introduced into the market — there are definite advantages to having a precise indication of the potential niche for a drug as early as possible in the development process.

Phase III Study

A randomised controlled trial: used to compare a new treatment with the previously accepted standard therapy.

Phase IV Study

A term used, mainly by the pharmaceutical industry, to describe post-marketing surveillance: efficacy is assumed, and the purpose of the study is to seek out uncommon but important side-effects.

Philadelphia Chromosome

This chromosomal abnormality, one of the first to be demonstrated in association with malignant disease, is associated with chronic myeloid leukaemia (CML). There is a reciprocal translocation between the distal part of chromosomes 22 and 9q, which results in the Abelson oncogene (*abl*) being moved from 9q to 22 and the *sis* oncogene moving from 22 to 9. There are a series of variant translocations in CML and all appear to involve the region at 22q11.21. The high frequency of breaks at this site has led to its designation as the 'breakpoint cluster region' (BCR). This region codes for a 160 kD protein with kinase activity and the C-terminal **domain** (**p. 102**) has GTPase activity that can activate *p21*. It is easy to envisage the potential for malignant transformation that will occur when an oncogene (*abl*) is translocated and placed immediately adjacent to a gene (*bcr*) with the ability to activate *p21*.

Phosphatase

A group of enzymes that removes phosphate groups from their substrates.

Acid phosphatases remove phosphates with a single negative charge. Alkaline phosphatases remove phosphates with a double negative charge.

Phosphorylases

These enzymes are involved in energy metabolism and are responsible for the regulation of glycogen breakdown: they add phosphate molecules to produce molecules such as glucose 1-phosphate. They are controlled by phosphorylase **phosphatases** (**p. 233**) that remove their phosphate groups, e.g. the conversion of phosphorylase a to phosphorylase b plus orthophosphate.

Photobleaching

An effect encountered in **photodynamic therapy** (PDT). Light, acting on the photosensitising chemical, causes it to be less efficient as a photosensitiser. This process is mediated by singlet oxygen, which causes chemical degradation of the sensitiser. The effect might improve the **therapeutic ratio** (**p. 307**) for PDT. If there is a differential concentration of sensitiser between the tumour and the normal tissues, then photobleaching could lead to a reduction of the effective sensitiser in the normal tissues to a level below that at which sensitisation occurs. If the levels of sensitiser within the tumour remained above the threshold for activity, then continuing the application of light would cause more damage to the tumour than to the normal tissues.

Photodynamic Therapy (PDT)

Photodynamic therapy is based on the principle that some tumours will specifically take up light-sensitising chemicals. If light of a wavelength that will specifically excite the photochemical is shone on the tumours and its immediate environment, then only the tumour cells will be exposed to the toxic effects of singlet oxygen produced by the photochemical reaction.

In practice, photodynamic therapy works as follows. Before exposing the tumour to light, the patient is treated with a photosensitising chemical. The drug is concentrated within the tumour and eliminated from the normal tissues. The tumour is treated using a coherent source of laser light of appropriate wavelength. The approach is only feasible for tumours that are endoscopically accessible and relatively superficial. As technology develops, more tumours will be accessible using fibreoptic sources and higher light energy will enable the light to penetrate more deeply into tumours. Currently the limit of penetration is between 5 and 10 mm.

The clinically useful photosensitisers can be summarised:

- Haematoporphyrin derivative (HpD):
  - — original photosensitiser
  - — complex mixture of a variety of porphyrins.
- Photofrin:
  - — purified version of HpD containing a mixture of various porphyrins together with their ether and ester derivatives
  - — several absorption peaks, maximal at 630 nm
  - — skin sensitisation lasts 6–12 weeks.
- 5-Aminolaevulinic acid (5-ALA):
  - — prodrug metabolically converted to the active agent, protoporphyrin IX (PpIX)
  - — can be used topically or systemically.

- Meta-tetrahydroxyphenychlorin (mTHPC, Foscan):
  - — stronger absorption peak than photofrin, peak is at 652 nm
  - — skin toxicity lasts 3–6 weeks.
- Benzoporphyrin derivative (BPD):
  - — strong peak at 690 nm, mono-acid form (BPD-MA) is more active
  - — skin photosensitization lasts <5 days.
- Mono-aspartyl chlorin e6 (Npe6):
  - — peak is at 664 nm
  - — short duration of action, skin photosensitization lasts <1 week.
- Lutetium texaphyrin (Lu-Tex):
  - — third-generation sensitiser, preferential accumulation within tumours
  - — transient skin toxicity, rapidly excreted
  - — peak is at 732 nm
  - — can be used for repeated treatments, a possible way round the problems of photobleaching and oxygen depletion.

Tumours that have been experimentally treated using PDT include: lung cancer, bladder cancer, ovarian cancer and oesophageal cancer.

Phylogeny

A term initially used to describe the development of a species but now extended to include the evolutionary history of a nucleic acid or protein.

Physician-Assisted Suicide

A procedure related to, but distinct from, **euthanasia (p. 119)** in which a physician furnishes a patient with the means to take their own life in the full knowledge that this is what they intend to do. The physician is an accomplice in the suicide but does not, directly, administer the lethal agent.

Pied Piper Effect

A term introduced by Kostoff to define a mechanism by which the scientific literature could be systematically misleading. The **peer review (p. 229)** process is innately conservative: it tends to reinforce the status quo rather than to encourage innovative or tangential thinking. If the current approach, for example our incremental progress towards successful treatment for cancer by combining surgery, drugs and radiation, is the correct one, then there is no problem. If, however, this approach is incorrect and the cancer is ultimately shown to be curable mainly through application of biofeedback or other psychosomatic intervention, then research will have been systematically heading off in the wrong direction: following a pied piper. The best protection against this type of error is to maintain a broad perspective, to be open- rather than narrow-minded and to assess evidence on its merits, not on its provenance. Pluralism is always going to be closer to the truth than dogmatism.

Planning Target Volume (PTV)

This term, used in ICRU Report 50, defines a volume used in radiotherapy treatment planning. PTV allows for the vagaries of patient positioning, beam alignment and the practicalities involved in setting up a patient for treatment on a day-to-day basis. It is a geometric concept that produces a volume larger than the clinical target volume, but smaller than the **treated volume (p. 315)** (figure 1).

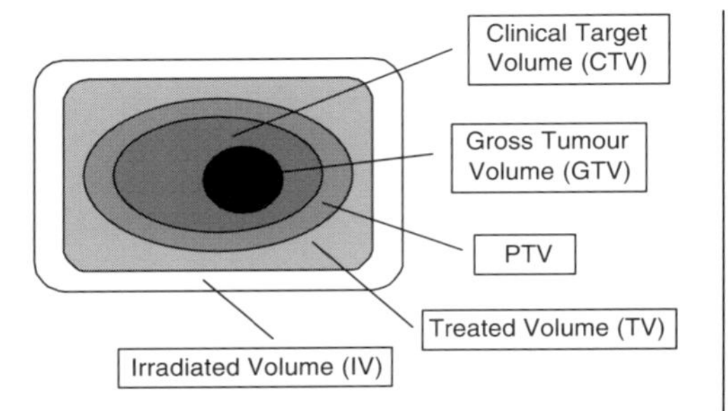

Figure 1.

Plating Efficiency

An essential step in calculating the surviving fraction of cells after exposure to cytotoxic agent:

$$\text{Plating efficiency} = \frac{\text{number of colonies formed}}{\text{number of cells plated out}},$$

where a colony, by definition, must contain at least 50 cells.

Point Prevalence

Indicates the number of people with a particular disease at a particular time point (e.g. on a certain date). Prevalence includes both the incident cases as well as survivors whose disease was diagnosed before the period of assessment.

Poisson Distribution

The Poisson distribution describes events the distribution of which is scattered at random: although the distribution may be random, the rate at which the events occur can be strictly defined. The essential features of events that obey a Poisson distribution are:

- Events occur at random.
- Events do not occur simultaneously.
- Events are independent (the occurrence of an event does not make the occurrence of a subsequent event either more, or less, likely).
- Events occur at a constant rate over time.

The approach is versatile: it was used in 1898 to study the pattern of incidence of soldiers in the Prussian army being kicked to death by horses. Radioactive decay is a process that obeys Poisson statistics: the rate at which an isotope decays is immutable and characteristic for that particular isotope, but whether an individual atom decays is a completely unpredictable process.

A similar argument applies in radiobiology: the distribution of ionising events within an irradiated biological target is a random process. Some targets will receive no hits, some will receive single hits, some will receive multiple hits. The distribution of damage is governed by Poisson statistics. The easiest way to visualise this is to consider an area of 100 square paving stones: 100 pebbles scattered at random will not distribute themselves evenly with one pebble landing in each square. Some squares will have no pebbles, others may have two or three. The distribution of pebbles is governed by the Poisson distribution formula:

$$Y = N\lambda^x/(2.7183)^\lambda x!,$$

where N is the number of squares (targets), λ is the average number of pebbles per square (hits per target); x is the number of pebbles per square (hits per target), $x!$ is factorial x, and Y is the number of times that there are x pebbles per square (hits per target). e is 2.7183, i.e. the base of the natural logarithms.

When the number of targets is set to 1, then Y is simply the **probability** (**p. 246**) of a specified number of hits per target. Figure 1 illustrates this for three different λ (average number of hits per target): the white bars are for an average of one hit per target (the probability of a target not being hit is 0.37);

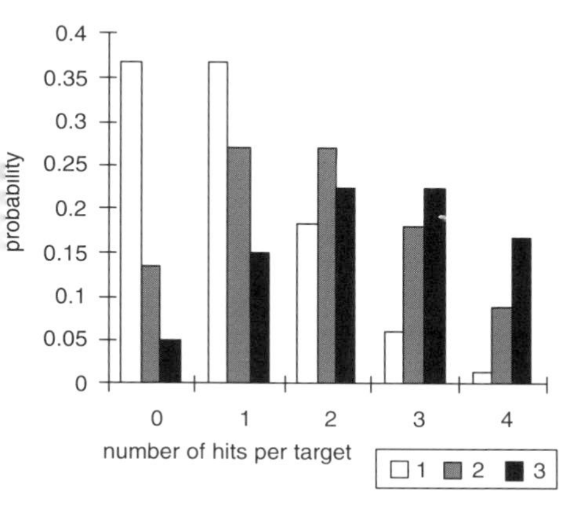

Figure 1.

he grey bars are for an average of two hits ›er target (the probability of a target not ›eing hit is 0.135). Even when there are hree hits per target (black bars), the ›robability that a target will not be hit is till 0.05.

Vhen the number of hits per target (pebbles ›er square) of interest is zero, then the ›oisson distribution simplifies to:

$$Y = N/e^{\lambda} = N \cdot e^{-\lambda}.$$

The more pebbles there are, the more hits ›er target there will be. Using Poisson tatistics in clinical radiobiology, we could imply equate dose with the number of ›ebbles. This can then be translated into a :onsideration of tumour control probability. f it takes only one surviving clonogen to :ause tumour recurrence, then all clonogens nust be inactivated, hit at least once, for a umour to be controlled. If the average number of clonogens that survive treatment s λ, then the probability of a tumour ›eing controlled is proportional to $e^{-\lambda}$. This dictates the classic sigmoid shape for he **dose–response curve** (**p. 103**) (figure 2).

The sigmoid dose–response relationship s governed by the expression $e^{-\lambda}$. Tumour control probability $= t_{cp}$, dose s expressed as BED (Gy).

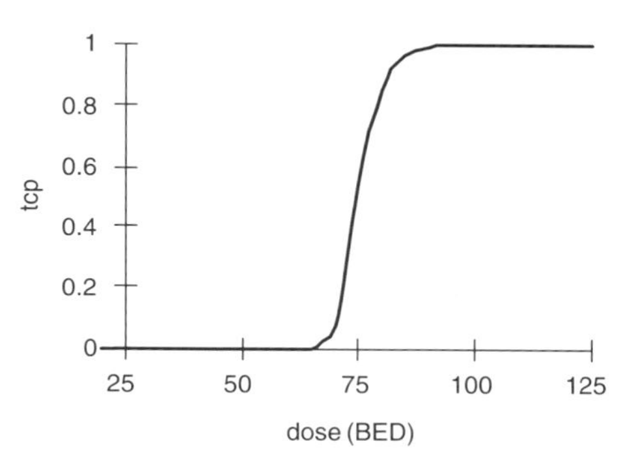

Figure 2.

Simeon Poisson (1781–1840) was a mathematician: he failed to fulfil his family's wish — that he became a surgeon. Since he was obviously fit for nothing better, they then decided that he should become a lawyer. Ultimately, he found his true vocation: mathematics. Another of his contributions was to expand **Bernoulli's Theorem** (**p. 27**), a cornerstone of probability theory: he rechristened it as the **Law of Large Numbers** (**p. 176**).

Poisson Heterogeneity Test

The test can assess whether the variation in counted colonies can be accounted for by random variation or whether, perhaps because of sloppy technique or some other systematic interference with the assay system, there is significant **heterogeneity** (**p. 150**) present. The null hypothesis is that the colony-forming cells are all drawn from a homogeneous population — a statistically significant result indicates heterogeneity.

An example is given here from everyday life: watching a toddler eat yoghurt from a pot. Are the hits (on the mouth) occurring purely at random or is there a method involved, i.e. does the toddler direct the spoon towards the mouth. If the Poisson heterogeneity test

were positive, then this would suggest that the child was in fact trying to convey the yoghurt to the mouth rather than splattering it randomly about. I am indebted to Max Hannam for this observation.

Polyglutamation

Polyglutamation is a process by which glutamate side-chains of a molecule polymerise, forming a large molecule from several smaller ones. Polyglutamation is an important feature of **methotrexate** (**p. 199**) metabolism: not only does the polyglutamated methotrexate remain in the cell for longer, but also the larger molecule is a more effective inhibitor of dihydrofolate reductase than methotrexate itself.

Polymerase Chain Reaction (PCR)

A technique for amplifying short stretches of DNA from small DNA samples. The crucial point is that, relatively rapidly and starting with only a small sample of the DNA in which you are interested, it is possible to increase the amount available by a factor of several million. The large amount of DNA produced means that we do not have to use the radiolabelled probes required to detect and identify small amounts of DNA. A reluctance to work with radioactive isotopes was in part responsible for motivating Kary Mullis to invent PCR: an innovation for which he received the Nobel Prize in 1993.

PCR does not require knowledge of the base sequence to be amplified. The principle of the technique is as follows. Complementary primers, one at each end, are used to define the single-stranded denatured DNA to be amplified. Heating denatures the DNA and the primers bind to the DNA when the solution is allowed to cool. The DNA, with its attached primers, is incubated with DNA polymerase, which synthesises DNA complementary to the single-stranded sequence defined by the primers. Heating the solution again denatures the DNA, but this time there is more of it, and, once again, the annealed primers define sequences upon which the DNA polymerase can act. With each cycle of heating and cooling, the DNA will double in amount, there is, therefore, over a number of cycles an exponential increase in the amount of DNA present. Since each cycle only takes ~5 min, DNA can be rapidly amplified. A series of replicatory cycles will produce an exponential expansion of the amount of DNA present, just as only 30 generations of clonogenic cells can produce a vast number of progeny so can 30 cycles of PCR produce enormous amplification of a specific DNA sequence.

The ability of PCR to amplify sequences of DNA is being exploited in an attempt to detect occult colorectal tumours. Mutant DNA, with patterns typical of colorectal carcinoma, is sought in exfoliated gastrointestinal lining cells obtained from samples of faeces — life at the cutting edge is sometimes a little messy.

PCR on peritoneal washings obtained during surgery for colorectal cancer may be useful for staging and prognosis. If free-floating cells are present, and these cells contain sequences known to be associated with malignancy, then this may be associated with adverse outcome. This is a more aesthetically acceptable use of PCR than using it to analyse faecal samples.

The invention, or discovery, of PCR is a fascinating example of how progress in science is made, and how it is handled. The idea of exponential expansion through repeated cycles of heating and cooling came from Mullis's background in computer programming. The concept is similar to that of a recursive loop. The original idea was sketched out during a long weekend. Putting the concept into practise took considerably longer. Disputes arose about scientific priority (it is one thing to have a bright idea, it is quite another to make it work). Disputes also arose about commercialisation, patents and intellectual property. Something that was thought up in a few days kept lawyers busy for years. The invitation for the Nobel Prize ceremony specified 'white tie and tails': Mullis had a suit made to obey the letter of this injunction — it was a black suit with white tails.

Population at Risk

The population susceptible to a particular problem. It is defined on the basis of demographic data, such as place of residence, sex, age group, etc.

Population Doubling Time

The time taken for the clonogenic cells in a tumour to double in number. The volume **doubling time** (**p. 107**) is usually less than the population doubling time because not all of a tumour's volume is accounted for by clonogenic cells.

POPUMET Regulations

A term used to describe IRR'88 — the Ionising Radiation Protection of Persons Undergoing Medical Examination or Treatment Regulations 1988 (UK). The regulations have their own vocabulary: radiographers are the physical directors of **exposure** (**p. 124**) while radiologists and radiotherapists are the clinical directors of exposure.

Positional Cloning

A form of **reverse genetics** (**p. 272**) in which a gene is isolated starting simply from knowledge of its position on the chromosome and in the absence of any knowledge about the product produced by that gene. Methods such as **chromosome walking** (**p. 64**) and chromosome jumping are used to generate overlapping DNA clones. Conserved sequences are then sought out and looked at more closely, in particular those conserved sequences whose mRNA is found in cells that exhibit the abnormality associated with the genetic disease. Probes based on the DNA sequence complementary to that of the mRNAs are used to look for mutations in the chromosomal DNA. Deletions are relatively easily identified using **Southern blotting** (**p. 292**) — other mutations are harder to pin down. The procedure is currently time-consuming; it takes between 10 and 100 person-years to isolate a gene. Technological developments in sequencing and the information from the Human Genome Project will speed up the process. The approach is expensive: it cost between US$5 and $50 million to find the gene for cystic fibrosis. The imprecision of the estimate is an interesting testament to the way that the promise of molecular biology has attracted funds without much in the way of **cost–benefit analysis** (**p. 76**). The gene was identified in 1989 but there is still no 'cure' and genetic counselling is still imprecise.

Positron Emission Tomography (PET Scanning)

This imaging technique is based on the same tomographic principles used in CT scanning, but with one important additional feature. Positrons interact with electrons to produce an annihilation reaction in which the mutual destruction that occurs produces two γ-rays. These depart the scene of the interaction at 180° to each other. A circular detection device based around the patient can detect simultaneous scintillations at 180°, i.e. directly opposite each other, coincidence

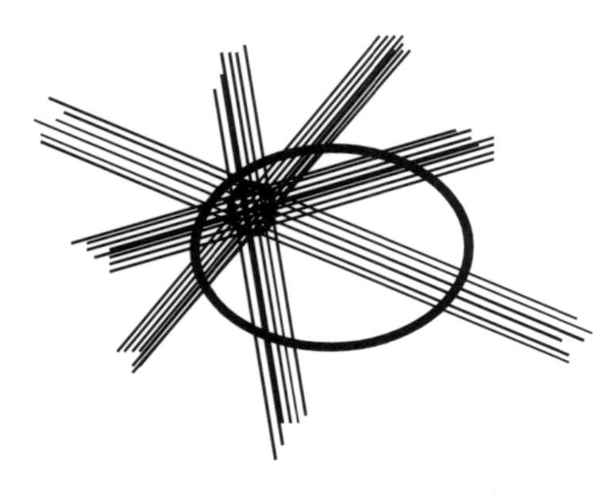

Figure 1.

scanning. The software in the scanner can use this information, based on multiple coincident emissions, to localise the source of the isotopic emissions in three-dimensional space (figure 1). The techniques can be used with a variety of isotopes:

- ^{11}Carbon: $t_{1/2}=20$ min.
- ^{15}Oxygen: $t_{1/2}=2$ min.
- ^{13}Nitrogen: $t_{1/2}=9.96$ min.
- ^{82}Rubidium: $t_{1/2}=1.3$ min.
- ^{18}Fluorine: $t_{1/2}=110$ min.

These isotopes are usually produced in a cyclotron and, given their short half-lives, PET scanning is only feasible if the scanner is within 2 h travelling time of a cyclotron. The isotopes of oxygen can be used to image the blood pool and blood flow; rubidium and nitrogen can demonstrate blood flow and the fluorine, as fluorodeoxyglucose (FDG), can investigate tumour metabolism. The crucial difference between PET and CT scanning is that PET scanning images function, it can even image thought, whereas CT scanning can only image structure.

Potential uses of PET in medicine include:

- Blood flow.
- Blood volume.
- Oxygen utilisation.
- Glucose utilisation.
- Amino acid uptake.
- Polyamine uptake.
- Permeability of blood–brain barrier.
- Regional pH+.
- Drug uptake studies.
- Receptor binding studies.

Recently ^{18}F-labelled **5-fluorouracil (5FU)** (**p. 130**) has been used to demonstrate 5FU metabolism in the liver, first-pass, and tumours.

The disadvantages of PET scanning are that spatial resolution is poor, the equipment is expensive and scanning procedures can be cumbersome. There is also the rather restrictive requirement of proximity to a cyclotron or other source of short-lived isotopes.

Post-Translational Modification

The polypeptide chain that emerges as a result of the DNA to RNA to ribosome sequence is not necessarily the finished, fully functional, product. Other processes, collectively referred to as post-translational modification, may be necessary for the formation of a functional molecule.
These processes include: modification of the N-terminus by removal of methionine; acylation of the N-terminus; addition of **ubiquitin** (**p. 319**); phosphorylation and/or dephosphorylation; addition of ribosyl groups; and cleavage of an inactive precursor, such as pro-insulin, into its active component. The glycosylation of haemoglobin that occurs in a glucose-rich environment is not, in contrast to the processes described above, dependent on enzymatic activity. It is an

enzyme-independent form of post-translational modification that can be used as a marker for diabetic control.

Potential Doubling Time (T_{pot})

A calculation, based on knowledge of **growth fraction** (**p. 144**) and **cell cycle** (**p. 52**) time, of the time it should take a population of tumour cells to double in number — provided there were no cell loss:

$$T_{pot} = \lambda\,(T_s/\mathrm{LI}),$$

where λ is a constant in the range 0.5–1.0 used to define the age distribution of the cell population.

The discrepancy between T_{pot} and the observed **volume doubling time** (T_d) (**p. 325**) can, in part, be explained by cell loss. T_{pot} is a measure of the time it would take a population of tumour cells to double in number were all the cells in the tumour able to proliferate at the same rate as that subpopulation of cells which constitutes the growth fraction. Unmasking of the latent T_{pot} of tumour clonogens has been invoked as a mechanism for explaining the **accelerated repopulation** (**p. 2**) of tumour cells observed during radiotherapy treatment. However, other mechanisms, such as shortening of cell cycle time, may also be involved.

Potentially Lethal Damage (PLD) Repair

If cells are not called on to divide immediately after a fraction of radiation treatment, then some of the radiation damage can be repaired. The extent of repair will, therefore, be influenced by the environmental conditions after irradiation. Single fraction experiments, with appropriate manipulation of the post-irradiation milieu, are used to measure PLD repair. The amount of PLD repair in a given experimental system is independent of dose.

Power

The power of a scientific study is a measure of the probability that, at a given level of statistical significance, the investigator will correctly conclude that there is a genuine difference between the two arms of the study.

Imagine that 100 clinical trials are performed. The sample size each time is the same, and the following results are obtained. In an Olympian fashion, it is known with absolute certainty that there is a real difference between the two arms of the studies:

- Trial concludes 'positive': 80 trials: rate $80/100 = 0.8 =$ power.
- Trial concludes 'negative': 20 trials: rate $20/100 = 0.2 = \beta$.
- Power $= (1 - \beta)$.

β is the probability of drawing a false-negative conclusion, which, in the jargon, is termed a Type II statistical error.

Imagine a similar study, but this time the absolute knowledge is that there truly is no difference between the two options under investigation:

- Trial concludes 'positive': five trials: rate $= 5/100 = 0.05 = a$.
- Trial concludes 'negative': 95 trials: rate $= 95/100 = 0.95$.

α indicates the probability that a 'positive' trial result is a false-positive: it is interpreted as p for statistical significance; it corresponds to the probability, again in the jargon, of making a Type I statistical error.

This whole formalism is reminiscent of that used in assessing a diagnostic test, i.e.

- β is equivalent to the false-negative rate (FNR).
- α is equivalent to the false-positive rate (FPR).
- Power = $(1 - \beta)$ is equivalent to sensitivity.

When calculating the required sample size for a planned clinical trial, both power and significance level (α) are among the parameters that need to be specified. As the power demanded increases, then, for any given level of statistical significance, the sample size will increase. The relationship between increased sample size and power is part of the rationale for pooling the results of clinical trials in a **meta-analysis** (**p. 195**). The width of the **confidence interval** (**p. 70**) is a rough visual guide to the power of an analysis — the narrower the interval, the more powerful the study. This is scarcely surprising since total sample size (N) appears as a denominator in the equations for calculating confidence intervals (CI): any increase in N will make the interval smaller.

PPAR (Peroxisome Proliferator-Activated Receptors)

These are nuclear hormone receptors discovered in the early 1990s with a wide variety of effects: lowering blood glucose through increased differentiation of adipocytes and improved peripheral utilisation of glucose; and differentiation of cancer cells. A synthetic ligand, **troglitazone** (**p. 315**), has been developed. Unfortunately, it causes liver failure and has, therefore, been taken off the market as a treatment for diabetes. Data from animal models on the use of troglitazone as a potential differentiating agent for cancer are conflicting: it may shrink some types of colon cancer but it stimulates others. PPARα is a potential mediator of carcinogenesis in rodents, but it probably has little role in human carcinogenesis. PPARγ ligands may induce differentiation in tumours of fat cells, the breast and in mouse models of prostate cancer. PPARβ may be an important component of APC-mediated carcinogenesis in the colon. It is antagonised by some non-steroidal anti-inflammatory drugs (NSAID) such as sulindac. These drugs can suppress tumour formation in the colon and it is possible that this effect is mediated by inhibition of PPARβ.

PPV (Positive Predictive Value)

The PPV of a **diagnostic** (**p. 95**) test simply indicates the probability that an individual, whose test result is abnormal, has the condition under investigation. It depends not only on the performance of the diagnostic test, in terms of sensitivity and specificity, but also on the prevalence (or pretest likelihood) of the condition being present. A test that is 85% sensitive and 92% specific will have, when prevalence is 100 per 1000, a PPV = 58%; when prevalence is 100 per 100 000, PPV = 1.2%.

Pragmatic Trial

A term used to describe a clinical trial that is less concerned with the details of how a treatment might work and how, in a detailed sense, it might compare with existing therapy, but is more concerned with the clinical usefulness of the new treatment. When using simple outcome measures, such as survival or relapse rate, how does it compare to standard therapy? Pragmatic trials usually have non-stringent entry criteria — the aim is to discover the usefulness, or lack of it, in a relatively unselected group of appropriate patients. The aim of a pragmatic trial is to assess the usefulness of an intervention in the real world, as opposed to the narrower, more restricted world of the explanatory clinical trial (table 1).

Explanatory	Pragmatic
Restrictive entry criteria to maximise **homogeneity (p. 152)** study population	broad-entry criteria to embrace all-comers: since these form the population to which the trial results will be applied
Careful data collection and analysis, study intermediate outcomes as well as simply assessing survival, pristine approach to measurement	simple data collection, e.g. survival or relapse and not much else; non-obsessional approach to data gathering
Ensure that subjects comply with allocated treatment	relaxed attitude to compliance — what will happen in the real world when one tries to use this intervention?
Small sample size	large sample size

Precision

A term is used ambiguously, indeed imprecisely, in science. Statisticians will often use it to indicate agreement, e.g. estimates of 5.1 and 5.0 would, in this specialised statistical sense, be 'precise', whereas estimates of 4.2 and 6.8 would not be so. The other use of 'precision' has to do with the amount of detail provided by an observation: 3.1 is a less precise estimate than 3.14159. By this criterion of precision, 5.099998752 and 5.100087253 would be equally precise estimates of 5.100000000. They, because of their close agreement, would also be precise in the statistical sense of the term. However, both figures, as estimates of π, would be completely wrong. Precision does not imply correctness. If systematic bias is present, estimates can be both precise and inaccurate.

Predictive Assays

It would be very useful in clinical radiotherapy to choose horses for courses: to know, before treatment, which tumours would be likely to respond to radiotherapy. By only treating those patients with sensitive tumours, the upset and distress caused by ineffective treatment might be avoided. Tumours will also vary in their suitability for different fractionation schemes: overall treatment time would be a relatively unimportant factor for tumours that lack the ability rapidly to repopulate. Conversely, protracted schedules would be totally inappropriate for rapidly repopulating tumours. Various predictive assays have been attempted but the results, so far at least, have been disappointing. The predictive value of the assays has been rather low and the assays themselves often take too long to deliver timely results for making decisions for individual patients.

Predictive assays can be applied to tumours or normal tissues. When applied to normal tissues they may suggest dose modifications that avoid excess damage to normal tissues in sensitive individuals. Predictive assays can be divided into those, such as both the **Courtenay–Mills and Brock assays (p. 37)**, that involve growing clonogenic cells and those, such as the Comet assay or using **Ki67 (p. 171)**, to measure proliferative index, which rely on non-clonogenic methods. The latter have more clinical potential than the former since they tend to be more rapid and, therefore, do not involve delaying treatment until assay results become available.

The future of predictive testing is bound to be shaped by recent advances in molecular biology. The development of DNA micro-arrays, and the consequent ability to perform gene expression profiling rapidly, will almost certainly have a major impact on predicting whether certain drugs should be used in certain patients with certain tumours. The dream of totally individualised therapy is in danger of becoming a reality — once the information-processing capability to keep up with the knowledge being produced is developed.

Predictive Factor

A predictive factor is one that can be used to indicate whether a disease will respond to treatment. Examples would include the expression of thymidylate synthase as an indictor that a tumour would probably respond to treatment with 5-fluorouracil (5FU); the expression of the *mdr* gene would suggest that a tumour would be unlikely to respond to **anthracycline** (**p. 16**). Predictive factors are distinguished from prognostic factors. A **prognostic factor** (**p. 248**) is one that, unrelated to any treatment given, affects outcome.

A predictive factor is a specialised form of prognostic factor — one that indicates the outcome of a therapeutic intervention. A prognostic factor is prognostic whether or not treatment is given; a predictive factor simply gives some indication of whether a treatment is going to be effective.

Pre-Randomization Consent

A procedure devised by **Zelen** (**p. 331**) to circumvent the problems arising from obtaining informed consent to randomised allocation of experimental treatment and thereby increasing recruitment to clinical trials.

Patients eligible for a trial are randomised to either the experimental treatment or conventional management. Those who draw standard therapy are treated conventionally and are never informed of their participation in the study. Patients allocated to the experimental treatment are approached and asked, using the appropriate methods for obtaining informed consent, if they wish to participate in the study. The design has pragmatic attractions but, ethically, appears slightly dubious. Control patients have been unwittingly entered into a trial and are never told that they might have been allocated an experimental therapy.

Prevalence

The prevalence of cancer is the number of cancer cases in a given population at a specified point in time. It depends on the incidence and duration of the disease, i.e. on survival.

Prevention

The prevention of cancer is an extremely important component of **cancer management** (**p. 40**) and control. In both human and economic terms it makes excellent sense to try to prevent cancer ever developing. It has been calculated that up to 80% of cancer in humans may be related to environmental factors and so, in theory at least, the opportunities for prevention are enormous. Translating theory into practice has proved extremely difficult. People continue to smoke, even though smoking is a major cause of cancer and accounts for up to one-third of all cancer deaths. This corresponds to ~50 000 deaths in the UK, and 250 000 deaths in the EU, each year as a result of smoking-induced cancers. There is little incentive in Western economies for governments to discourage smoking. By paying tax on the cigarettes they smoke, and having the good grace to die prematurely, smokers are net contributors to the wealth of the nation: the popular belief that smokers

re a drain on the healthcare budget is a
nisconception. There are few economic
rguments to make against smoking, only
umanitarian ones.

n 1990, the most recent year for which
vorld-wide data are available, 4.25 million
eople developed cancer and 2.95 million
eople died from cancer. By 2020 there will
e, world-wide, 20 million new cases of
ancer in the world each year. Of these,
4 million will occur in the developing
vorld. Currently, ~25% of cancers in the
eveloping world are related to infection,
ome of which are potentially preventable.
Iepatitis B and C are associated with
epatoma, and there is an effective vaccine
vailable for preventing hepatitis B. The
uman papilloma virus (HPV) (p. 155)
related to the development of cancer of
e cervix. Safer sexual practices will decrease
e rate of transmission of HPV and should
elp to reduce both cancer of the cervix and
IIV-related tumours (p. 151). In some
arts of Sub-Saharan Africa, >50% of patients
vith cancer of the cervix are HIV-positive.
'he proportion of cancers related to **tobacco**
se (p. 309) can be expected to rise in the
eveloping world, particularly in China.
'his is in some measure due to aggressive
arketing by tobacco companies. As they
nd it more difficult to maintain their
arket in the developed world, the
ompanies look to the developing world
s a source of future profit.

trategies for preventing cancer can be
ivided into primary prevention, preventing
ndividuals from ever developing cancer, and
econdary prevention (p. 281), preventing
ubsequent tumours in patients who have
lready been treated for one cancer.
dvising people to stop smoking and
iving **tamoxifen (p. 304)** to women at
igh risk of developing primary breast cancer
vould be examples of primary prevention.
'he use of retinoids to prevent second
tumours in patients treated for cancers of the
head and neck is an example of secondary
prevention.

Primary Prevention

The prevention of cancer is classified as primary, where the intention is to prevent individuals from ever developing cancer, and secondary, where the aim is to prevent second primary tumours in patients who have already been successfully treated for cancer. Primary prevention can be classified as:

- Strategies aimed at increasing awareness of individual risk, to both patients and their families, of developing cancer: genetic counselling (prevention is but one goal of this complex discipline).
- Strategies aimed at preventing the exposure of susceptible normal cells to carcinogens:
  - — health education: advice about diet (fibre, vitamins, avoiding excessive animal fat); advice about tobacco use
  - — occupational exposure: avoidance of excessive risk in the workplace (e.g. prevention of exposure to asbestos fibres, aniline dyes, excessive radiation, etc.).
- Strategies involving the removal of normal tissues known to be at risk of malignant transformation:
  - — colectomy in patients with familial **polyposis coli (p. 125)**
  - — prophylactic mastectomy in patients at high risk of breast cancer.
- Strategies aimed at preventing malignant transformation: **tamoxifen (p. 304)** as a chemopreventative agent in individuals at high risk of breast cancer.

Primer

A short synthetic sequence of nucleotides used in the **polymerase chain reaction (PCR) (p. 238)** to mark out the DNA sequence to be amplified.

Prions

These agents are implicated in the transmission of spongiform encephalopathies, including the bovine form (BSE) and new variant Creutzfeldt–Jacob disease (nvCJD). They are related to a normal sialoglycoprotein found in cell membranes: indeed the amino acid sequences of the normal protein (PrP^{c}) and the prion (PrP^{Sc}) are identical. The pathogenicity of prions is consequently believed to be due to their **secondary structure (p. 281)** — prion protein is predominantly a β-pleated sheet whereas the normal protein exists mainly as α-helices. A simplified view of the pathogenesis of the spongiform encephalopathies would regard the abnormal prion protein as acting like the seed of a crystal and converting the host PrP^{c} into insoluble aggregates, which then produce the classical changes in histopathology and function. In one sense, prions could be regarded as **chaperones (p. 62)** gone wrong.

Recent evidence suggests that the varied incubation periods of prion-related diseases are due to variations in the sensitivity to protease of the various prion 'strains'. There are eight such strains so far identified, and the biological properties of each are related to the conformation of the protein. So far, 27 people in the UK have died from nvCJD.

Probability Density

Simply, it is the height of the normal distribution curve. Probability density has no intrinsic value or meaning — it is the area under the curve (AUC) that is important. AUC is always 100%: 95% of values will lie within the range:

$$(\text{mean} - 1.96 \times \text{SD}) \text{ to } (\text{mean} + 1.96 \times \text{SD})$$

where SD is the **standard deviation (p. 296)**.

Procrustean Method

The Procrustean method is a necessary evil in scientific measurement whereby, for purposes of counting, computation and analysis, that which is complex is oversimplified. **Quality of life (p. 253)** is reduced to a single number: the results of examining a trainee's knowledge are reduced to P+, P, P− or F. Other examples are less obviously crude, but nevertheless represent the underlying pragmatic approach that lies at the root of the Procrustean method. Patients with lung cancer who reply 'yes' to the question 'Are you short of breath on climbing stairs?' may have varying degrees of respiratory impairment — from a mild subjective feeling of breathlessness to severe physical difficulty requiring frequent stops. Another unasked and, therefore, unanswered question in this example is 'How many stairs? Five stairs between the basement kitchen and the living room, or three flights to the top the house?' By lumping all these various levels of breathlessness within the answer to a single question we might gain, in terms of simplicity and convenience, but we also lose: ambiguity increases while precision is lost. Procrustes was a giant robber who amputated or stretched his victims so that they would fit the size of his bed exactly.

Prodrug

A prodrug is a drug that is, in its unmodified form, inactive. The drug has to be activated to have its pharmacological activity. Usually this involves enzymatic conversion to an active metabolite. Examples of relevance to oncology include:

- Thotepa: release of an aziridine moiety produces ethyleneimines that act as monofunctional alkylating agents.
- Tetraplatin (Platinum IV): reduction via a glutathione-dependent pathway produces Platinum II, the active drug.
- Azathiaprine: activated to the antimetabolite 6-mercaptopurine.
- **Cyclophosphamide** (**p. 83**): multistep activation, both in the liver and in the periphery, produes the toxic compounds phosphoramide mustard and acreolin.
- **5-Fluorouracil** (**5FU**) (**p. 130**): activated to FdUMP, a potent inhibitor of thymidylate synthetase.
- **Fludarabine** monophosphate (**p. 129**): initially dephosphorylated, then transported into the cell and rephosphorylated to its active, triphosphate, form.

The existence of prodrugs has been exploited in the **ADEPT** (**p. 7**) and VDEPT concepts. A vector, e.g. an antibody to a tumour-associated antigen, is used to carry the enzyme which activates the prodrug to a specific target. The unbound enzyme is excreted leaving only active enzyme at the target. The prodrug is then given. It will only be activated at sites where enzyme is present — in this way, the action of the drug can be specifically localised to the tumour.

Product Limit Estimate

An alternative term for the **Kaplan–Meier** (**p. 170**) method of survival analysis.

Profile-Based Methods (syn. Psychometric Methods)

These terms describe an approach to assessing **quality of life** (**p. 253**) based on looking at a series of items, for each patient, that might contribute to their **quality of life** (**p. 253**). By assessing a variety of items for a wide variety of patients, it is possible, using techniques such as **factor analysis** (**p. 125**), to identify a core set of items important in influencing the quality of life. **Factor analysis** (**p. 125**) usually identifies items that accurately reflect quality within a series of distinct domains such as physical function, cognitive function, emotional function, social function, etc.

These items can be selected for use in an instrument used to assess quality of life. Such instruments include: EORTC QLQ C-30; **Nottingham Health Profile** (**p. 215**); **SF-36** (**p. 285**); Functional Living Index-Cancer (FLIC); and Sickness Impact Profile (SIP). The problem with profile methods is that when the results are collapsed into a global score, much of the meaning is jettisoned. For example, one treatment may have a favourable impact on physical function but be associated with impaired psychological function: surveillance after orchiectomy for seminoma is without physical side-effects (unlike radiotherapy) but may be associated with increased anxiety. What is gained on the swings may be lost on the roundabouts, and this detail is lost if only collapsed, global, scores are evaluated. An alternative approach to assessing general health and quality of life is based on decision theory. These are the so-called utility-based measures.

Progestagens/ Anti-progestagens

The main progestagens used clinically are **medroxyprogesterone** (**p. 193**) acetate and **megestrol** (**p. 193**) acetate. Both can be used for treating metastatic breast cancer. The main side-effects are weight gain, through a combination of fluid retention and increased appetite mood swings, and, in women with a uterus, spotty vaginal bleeding. The side-effect of increased appetite can be used therapeutically in the supportive care of patients with AIDS or cancer-related cachexia. **Medroxyprogesterone (p. 193)** acetate has been widely used in the treatment of metastatic renal cancer. The reasons for this are obscure as there is little evidence that it does any good.

Breast cancers can express not only just oestrogen receptors, but also progesterone receptors. This has led to the development of anti-progestagens such as onapristone and mifepristone. Clinical trials are under way to establish whether these agents have any potential in the treatment of breast cancer.

Prognostic Factor

A prognostic factor may be defined as a factor the presence or absence of which affects the outcome of an illness. The factor may be a feature of the illness itself (axillary node involvement, differentiation, tumour size), of the patient (menopausal status, haemoglobin level, **body mass index, p. 33**) or of the patient's environment (social class, income, education). Interactions can be complex: are fat patients poorer than thin patients and is it easier to detect a small breast lump in a thin patient? Treatment itself can be regarded as a prognostic factor: sometimes beneficial, sometimes adverse. **Predictive factors (p. 244)** are often distinguished from prognostic factors: a predictive factor indicates whether a disease will respond to treatment.

Progression-Free Survival

A term used in survival analysis. The event of interest is a recurrence or increase in size of the local tumour or the development of new lesions — it does not imply that complete local control has ever been achieved by treatment. It is useful in the assessment of therapies, such as hormone manipulation for advanced disease, where complete remission is unlikely and, were it ever to occur, it would be almost impossible to prove.

Progressive Disease

A term used in the evaluation of cancer treatment. Progressive disease is characterised by:

- An increase in the size of assessable lesions. Assessment can be clinical or radiological.
- Development of new lesions.

The presence of progressive disease does not imply that a treatment is totally ineffective. A treatment that slows the rate of disease progression may be therapeutically useful: hormone manipulation in the treatment of metastatic cancer is an example of a clinically useful treatment in which the best available response might well be progressive disease. For a slow-growing tumour, anything that slows the rate at which progression occurs might be useful.

Prometaphase

The second phase of **mitosis (p. 201)**. It lasts ~5 min. The nuclear envelope breaks down, **kinetochores (p. 171)** form and the chromosomes start jiggling about.

Proof of Principle

If a treatment ought to work because it is thought to have an effect, X, then proof of principle demands that we demonstrate that, indeed, the treatment truly does have effect X. Demonstration that X happens is, of course, no proof that the treatment will be of benefit. Satisfying the concept of proof of principle is an essential first step in the evaluation of a new approach — it is, y itself, not a sufficient justification or adopting a new treatment. The emonstration that a given vector–construct ystem was capable of restoring expression of normal ***p53*** (**p. 224**) to a bronchogenic arcinoma would provide important proof of principle: it would not, however, establish uch an approach as therapeutically effective.

Prophase

The first phase of **mitosis** (**p. 201**). It lasts ~17 min. The **chromosomes** (**p. 64**) ondense and the centrosomes separate.

Cox's Proportional Hazards Model

The technique is used in the analysis of urvival data and can assess potential **prognostic factors** (**p. 248**). It is possible o define for a group of subjects an average hazard: that is the average instantaneous event ate for the whole group. It is then possible to calculate, for specific subgroups of patients, the average hazard associated with membership of hat subgroup and to compare it with the average for the whole group. Provided that the hazards in the various groups are in constant proportion over time, and this can be checked using **complementary log transformation** (**p. 69**), there is no requirement that the **hazard rates** (**p. 148**) be absolutely constant over time.

In its simplest form the Cox model looks at one variable (e.g. treatment) to see whether it has a significant effect on the average hazard rate. The result is usually expressed as the hazard ratio (HR):

$$\mathrm{HR} = \frac{\text{hazard rate with treatment}}{\text{hazard rate without treatment}}.$$

If HR for death is <1.0, then it suggests that treatment is beneficial; HR >1.0 indicates that treatment is harmful. The Cox model can be extended to multiple variables and can be adapted to accommodate both categorical and binary variables as well as continuous variables. The influence of multiple variables on outcome can be considered simultaneously and, used in this way, the Cox model is an example of multivariate analysis.

The temptation is to use the Cox model to look at many of potential **prognostic factors** (**p. 248**) simultaneously. The problem is that the data can only support so much interpretation. There are two approximations that can be used to define the appropriate number of variables for a Cox model, and these methods do not necessarily agree. The first method is simply to use the fourth root of the number of events as the maximum number of permissible variables: if there were 1000 events, then 10 variables could be evaluated; if there were only 75 events, then no more than three variables should be investigated. The other approach is to demand that each variable be associated with at least 15 events before it could be considered for inclusion in a Cox model. Thus, if there were only five deaths in patients >60 years of age, it would be inappropriate to use age >60 as an independent variable in a Cox model.

The main use of the Cox model is in assessing factors, including allocated treatment, that can influence outcome. The model generates a term (usually notated '*b*') for each variable that indicates the amount of effect that variable has on outcome: variables with high values of *b* are highly influential. The 95% **confidence interval** (**CI**) (**p. 70**) around HR will give an estimate of whether the influence of a putative prognostic factor is statistically significant.

Proteasome

A proteasome is a complex made up from several subunits and named according to its sedimentation characteristics. Proteasomes are involved in the degradation of protein complexes and are also important in antigen processing in dendritic cells. The 20S proteasome is involved in the ATP **ubiquitin (p. 319)** pathway; the 26S proteasome is involved in the degradation of the ***MYC* oncogene (p. 211)** product. Inhibition of the activity of the 26S proteasome may be a route through which the *RAS* oncogene product acts.

Protein Folding Problem

The amino acid sequence of a protein dictates its higher structure. The ultimate form a protein adopts is crucial to its function: if we could deduce the form then the function might be deduced. Although the shape of a protein is implicit within in its amino acid sequence, not enough is known yet about how and why proteins adopt the conformations they do. This places a major limitation on approaches that involve trying to deduce the function of a gene product from the nucleotide sequence of that particular gene. RNA splicing, and other forms of **post-transcriptional modification (p. 240)**, place similar constraints on trying to deduce protein function from knowledge of a particular DNA nucleotide sequence.

Proteomics

Proteinomics is the study of the levels of proteins within a cell. Its further expression is **functional proteinomics**, which involves not only investigating whether a protein is produced, but also studying whether the proteins are functionally active. An immediate problem is that the amounts of detectable mRNA do not always correlate with the extent of protein expression. Given that there are only four paired nucleotides in DNA and that proteins are made from 20 amino acids, proteomics is inevitably going to be more complex than **genomics (p. 140)**.

The main direction of proteomics research a present is to study the expression of proteins by normal cells and to contrast the pattern with that found in malignant cells. By looking for specific differences, the aim is to define markers for diagnosis and prognosis and targets for specific treatment.

Proton

A proton is a positively charged subatomic particle. A single proton forms the nucleus o hydrogen. In other isotopes and elements both neutrons and protons are present in the nucleus. A proton has a mass of 1.6734×10^{-27} kg (1.00727 amu) and a charge of $+1.6 \times 10^{-19}$ Coulomb.
The number of protons in the nucleus, or th number of electrons orbiting in an electricall neutral atom, defines the atomic number (Z). The number of nucleons (protons and neutrons) in the nucleus defines the mass number (A) of the atom. The number of neutrons in the nucleus is given by:

$A - Z = N$ (neutron number).

Psychometric Testing

The use of interviews, questionnaires and scales to find out about an individual's psychological or emotional state. Areas of interest to oncology include anxiety, depression, emotions, attitudes (optimism, pessimism, fighting spirit), cognitive function, etc. The theoretical basis of psychometric testing is of fundamental relevance to the design of instruments intended to measure **quality of life (p. 253)**. This includes concepts such as **validity (p. 322)**, reliability responsiveness as well as techniques such as **factor analysis (p. 125)**.

Psychosocial Oncology

That branch of oncology concerned with the interactions between cancer, psychology and well-being (physical, social and emotional), can be regarded as psychosocial oncology. The traffic is two-way: the study of how pre-existing social factors and psychological traits might influence the incidence and prognosis of cancer; the effects of cancer on health and well being.

The main areas of investigation include:

- Quality of life research.
- Research on psychological interventions intended to improve the outcome of treatment.
- Research on the association between personality type and outcome of cancer treatment.
- Relationship between social circumstances and the incidence and prognosis of cancer.
- Both the effects of stress on cancer and of cancer on stress.

Up until now, the emphasis has been mainly on the psycho and less on the social. This is despite the fact that there are major gradients in cancer survival according to social class or ethnic background. The emphasis on psychology reflects the background of most psychosocial oncologists; it also reflects an general unwillingness to ask awkward questions about why some patients with cancer have had a raw deal. Why should the majority of studies in psychosocial oncology concern breast cancer? Recently, there has been a welcome increase in research into cancer in ethnic and economic minorities.

PTCH Gene

This gene is the name for *Drosophila*'s 'patched' gene. It is a transmembrane protein that inhibits expression of genes which encode proteins related to the TGF-β family. In humans, it is on chromosome 9q22.3 and it may function as a tumour suppressor gene.

Loss of wild-type *PTCH* can be associated with the Gorlin (naevoid basal cell carcinoma, NBCC) syndrome: multiple basal cell naevi; CNS tumours, etc.

PTEN (Phosphatase and Tensin Homologue on Chromosome 10)

This tumour suppressor gene is on chromosome 10q23.3. Mutations of the gene have been found in a variety of different cell lines derived from human tumours: prostate, glioblastoma, breast cancer. Abnormalities of *PTEN* are frequently identified in the endometrioid subtype of endometrial carcinoma and may serve as useful markers of premalignant change.

The gene product may act by inhibiting tyrosine **kinase** (**p. 171**) and preventing progression through G_1-phase. Abnormalities have been identified in several families with the **Cowden syndrome** (**p. 79**) (multiple hamartomas, thryoid cancer, breast cancer) and also in **Bannayan–Zonana syndrome** (**p. 26**) (large head, lipomas, vascular abnormalities, freckled penis, Hashimoto's thyroiditis).

Pulsed-Field Gel Electrophoresis (PFGE)

A technique used to measure double- and single-strand breaks after the DNA has been exposed to a potentially damaging agent such as radiation or a mutagenic chemical.

After exposure to the toxic agent the DNA is lysed. Alkaline lysis is used for measurement of single-strand breaks, and double-strand breaks are detected after lysis at neutral pH.

DNA fragments are negatively charged. The rate of movement in electric field applied to a gel will be inversely proportional to fragment size and this effect can be exaggerated by using pulsed charges at an angle (30–45°). The DNA is labelled either with an isotope or fluorescent chemical before running the electrophoresis and the system is calibrated using DNA fragments of known size. Variants include: orthogonal-field gel electrophoresis (OFGE) and field-inversion gel electrophoresis (FIGE).

Purines

The purines are a class of organic nitrogenous bases. They share a common structure: the purine ring. Guanine and adenine are purines and are important components of DNA and RNA. Caffeine and uric acid are also purines.

Pyrimidine

Pyrimidines are nitrogenous bases. In RNA, the pyrimidines are cytosine and uracil; in DNA, they are cytosine and thymine. The base pairing between purines and pyrimidines is the basis of the complementarity that enables DNA to act as a template, both for DNA replication and for transcription. Cytosine pairs with guanine; adenine with uracil (RNA) or thymine (DNA).

Quality-Adjusted Life-Year (QALY)

A measure of utility used in the assessment of health states and in economic analysis. Mathematically, it is simple; conceptually and ethically it is suspect. QALY is calculated as:

years of survival × utility factor.

The utility factor applies an adjustment for the quality of survival: perfect health = 1; death = 0. Intermediate states are between 0 and 1. Two years in a state scored as 0.75 would be: 2 × 0.75 = 1.5 QALY. Four years in a state scored as 0.375 would be: 1.5 QALY. The assumption here is that the value placed on 1.5 QALY is the same: whether acquired as 2 years of moderate misery or 4 years of considerable misery. This assumption is possibly unwarranted.

Another problem with the use of QALY is the tyranny of youth: this particularly applies to its use in economic evaluations. If, after treatment for cancer costing £10 000, an individual is cured and left with a quality of life that is 90% of normal, the utility factor is 0.9. If they are 45 years of age and have a life expectancy of 37 years, the QALY gained by treatment is:

37 × 0.9 = 33.3 years.

If the patient is 75 years of age and has a life expectancy of 7 years, the QALY gained by treatment is:

7 × 0.9 = 6.3 years.

The cost per QALY is £10 000/33 (£303) for the 45-year-old, and £10 000/6.3 (£1587) for the 75-year-old. If cost per QALY is then used as a criterion for allocating resources, the 45-year-old will be given a higher priority than the 75-year-old. This is fundamentally unjust — particularly since the 75-year-old is likely to have paid taxes for longer than the 45-year-old.

Q-TWiST

Q-TWiST is a variant of **TWiST** (**p. 317**). It incorporates a quality factor into the calculation of survival time.

Quality

This term is an extremely popular component of many of the expressions used in health services management: quality circles, quality assurance programme, director of quality. The term is also used in assessing health status, as in **quality of life** (**p. 253**). Quality, in its current managerial sense, can be easily defined: excellence, good workmanship and performance, possessing true worth and value. Quality can also simply be used to describe an attribute: 'the quality of mercy'.

Quality Factor

A factor (Q) used in the calculation of dose-equivalent for different types of radiation. Its value primarily depends upon the linear energy transfer (LET) of the radiation. For X-rays, Q = 1.0; it is ~10 for fast neutrons and approaches 20 for α-particles. The main use for *Q* is in calculations for radiation protection since the dose-equivalent (sieverts) is:

$$H = DQN,$$

where H is the dose-equivalent (sieverts), D is the dose (Gy), Q is the quality factor and N is a factor introduced to allow for differences in dose-rate or fractionation. Conventionally, 1 is substituted for N.

Quality of Life

Quality of life is a concept that everyone understands, in their own personal interpretation, but which no one can satisfactorily define. The closest approach

to a sensible definition is that of Calman. This defines quality of life in terms of an individual's ability to achieve those goals that they wish, and should have a reasonable expectation of being able, to achieve. It reflects the extent to which an individual's hopes and expectations are matched by reality. Impairment of quality of life, by this definition, is equivalent to T. S. Eliot's shadow:

> Between the idea
> And the reality
> Between the motion
> And the act
> Falls the shadow.
>
> (*The Hollow Men*, 1925)

One way through the philosophical minefield involved in trying to define quality of life is simply to avoid the issue altogether. Just as intelligence can be defined as that which is measured by intelligence tests, so quality of life can be defined as that which is measured by instruments that measure quality of life. This argument is both circular and of little intellectual merit: it is, however, a reasonable reflection of the current state of the art. There are several issues that need to be considered when assessing quality of life, whatever it may be, and these can be translated into requirements that should be fulfilled by any instrument purporting to measure quality of life.

Any quality of life (QOL) test should produce meaningful data based on the opinions of patients, or their carers, rather than those of doctors or nurses. These opinions should be elicited using a validated assessment technique, such as a questionnaire or interview. The instrument should be reliable and should satisfy the various, and opaque, criteria for psychometric validity. The data should be expressed in a numerical form suitable for statistical manipulation and analysis. The assessment should cover several key areas of the human condition: physical ability and capacity for playing a constructive role in society, emotional and psychological state; social interactions and activities and support; physical symptoms and distress; an overall global assessment of where the individual feels that they are in the world, the extent to which their aims are matched by reality. These points cover the core issues in any quality of life assessment. Features related to specific diseases also need to be included. Patients with head and neck cancer, for example, may have particular problems with eating and communication; patients with testicular tumours may have concerns about sexuality and fatherhood.

The QOL instrument must also be practical. it should be feasible, should not take too long to complete, should present questions in comprehensible language; and it should also be acceptable, the sensibilities of the potential respondents should be taken into account.

There is no ideal instrument. In the USA, the FACT scale is popular, whereas in Europe the various disease-specific modules and the EORTC QLQ C-30 core questionnaire are more widely used. Instruments that pay heed more explicitly to the mismatch definition of quality of life, described above, include SEIQoL and PGI. SF-36 is widely used as a measure of general health. The problem with choosing the 'best' is that there is no such thing. There is no **Gold Standard** (**p. 141**): the instruments have to be assessed on their relative, as opposed to their absolute, merits.

Quasi-Threshold Dose (D_q)

This is a measure of the width of the shoulder of the cell-survival curve for irradiated cells. It may be defined as the dose at which the backward extrapolation of the linear portion of the curve intersects with a surviving fraction = 1.0.

There is a mathematical relationship between D_0, D_q and n:

$$D_q = D_0 \log_e n,$$

where D_0 is the slope of the linear portion of the curve and n is the extrapolation number.

Quaternary Structure

The term used to describe the arrangement of subunits found in a complex protein.

Quetelet and His Index

A useful name to know if ever challenged to name six famous Belgians (along with Tintin, Eddy Mercx, Hercule Poirot, René Magritte and…). Quetelet was a Belgian mathematician and statistician. He had a mania for measuring; he derived an index of body mass that is still used in epidemiological surveys relating bodily habitus to incidence of disease. He measured the chests of 5738 Scottish soldiers and found that the measurements conformed to a normal distribution. This finding was unlike those found in French conscripts who showed an excess of shorter men, who, he felt, might be lying about their height to avoid military service.

Quetelet's index of body mass is calculated as:

$$\frac{\text{Weight}}{(\text{height})^2}.$$

Rad51 and its Protein

Rad51 is a gene implicated in the regulation of recombination and DNA repair in mammalian systems. The *BRCA*1 and *BRCA*2 genes interact with Rad51 protein, suggesting that these genes, which function as tumour suppressor genes, are involved with DNA repair. If repair fails, then ***p53*** **(p. 224)**-dependent mechanisms come into play: **cell cycle (p. 52)** arrest, via *p21*; **apoptosis (p. 19)**, via ***Bax*** **(p. 26)**. The *Rad51* knockout mouse fails to develop, presumably due to a *p53*-dependent mechanism, since the arrest can be partially overcome by mutant *p53*. *Rad51*-deficient embryos also exhibit increased radiosensitivity.

Radiation Cell Sensitivity and the Cell Cycle

In synchronously growing cell populations, the sensitivity of cells to radiation varies according to the phase of the **cell cycle (p. 52)**: sensitive in G_2 and M, relatively resistant in S-phase.

In theory, it might be possible to synchronise the cycling cells in a tumour and then give the next fraction of radiotherapy when the cycling cells were in a relatively sensitive phase of the cycle (G_2 or M). It would be even more desirable to synchronise cells of the normal tissues so that when tumour cells were in G_2 or M the normal tissue cells were in late S-phase. Although the concept of cell synchronisation is theoretically attractive, attempts to implement it clinically have been disappointing.

Radiation-Induced Cell Killing

Until fairly recently, the dominant factor in radiation-induced cell killing was thought to be biochemical damage to DNA caused by ionising events occurring in the immediate vicinity of the DNA molecule. It is now known that the picture is much subtler. Radiation oncologists, just as Monsieur Jordain, who discovered that, much to his astonishment, he had been speaking prose all his life, are somewhat surprised to find that unknowingly they have been using a powerful tool in the regulation of gene expression. There are now scores of genes known to be affected, either directly or indirectly, by radiation. It is now realised that not all radiation-induced cell killing is the result of damage to DNA and subsequent loss of reproductive integrity. Radiation can act directly on genes, radiation responsive elements, and can directly trigger important cellular mechanisms such as apoptosis. A dry mouth occurs within a week of irradiating the parotid. Explanations based on reproductive integrity cannot explain the rapidity of this phenomenon: the explanation lies in radiation-induced apoptosis of parotid cells. The recognition of alternative pathways for radiation action opens up whole new areas of radiobiological and clinical research.

Radiation Myelopathy

This clinical syndrome was first described in 1941. It describes damage to the spinal cord arising as a result of previous **exposure (p. 124)** to ionising radiation. The diagnosis requires that there be evidence of damage to the cord at a level that has previously been irradiated. This definition is not as straightforward as it seems. Recurrent tumour at the irradiated site may mimic radiation-induced damage; it may not always be possible to reconstruct the previous treatment so as to be entirely certain that a damaged segment is in fact within a previously irradiated area. There are no pathognomonic features, it is a diagnosis of exclusion.

Tolerance doses were first defined, in 1948, as 33 Gy in 17 days for large fields and 43 G in 17 days for small fields. The issue of defining tolerance has been controversial ever since. Several factors are known to be important: fraction size, volume effect, site effect, concomitant factors and time since any previous radiation treatment.

Dose (Gy)	Incidence of myelopathy (%)
45	0.2
57–61	5
68–73	50

Lower total doses are recommended if the patient is <16 years old, synchronous chemotherapy is being used, and there is an underlying predisposing factor, such as severe hypertension, vasculitis or diabetes. Keeping the fraction size as small as possible will, since the α/β ratio for spinal cord is between 2 and 3 Gy, selectively spare the cord. However, there may be problems with multiple fractions per day as repair half-times for the cord may be longer than previously supposed. Even an interval of 8 h between fractions may not be sufficient to allow for repair of all the damage inflicted by the preceding fraction. When retreatment is being considered, a useful estimate is that after 2 years ~50% of the original damage is recovered.

The pathology and pathogenesis are complex. Selective injury to grey matter is incompatible with diagnosis of radiation myelopathy — any selective damage is to white matter (demyelination, necrosis). This may be due to primary depletion of glial precursors such as oligodendrocytes or perhaps to primary obliterative vascular changes. Most probably it is due to some combination of the two.

Radiation Recall

Radiation recall is a phenomenon encountered with certain **cytotoxic drugs** (**p. 84**), particularly the anthracyclines, **taxanes** (**p. 305**), actinomycin D and **bleomycin** (**p. 32**). In patients who have previously been irradiated, subsequent administration of the drug causes the radiation-induced skin reaction to reappear. The clue to the nature of the reaction is that, in contrast to a generalised drug eruption, the skin changes are confined to the previously irradiated fields.

Radiation and its Interactions with Living Tissues: Scale and Time

A question of time

There are three overlapping phases in the response of cells and tissues to radiation

- Physical ($10^{-20} - 10^{-9}$ s).
- Chemical ($10^{-12} - 10^{2}$ s).
- Biological ($10^{0} - 10^{9}$ s).

The yard was, originally, the distance between the tip of the king's nose and the fingertips of his outstretched arm. Now the metric (and the metre) is used.
If the metre is taken as the equivalent to the human life-span (~10^9 s), then the relative time intervals can be converted to length:

- Physical: smaller than a subatomic particle.
- Chemical: about the size of an erythrocyte.
- Biological: the whole metre.

As in cosmology, the events of the first minuscule intervals of time have effects of comparatively immense duration.

A question of scale

The physical phase produces fast electrons and ionisation and these can be related to the dose and the type of radiation (table 1).

Table 1

Type of ionisation	Number Gy^{-1}	Energy (eV)	Target size (nm)
Sparse	1000	10–40	2
Moderate	20–100	100	2
Large	4–100	400	5–10
Very large	0–4	800	5–10

The amount of **DNA damage** (**p. 98**) Gy^{-1} per cell can be quantified approximately:

- 10 000 damaged bases.
- 1000 damaged sugars.
- 1000 single-strand breaks.
- 40 double-strand breaks.
- 150 DNA–protein cross-links.
- 30 DNA–DNA cross-links.

Imagine that a DNA molecule could be magnified so that a base pair occurred every yard:

- Human genome is 5.5 million miles long.
- DNA molecule is 6 yards wide.
- Smallest radiation cluster is the size of a large room.
- Largest radiation cluster is the size of a large house.
- Nucleosome is the size of a country mansion and contains 200 yards of DNA.

After 1 Gy radiation:

- Small or moderate cluster occurs every 5000 miles.
- Large or very large cluster occurs every 50 000 miles.
- Double-strand break occurs every 120 000 miles.
- Single-strand break occurs every 4800 miles.
- Damaged base occurs every 550 miles.
- Damaged sugar occurs every 5500 miles.
- DNA–protein cross-link occurs every 37 000 miles.
- DNA–DNA cross-link occurs every 180 000 miles.

Single-hit killing is easy to imagine; multiple hits adjacent to a single target are less easy to envisage.

The chemical phase of the interaction between radiation and living tissue involves damage to critical molecules:

- Direct damage to DNA via molecular excitation.
- Indirect damage to DNA via the radiolysis of water into **free radicals** (**p. 132**) and hydroxyl ions, etc.

The relative importance of each process is undefined.

The chemical interaction between radiation and tissues can be modified in a number of ways:

- Mop up **free radicals** (**p. 132**) (scavengers), e.g. 10% EtOH, DMS.
- Donate H to DNA, e.g. glutathione, cysteamine.
- Rapid enzymatic degradation of radicals, etc., e.g. catalase, peroxidase, **superoxide dismutase** (**p. 300**).
- Increased vulnerability of certain sites within the nucleus (e.g. peripheral).

Quite apart from these physicochemical effects, radiation also affects gene expression. The study of the molecular biology of the interaction between radiation and tissues is in its infancy. Already tantalising clues are emerging: ionisation can have chemical effects on surface membranes that lead to signal transduction and gene expression.

Sphingomyelinase can hydrolyse membrane sphingomyelin and produce ceramide, which can activate an SAPK (stress-activated protein **kinase**, **p. 171**) system, which then triggers **apoptosis** (**p. 19**). Radiotherapy may act on cell membrane to trigger apoptosis by a similar pathway. This implies an entirely new set of possibilities for manipulating radiation action, particularly since a great deal is known about the differences between the surfaces of malignant cells and normal cells.

Radical Treatment

Radical treatment is given in the hope that it might be possible to cure the patient. It is accepted that some normal tissue damage will inevitably occur — it is a matter of experience and judgement to choose a treatment that will have a good chance of curing the patient without producing toxicity that, either in the short- or long-term, is unacceptable. It is interesting that studies of patients' attitudes suggest that they might be prepared to accept greater toxicity, for lower chances of cure, than their doctors might assume. This observation emphasises that it is vital to involve the patient in any decision-making about treatment. The doctor has the duty to explain, and perhaps to recommend, but the right to decide must always rest with the patient.

Radical treatments can be divided into definitive and adjuvant. Definitive treatment is intended to extirpate the known tumour: mastectomy for breast cancer and chemotherapy for leukaemia. **Adjuvant treatment** (**p. 8**) is treatment given in addition to definitive treatment in an attempt to lower the risk of relapse. Implicit within the philosophy behind adjuvant treatment is the realisation that many patients will be treated unnecessarily having already been cured by the definitive therapy.

Radiobiological Hypoxia

Radiobiological hypoxia is defined as cellular behaviour in response to irradiation that mimics that of cells which are hypoxic at the time of irradiation. Although lack of oxygen is assumed to be the cause of that behaviour described as radiobiological hypoxia, the assumption is hard to prove. The proportion of radiobiologically hypoxic cells within a tumour can be estimated in the human, *in vivo*, using the comet assay. It should be clearly distinguished from physiological hypoxia: in which the low pO_2 may not be sufficient to produce a hypoxic-type cell-survival curve, but may be sufficient to turn on genes, such as *HIF*-1, which can have profound effects on cellular behaviour.

Radiographer

A person not medically trained but who has been trained in the use of ionising radiation for either diagnosis or treatment. Although working closely with physicians, radiographers have their own professional body and are in no way simply the doctors' handmaidens. Radiographers directly administer most of the imaging procedures performed and radiation treatments given in the UK. They have their own professional body — the Society of Radiographers.

Other related terms for 'radiographer' used in North America and Europe include radiation technologist and radiation technician.

Radioprotector Genes

These are genes that protect cells against the adverse effects of radiation. A variety of mechanisms can be involved: facilitating recombinatorial repair and modification of damage caused by **free radicals** (**p. 132**). Four main radioprotector genes have so far been sequenced: superoxide dismutase gene;

ATM (**p. 23**); *XRCC2*; *Ku*80. *Ku*80 confers considerable radioprotection, the **dose-modification factor** (**p. 103**) may be as much as five or six. *ATM* and *XRCCC2* are less potent: dose-modification is about twofold. The superoxide dismutase gene is the least active — its protective effect is probably only ~1 log cell kill.

The identification of radioprotector genes offers intriguing possibilities for combining **gene therapy** (**p. 137**) with radiotherapy. If a construct, based perhaps on antisense RNA or ribozymes, that specifically inactivated radioprotective genes in tumours could be administered at and around the time of radiation treatment, then this could substantially improve the effectiveness of radiotherapy. The reverse approach, also feasible, is to insert into tumour genes that confer increased radiosensitivity.

Radiosensitivity

Not all cells or cell lineages are equally sensitive to radiation. Although it is often difficult to disentangle environmental factors from innate factors, there is evidence from low dose-rate experiments, using cultured cells derived from human tumours, that some human tumours are more resistant to radiation than others. This corresponds to what might be better termed intrinsic radiosensitivity: that component of response to radiation that is innate to the cell itself and which is uninfluenced by environmental manipulation.

RAGE

The RAGE (receptor for advanced glycation end-products) molecule is part of the immunoglobulin superfamily. It is a cell surface molecule that can bind a multiple **ligand** (**p. 180**). It participates in the cellular response to stress causing increased expression of NF-κB and its target genes including interleukin 6 and monocyte/macrophage colony stimulating factor. It may play a pivotal role in the development of amyloidosis. It may also have a role in **invasion** (**p. 167**) and **metastasis** (**p. 198**). Amphoterin can bind to RAGE and this action activates MAP-kinase and a signalling pathway which leads to an increased expression of tissue metalloproteinases. Blocking the RAGE/amphoterin interaction can, in a mouse model, cause tumour regression.

Raltitrexed

This drug is a thymidylate synthase inhibitor that has been extensively used in the treatment of colorectal cancer. It has a long plasma **half-life** (**p. 147**) and can be given by bolus injection on a 3–4-weekly schedule. Its **clearance** (**p. 68**) is markedly altered by changes in renal function, and dose reductions or attenuations are necessary in patients with elevated creatinine levels.

Randomisation

Randomisation is, in the design of a study, the major protection against the introduction of **bias** (**p. 28**), whether intentional or unintentional. The rationale is simple: there are factors, only some of which are known, that can, independent of any intervention under investigation, affect outcome. If subjects are allocated at random to the interventions under investigation, then the play of chance will ensure that the various factors that can influence outcome will be evenly distributed between the groups of subjects. There will not be a preponderance of well-differentiated tumours in patients with high-performance status allocated to one treatment, while the other group of treated patients contains a majority of patients with poor performance status and poorly differentiated tumours.

Randomisation can be achieved in a variety of ways, but some 'randomisation' procedures are in fact flawed. The **Gold Standard (p. 141)** is randomisation performed at a distant centre using telephone, fax or e-mail contacts and with groups allocated according to a sequence of computer-generated random numbers. The investigator has no control over the allocation process and no way of predicting the group to which the next patient entered into the study will be allocated. The centralised procedure means that **stratification (p. 298)** and blocking can be used appropriately. Allocation to a particular group according to whether a subject's birth date is odd ('assign to Group A') or even ('assign to Group B') appears, at first sight, to be a random process but is, in fact, non-random. The investigator will know the subject's birth date, and by extension their treatment allocation. Knowing the allocation they may decide not to enter them into the study, thereby introducing bias. Randomisation is a potentially subvertible process and, in a good trial design, explicit steps are taken to prevent investigators breaking codes or in any other way interfering with the ability of the process to minimise **bias (p. 28)**.

Randomised Controlled Clinical Trial (RCT)

A well-designed and fully reported randomised controlled trial is, provided it deals with patients similar to those for whom decisions have to be made, the best evidence on which to base a management decision in oncology. Subjects are assigned at random to the interventions under investigation. There is a control group, either untreated or receiving a conventional, as opposed to an experimental, treatment. The interventions are administered to a group of prospectively defined patients. The interventions themselves are carried out according to a protocol specified in the design of the study. Similarly, follow-up procedures and the evaluation of **endpoints (p. 114)** are performed according to a predefined schedule. The endpoints are unambiguously defined and the results reported according to their significance, both clinical and statistical. The trial should be planned to be large enough to detect, with a low probability of error, a clinically significant difference between the interventions being compared. The design should, therefore, minimise the likelihood of either Type I or II statistical errors. The **CONSORT statement (p. 74)** concerns the design and reporting of a randomised controlled clinical trial and establishes what should be regarded as the basic standards for this type of research.

REAL Classification

A classification of lymphomas that attempts to unite two previously distinct systems operating on different sides of the Atlantic. REAL stands for Revised European–American Lymphoma.

Lymphomas are divided into the B-cell types and the T-cell and putative natural killer (NK) types, and within each type are both the precursor and peripheral neoplasms. There is also a catch-all category of unclassifiable neoplasms. The REAL classification can be used to group tumours according to their clinical characteristics (table 1).

Reciprocal Iso-effect Plot

It is difficult to measure α/β ratios directly for tissues *in vivo*. One indirect method that can be used to provide an estimate of the ratio is to use the **intercept (p. 163)**: slope ratio of the reciprocal iso-effect plot. The basic linear–quadratic equation can be rearranged to give:

$$\log_e S/n\ d = \alpha + \beta d,$$

Table 1

B-cell tumours	T-cell tumours
Indolent	
Chronic lymphatic leukaemia/small lymphocytic lymphoma	large granular lymphocytic leukaemia T-cell and NK types
Lymphoplasmacytic, e.g. Waldenstrom's macroglobulinaemia	*Mycosis fungoides*/Sezary syndrome
Hairy cell leukaemia	smouldering and chronic adult T-cell lymphoma/leukaemia
Splenic marginal zone lymphoma	
Marginal zone B-cell lymphoma: extranodal (MALT); nodal (monocytoid)	
Follicle centre lymphoma follicular grade I (small cell)	
Follicle centre lymphoma follicular grade II (mixed small cell and large cell)	
Aggressive lymphomas (intermediate risk)	
Prolymphocytic leukaemia	prolymphocytic leukaemia
Plasmacytoma/multiple myeloma	peripheral T-cell lymphoma unspecified
Mantle cell lymphoma	angio-immunoblastic lymphoma
Follicle centre lymphoma follicular grade III (large cell)	angiocentric lymphoma
Diffuse large B-cell lymphoma (immunoblastic and centroblastic)	intestinal T-cell lymphoma
Primary mediastinal (thymic) large B-cell lymphoma	anaplastic large cell lymphoma T-cell and null cell types
High-grade B-cell lymphoma, Burkitt like	
Very aggressive lymphomas (high risk)	
Precursor B-lymphoblastic lymphoma/leukaemia	precursor T-lymphoblastic lymphoma/leukaemia
Burkitt's lymphoma/B-cell acute leukaemia	adult T-cell lymphoma/leukaemia
Plasma cell leukaemia	

where S is the number of surviving clonogens, n is the number of fractions and d is the dose per fraction. It is reasonable to assume that, for any given iso-effect, i.e. equal degree of damage, the surviving cell number will be equal. Total dose $= n\,d$ and, if S is held constant as in an iso-effect plot, then the above equation is that of a straight line: when reciprocal dose is plotted against dose per fraction, the slope of the line is given by β and the intercept is given by α (figure 1). The intercept to slope ratio is, therefore, the same as the α/β ratio. This statement is only true if repair is complete between fractions; if repair is incomplete, then the data will follow a curved line that approximates to the expected line but which is concave downwards.

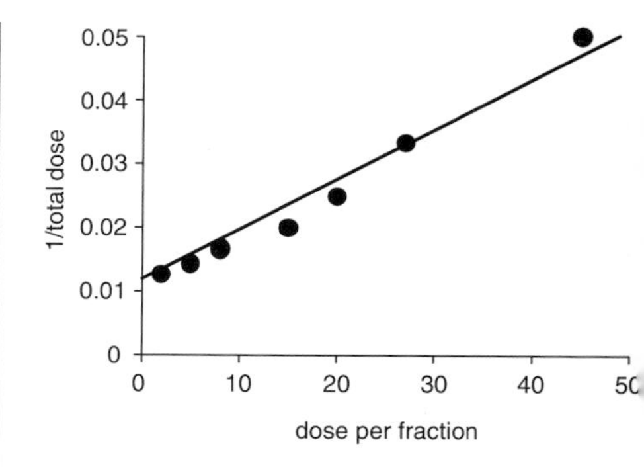

Figure 1 — Reciprocal iso-effect plot.

This example, of a reciprocal iso-effect plot, shows points corresponding to different combinations of dose per fraction and total dose at which the biological effect of the radiation is equal, e.g. the **endpoint** (**p. 114**) could be moist desquamation of the irradiated skin:

- Slope of the line $= \beta/\log_e S = 0.00872\ Gy^{-2}$.
- Intercept on the y-axis $= \alpha/\log_e S = 0.011\ Gy^{-1}$.
- Ratio of the intercept to the slope $= 0.011\ Gy^{-1} = /0.00872\ Gy^{-2} = 12.61$ (an α/β ratio characteristic of an acute response to radiation).

Recognition Site

A recognition site is simply that sequence of DNA recognised by a specific restriction enzyme.

Recommended Dose

In cancer chemotherapy this may be defined as a dose that causes moderate, reversible, toxicity in the majority of patients.

Recursive Logic

A fallacious process whereby truth is obscured by self-reference. If the data used to generate a hypothesis are then used to assess the 'truth' of that hypothesis, then it is no surprise when the hypothesis is confirmed as 'true'.

The best examples are usually found in studies on **prognostic factors (p. 248)**. A putative prognostic factor is studied on a population of patients whose fate is known. The new test is applied to the patients and a cut-off point between normal and abnormal is chosen that best discriminates between patients with a good prognosis and those with a poor prognosis. Using this cut-off point, the test is applied to the same group of patients and, unsurprisingly, it shows an excellent prognostic discrimination. The correct procedure would be to derive cut-off values from one group of patients (the data-generating set) and then to test the performance of the prognostic factor on a completely different group of patients (the hypothesis-testing set).

Redistribution

Cells vary in their sensitivity to radiation according to their **cell cycle (p. 52)** phase. Resistance is least during late G_2 and **mitosis (p. 201)** and greatest during late S-phase. There may be up to a fivefold variation in cell killing according to cell cycle phase. In a fractionated course of treatment, each radiation treatment will, to some extent at least, produce cell cycle synchrony since there will be preferential sparing of some cycling cells and preferential killing of others. The net result could be a synchronised cohort progressing through the cycle. The effect of a subsequent dose of radiation on this cohort would depend on the phase of the cycle that the cohort had reached.

Reductionism

In biology, is that scientific approach whereby the explanation of a complex process is sought in terms of its component parts. The underlying assumption is that the simple components can produce the complex, the process, without the need to invoke outside influences. It is in contradistinction to concepts such as **vitalism (p. 325)**. **Reductionism (p. 263)** is related to **determinism (p. 94)**, except that the latter moves from the simple towards the complex: 'almost all aspects of life are engineered at the molecular level' (Francis Crick, *What Mad Pursuit*). Reductionism has its dangers: 'Like following life through creatures you dissect, You lose it in the moment you detect' (Alexander Pope, *Moral Essays*).

Reductionism implies a tendency to look at components of a complex mechanism in isolation. This may be misleading and, at present, seems particularly true for studies looking at *p53* as a prognostic factor. ***p53*** (**p. 224**) can have different roles according to the functional state of a complex system, with numerous interactions and feedback loops, involving *p53*, *p21* and *mdm2*. Simply looking at *p53*, and its mutations, in isolation may be grossly misleading — it is the function of the system as a whole that is going to be important. You cannot predict the performance of a car engine simply by looking at a single piston.

Regression (Regression Analysis)

Regression analysis is a repertoire of mathematical techniques used to define the best mathematical relationship between two variables. Unlike **correlation** (**p. 75**), there is no requirement that the relationship must be linear, although it could be (linear regression). The essence of regression is that one variable (the explanatory variable) may be used to predict another (the dependent variable). Regression techniques are invaluable in oncology. A putative **prognostic factor** (**p. 248**) can be explored as an explanatory variable in a regression analysis: the dependent variable would then be survival or other appropriate **endpoint** (**p. 114**). Similarly, in diagnostic testing, the test result can be used as the explanatory variable with the presence of disease being the dependent variable. Other types of relationship, such as proportional hazards (Cox model) or **logistic regression** (**p. 183**), may also be explored in regression analyses.

Relative Biological Effectiveness (RBE)

RBE is a measure of the difference in cell killing between the test radiation and X-rays: 250 kV X-rays are used as reference. For any given level of biological effect:

$$\text{RBE} = \frac{\text{dose of 250 kV X-rays}}{\text{dose of test radiation}}.$$

As with **oxygen enhancement ratio** (**OER**) (**p. 221**), RBE will depend on the endpoint chosen. The RBE for neutrons is usually taken as 3.0, but, this is an oversimplification. RBE rises at smaller doses; different tissues each have a different RBE. This has caused problems clinically, e.g. high rates of temporal lobe necrosis in the neutron therapy of tumours of the maxillary antrum. Neutron fraction sizes used clinically range from 0.9 to 1.3 Gy, and are, therefore, on the steepest and least predictable parts of the RBE curve. Another physical effect is also important: dose per fraction will vary in the penumbra of a beam and it is entirely possible that under some circumstances there will be regions outwith the treatment volume in which, as a result of the effect of fraction size on RBE, the fall in the physical dose of neutrons is offset by an increase in biological dose. RBE cannot be considered as an absolute value: different values will apply to different tissues under different circumstances.

Relative Effectiveness

This term is used to calculate the **biologically effective dose** (**BED**) (**p. 29**) of radiation. It is defined as:

$$(1 + [\text{dose per fraction}/\{\alpha/\beta\}])$$

used in the BED equation. BED is the product of total dose and relative effectiveness.

Relative Risk

Relative risk (RR) is defined as the rate at which, over a given period, an event of interest occurs in an experimental group divided by the rate at which the event occurs in a control group. If there is no difference between the two groups in the frequency with which the event occurs, then RR = 1.

Exposed	Cancer	No cancer
Yes	a	c
No	b	d

$$\text{RR ratio} = \{a/(c+a)\}/\{b/(d+b)\} = a(d+b)\ /b(c+a).$$

The relative risk ratio in this context becomes somewhat arbitrary since it depends on how many patients with cancer (a + b) there are relative to the number of controls (d) studied.

The relative risk can only be accurately estimated from a prospective study. In a retrospective study the relative risk will alter depending on numbers chosen for both the experimental and control groups. For retrospective studies the **odds ratio (p. 217)**, which is independent of the size of the groups, is used: the odds ratio from data laid out as above would be: ad/bc. The relative risk is not the same as the **absolute risk (p. 1)**. A relative risk of 25 for dying from breast cancer within 5 years simply indicates that this particular woman, or group of women, has an absolute risk of dying from breast cancer within the next 5 years which is 25 times greater than the risk that applies to a suitable reference population. In terms of advising an individual patient, the relative risk is not particularly useful: its use begs the question, 'relative to what?' A measure of the absolute risk is more useful and understandable. To convert a relative risk to absolute risk, it is necessary to know the baseline absolute risk year by year, and the survival probability for each year for the reference population. This latter information is required to allow for deaths from causes other than the condition under investigation.

Because of the effects of competing causes of mortality, the relationship between absolute risk and age is complex, even though the relative risk remains constant. When absolute risk is low, then simply multiplying an individual's relative risk by the **absolute risk (p. 1)** for the reference population will give a reasonable approximation of their absolute risk. This relationship will not hold when the absolute risk is high.

Reliability

The term is used to define the extent to which a scale or scoring system produces consistent results when applied and reapplied to a subject, or group of subjects, in whom the condition being assessed remains stable. **Consistency (p. 74)** and reliability are synonymous: they indicate the extent to which the results of repeated measurements of the same phenomenon agree with each other. If the height of a patient is measured, on various occasions, as 160, 134 and 178 cm, then this shows a suspicious lack of reliability. Variation represents a lack of reliability. This lack of reliability can arise because different observers measure the same thing differently (interobserver variation) or because the measurements made by an individual observer vary over time (intra-observer variation). Reproducibility and repeatability are measures of reliability. The means should not be confused with the end: reliability is not the same as repeatability, repeatability is a quality possessed by a reliable measure.

Reliability refers, therefore, to the reproducibility of test results. In clinical measurement there are three main sources of impaired reliability:

- Interobserver variation: where different observers applying the same test to the same subject report different results.

- **Measurement error (p. 190)**: a random source of error in which the same observer applying the same test to the same subject reports different results each time, the intra-observer variation. This can be assessed using test–retest data, provided that there has been no change in the subject's state between the test and retest.

- Subject variation: where one observer applying the same test to a group of allegedly similar subjects obtains a different result for each subject.

Analysis of variance (p. 13) can be used to apportion the variance in a set of test results. The resulting coefficient of reliability is known as the intraclass correlation coefficient (ICC). Reliable tests should have a value of at least 0.8. Other statistical methods can assess reliability: the Pearson correlation coefficient provides a reasonable approximation for continuous variables but it tends to overestimate reliability. When categorical variables are used then **kappa κ (p. 170)**, or weighted κ, is used. This is a modified form of the contingency table with a correction applied for chance agreements. Weighted κ provides a method for allowing for partial agreements. When weights are applied using a quadratic relationship, then weighted κ and the ICC will produce identical results.

Reliability and consistency, when referring to scales and scores, are not the same. Consistency is used to indicate the **homogeneity (p. 152)** of the items within a test instrument that uses multiple items. For example, in an instrument designed to assess fatigue, we might wish to include separate items about feeling sleepy and about difficulty in concentrating. The extent to which these different items correlate with each other, and with the test instrument as a whole, is defined as consistency or homogeneity. If the test included an item on dietary fat intake, this would be inconsistent with the other items and we would probably discard it from the final version of our instrument. There are two main methods for assessing consistency: the item–total method and split-half studies. The Pearson correlation coefficient is adequate for item–total studies; for spilt-half studies either the Kuder–Richardson formula or Cronbach's α are used.

Reorganisation

This is a phenomenon to which health services seem particularly prone. The usual assumption appears to be that performance is bad and that the best way to improve performance is not to provide more resources but to change the structures used to organise and deliver services. It is as if, after a process of trial and error, the crew of a quinquereme (an ancient Roman galley) were correctly positioned according to their individual talents. The boat progresses through the water reasonably smoothly. Suddenly a whistle is blown, the oarsmens' benches are moved, some are completely removed, and there is a mad scramble for seats and position. The subsequent progress of the boat is not necessarily improved by such manoeuvres.

Reoxygenation

The efficiency of radiation-induced cell killing is dependent on the oxygen tension in the tissues at the time of irradiation. Hypoxic cells are relatively resistant to irradiation. After the first fraction of a course of fractionated treatment, many cells are killed and this reduces the competition for oxygen among the tumour cells. Cells that were previously hypoxic, and relatively radioresistant, may, by the time of the next fraction, be re-oxygenated and consequently be more effectively killed by radiation. It is entirely possible that appropriate fractionation might have made redundant many of the strategies (hyperbaric oxygen, neutron therapy, hypoxic cell sensitisers) used to deal with the so-called hypoxic cell problem.

Repair of Radiation-Induced Damage

Not all radiation-induced damage is lethal. Some of it can be repaired. Repair of this sublethal damage is a time-dependent process. The clinical results with schedules using multiple fractions per day suggest that 6–8 h are required for adequate repair. Shorter interfraction intervals may be associated with unexpectedly severe toxicity. The experimental data are consistent with this. Repair half-times are of the order of 2–3 h, but there is considerable **heterogeneity (p. 150)** among the various normal tissues. A slower repair component may be particularly important in determining CNS damage.

Fidelity is crucial to repair processes: it is not sufficient to cram all the components of an exploded engine back under the bonnet in no particular order. The engine has to be carefully reassembled, the original state of affairs being restored, for the car to work. In terms of **DNA damage (p. 98)**, rejoining and repair are not equivalent (rejoining just indicates that the components are back under the bonnet, it tells nothing of the function of the engine). Some sites are more rapidly repaired than others (e.g. preferential repair of transcribed regions). Repair will also depend on the time available: base damage, single-strand breaks need <1 h; double-strand breaks may need up to 4 h. DNA repair requires gene expression and in conditions such as **ataxia-telangiectasia (p. 23)** and **xeroderma pigmentosa (p. 329)** abnormal genes cause ineffective repair. There are several chemical inhibitors of DNA repair, e.g. 3-aminobenzamide. Repair will also depend on cell cycle phase, peaks in radioresistance during G_1 and at S-phase/G_2 border may be related to particularly efficient repair occurring at these specific phases of the cycle. The phenomenon of **potentially lethal damage repair (p. 241)** is defined by that component of repair which is influenced by the conditions after, as opposed to before or during, radiation.

The results of attempts at repair can be summarised as: complete repair with fidelity and unimpaired survival of clonogens; misrepair, in which mutation or incomplete repair result in a change in clonogenicity; and no repair, causing a total loss of clonogenicity.

Replication

Replication is the process whereby the two strands of a DNA molecule can copy themselves, thus producing a second DNA molecule identical to the first. The process is complex and involves many checks for integrity and fidelity. The main enzymes involved are **DNA polymerases (p. 101)**.

Repopulation

If a course of radiation is sufficiently protracted, and the interval between treatments long enough, then it is possible for tumour cells to proliferate during a course of treatment. Some of the extra dose required with longer radiotherapeutic schedules might be necessary simply to counteract the proliferation that occurs during treatment. It is not clear whether radiotherapy induces an accelerated repopulation of clonogenic cells or if it simply unmasks the latent proliferative abilities of the cells. Nor is it clear whether there is a lag before repopulation starts. Whatever the causes and mechanisms, the clinical consequence of repopulation is relatively straightforward: unnecessary prolongation of a course of fractionated radiotherapy will compromise the chances of controlling the tumour. If repopulation is occurring, then protracted treatment is like pedalling a bicycle with a slipping chain: there is much effort with little gain.

Reproductive Integrity

The clonal/cellular basis of cancer implies that, to form a tumour, an individual cell must be capable of reproducing and that each of that cell's progeny must also have the same capacity — this is what is meant by reproductive integrity. This defines the ability of a cell to divide in a self-sustaining fashion over a number of generations — it is an essential prerequisite for clonogenicity. A cell can be viable, in the sense that it is alive, but can lack reproductive integrity. The most obvious analogy is with a mule: the animal is viable and looks not much different from a donkey. The donkey can, however, produce more donkeys. A mule, by contrast, cannot produce any descendants. Just as a mule can eat, breathe, excrete and carry tourists into the Grand Canyon — but cannot breed — so a thyroid follicular cell after a lethal (in the sense outlined above) dose of radiation can continue to metabolise and synthesise thyroid hormone. The lethal effects of the radiation will not be expressed until that cell is called on to divide; this may be months or years after the original dose of radiation. This is why hypothyroidism is an important, but late, complication of radiotherapy to the neck.

If a cell, no matter whether it is 'alive' or 'dead', cannot divide, then it cannot form a tumour. Conversely, if in treating a tumour cells are left alive but reproductively non-viable or sterile, then the tumour cannot develop further and, for all practical purposes, the patient is cured. There is a clear distinction to be drawn between viability in general and reproductive viability (or integrity) in particular. Cancer treatments are primarily aimed at destroying the reproductive integrity of tumour cells. The technical difficulties of growing tumour cells *in vitro*, and the practical difficulties associated with sequential biopsies of tumours studied *in situ*, mean that direct measurement of the loss of reproductive integrity is usually not possible and we are forced to rely on indirect evaluations.

Recent evidence using a *p21*-deficient cell line from a colorectal cancer cell line illustrates the problem of defining **cell death** (**p. 55**) and/or reproductive integrity in a way meaningful in terms of cancer treatment. *p21*(+/+) cells irradiated *in vivo* showed few surviving colonies when plated out; nor did *p21*(−/−) cells — the radiation survival curves were identical for the two cell lines. However, the *p21*(+/+) cells produced a fine lawn of cells in the background: the *p21*(−/−) cells did not. In a xenograft assay, the irradiated *p21*(−/−) tumours regressed completely whereas the *p21*(+/+) tumours continued to grow. Thus, the two assays, both apparently testing reproductive integrity, give different results. The explanation lies in the fact that *p21*-deficient cells cannot arrest their **cell cycle** (**p. 52**) after irradiation and proceed promptly to die by **apoptosis** (**p. 19**), and dead cells surround the few surviving clonogens. There is no background lawn of living but sterile cells. The *p21*(+/+) cells, in contrast, can arrest their cell cycle in G_1 following irradiation. These growth-arrested cells may act as a feeder layer for the few surviving clonogens, which go on fully to repopulate the tumour.

Re-replication Block

This prevents a cycling cell from producing more DNA than required for the two daughter cells. The DNA must double in amount: no less, no more. The mechanism f the re-replication block is not well nderstood. One manifestation of the block that G_2 cells, which should contain double ie original amount of DNA, are refractory) factors capable of driving a G_1 cell into)NA synthesis.

esiduals

ı linear regression, the residuals are the ifferences between the individual values and ie regression line. The least-squares method f regression fits the regression line to the ata so that the sum of the squares of the esiduals is the lowest possible number. ıspection of residuals is important as, y highlighting **exceptions (p. 122)**, it may raw attention to unexpected inconsistencies ı the data. These could arise as a result of an nexpected new finding or from entry of orrupt or inappropriate data. Figure 1 of ımour marker results for patients with on-seminomatous germ cell tumours plots ie residuals for the points with the largest eviations from the regression line. r for the egression line is 0.79; $r^2 = 0.62$. Point c ıows a particularly large residual. Review of ie histology showed that this patient had a eminoma and was, therefore, inappropriately ıcluded in the analysis. With point c liminated, the regression line fits more recisely the data points. Now $r = 0.98$; $^2 = 0.96$. In a one-way analysis of variance,

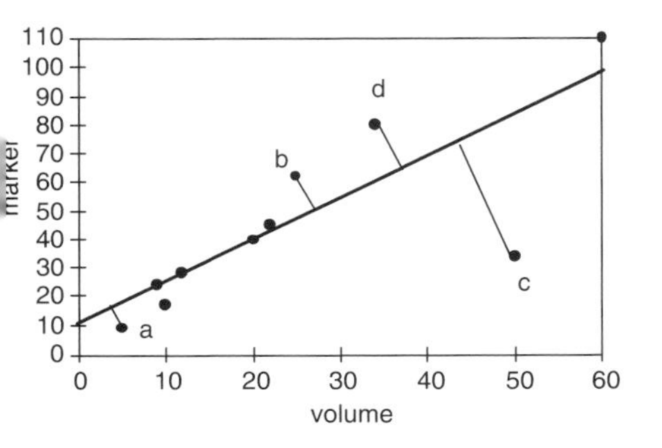

Figure 1.

the residuals are the differences between the observed values within the group and the value, fitted by the model, for the mean.

Response Criteria

Response categories for cancer chemotherapy are fairly strictly defined but are, in a scientific sense, rather primitive. They may be useful in a comparative sense, but in a biological or clinical sense they may be too rigid:

- **Complete response (p. 70)**: complete disappearance of all evidence of malignant disease for at least 4 weeks: response may be defined clinically, radiologically or pathologically — the latter requires re-biopsy.
- **Partial response (p. 228)**: decrease in tumour size of at least 50%. Tumour size is defined as the area for a single measurable lesion or as the products of the perpendicular diameters for a series of measurable lesions. The reduction should last at least 4 weeks and no lesion should grow, or new lesion develop, during this period.
- Stable disease: decrease in size of <50% or a <25% increase in tumour size. No new lesions develop.
- Progression: increase of >25% in tumour size or the development of new lesions.

Even superficial inspection of the above criteria indicates the problems:

- Lack of any incorporation of patient's symptoms or quality of life.
- Lack of uniformity of assessment: clinical, radiological, pathological.
- Heavy dependence on reproducible measurement.
- Equating area regression with volume regression.
- Inability to identify subtle shrinkage of tumour.

- Equating tumour size with number of viable cancer cells.
- Failure to exploit modern technology: CT or MRI reconstructions of tumour volume (ml); marker regressions.

One of the challenges for the future in oncology will be to devise more accurate methods for assessing the response of tumours to treatment. The use of surrogate markers is no solution, though it may help establish **proof of principle (p. 248)**. Given that some of the newer therapeutic approaches, such as inhibition of angiogenesis or restitution of the function of tumour suppressor genes, cannot reasonably be expected to produce the type of response defined by the above criteria, we will have a problem assessing the **clinical effectiveness (p. 69)** of some of the newer therapies for cancer.

Responses to Questions: A Simplified Taxonomy

The possible responses to items in a questionnaire can be obtained and classified as follows:

Categorical

- Yes/no with no intermediate values, e.g. do you have trouble climbing stairs, Y or N?

The problem here is that the approach is relatively coarse: how many stairs do there have to be before there is a problem? What, in any event, is the nature of the problem: breathlessness, fatigue, claudication?

Continuous

This is a preferable method since we can always dichotomise continuous data later, if we wish: it does not work the other way round.

- Direct estimation:
  — **visual analogue scale (p. 325)**
- Comparative methods: much more difficult to understand and perform, but may have statistical and interpretative advantages. Examples include: Thurston Guttman and paired-comparison metho
- Econometric:
  — **standard gamble (p. 296)**
  — **time trade-off (p. 308)**.

It is important that appropriate statistical techniques are used when analysing respons responses on a **Likert scale (p. 180)** canno even though they may be numerically expressed, be analysed as if they reflected a continuous variable. Visual analogue, or LA scales, can be treated as generating continuo variables provided no subcategorisation has been applied.

Responsiveness

The responsiveness of a score or scale is a reflection of its ability accurately to reflect changes over time. Medical interventions are intended to produce change for the better but sometimes, in certain respects, they produce change for the worse. The ability to capture information on change, before, during and after treatment, is an essential feature of a clinically useful tool for evaluating the effects of therapy. Unfortunately, it is not always easy to measure the responsiveness of a test instrument. One method is simply to exter the ANOVA approach and to estimate the variance introduced by entering time as a variable. In designing any instrument, there is an intrinsic tension between reliability and responsiveness since that which makes a test responsive may make it unreliable, and vice versa.

Restriction Digest

The collection of DNA fragments, of varying lengths, produced when DNA is subjected to the action of restriction enzymes. The preparation of a restrict digest is the first step in **Southern blotting** (**p. 292**).

Restriction Enzymes

These enzymes cut DNA at specific sites: hey are endonucleases. They evolved in bacteria as a protective mechanism against parasitisation by **phage** (**p. 232**). Each endonuclease recognises a particular phage-related DNA sequence and snips it out of the host genome. The specificity of he site of action is crucial to this defence — otherwise the endonuclease would snip estructively at the host's own DNA. Restriction enzymes are major tools in nolecular biology for both mapping and ngineering recombinant DNA. There are iree classes (I–III) and it is the Type II orms that are the most useful. For ecombinant techniques, those endonucleases ich as EcoR1 that have the ability to create icky ends are particularly useful. The sticky nds are formed of single-strand extensions f DNA: one chain has been cut slightly nger than the other. The single-strand quence can then be attached to another striction fragment with a complementary icky end.

estriction Point

hat point during the G_1-phase of a ammalian cell at which commitment DNA synthesis is made. The 'decision' influenced by nutritional state, cell mass, d all the other checks and balances at control progress through the **cell cle** (**p. 52**).

Ret Gene

The gene is on chromosome 10. Mutations are associated with **MEN Type 2 syndrome** (**p. 194**). There are interesting associations with the embryonic development of the kidneys and the nerve plexuses of the bowel. The knockout mouse without *Ret* has poorly developed or absent kidneys, hypoplasia or agenesis. Patients with Hirschsprung's disease have *ret* abnormalities.

Retinoblastoma

A dominantly inherited syndrome associated with retinal tumours and sarcomas, particularly radiation-induced osteosarcomas. The defective gene is designated *RB*1 and its gene product is intimately concerned with regulation of the **cell cycle** (**p. 52**).

Retinoic Acid Receptor (RAR)

A nuclear receptor activated when bound by retinoic acid. The gene is on chromosome 17q12. When activated by retinoic acid, the receptor acts by transactivation to control cellular differentiation. The role of retinoids in the chemoprevention of cancer is mediated via the RAR and the receptor is a promising target for approaches to cancer prevention or to inducing the differentiation of established tumours. The gene contains four regions designated A/B, C, D and E. C encodes the DNA binding site, while E codes for the ligand-binding **domain** (**p. 102**).

Acute promyelocytic leukaemia (AML3 or APL) is associated with a reciprocal translocation involving RAR: t(15,17)(q22;q11.2–q12). This translocation probably leads to the formation of a **fusion protein** (**p. 133**) that contains the activating portion of the RAR, from chromosome 17, and a gene, *PML*, from chromosome 15. The *PML* gene product is co-activator of

RAR: the fusion product is, therefore, a **transcription factor (p. 313)** joined to one of its own cofactors.

Reverse Genetics

A process in which, in the absence of any knowledge about the product of a gene, techniques such as **chromosome walking (p. 64)** are used to define the position of the gene and its DNA sequence. From the DNA sequence, the protein structure and possibly function can be deduced. Another method uses cDNA. A cell line with high expression of the protein of interest is used to obtain large amounts of the mRNA corresponding to the protein. Reverse transcriptase is used to create a **DNA sequence (p. 101)** complementary to the mRNA sequence. The DNA formed in this way is termed cDNA. It can be sequenced and once its sequence is known, this can be compared with DNA sequences whose chromosomal location is known.

Reverse Transcriptase

This is an enzyme that shattered a dogma and is destroying nations. It is a viral enzyme, an RNA-dependent **DNA polymerase (p. 101)**, capable of forming a DNA molecule from an RNA template. It shattered the central dogma (DNA to RNA to protein, and never in the reverse direction). It is part of the essential machinery of retroviruses including **HIV (p. 151)**, the virus which is currently laying waste to Sub-Saharan Africa.

rh

This is used as a prefix for various biological products, e.g. **interferons (p. 163)** and **growth factors (p. 144)**. It is an abbreviation for **r**ecombinant **h**uman and indicates that the product is of human type and has been prepared using recombinant genetic techniques.

Rho

This small molecule is concerned with cell adhesion and motility and acts by reorganising the cytoskeleton and affecting the contractility of actomyosin. It has GTPase activity and may interact with, among other molecules, a family of Rho-associated serine–threonine protein **kinases (p. 171)** (ROCK). Rho has a role in tumour invasion and **metastasis (p. 198)**. Inhibition of ROCK, its possible intracellular target, may have therapeutic potential. Such inhibitors are now available (e.g. Y-27632) and in an animal model have been shown to inhibit the intraperitoneal spread of a tumour.

Ribozyme

A ribozyme is RNA that acts as an enzyme with the ability to cleave or ligate RNA wit high specificity. Although only first describe in the mid-1980s, ribozymes are increasingly being used in cancer research and therapy. Their specificity, which is sufficient to discriminate between substitutions involving single nucleotides, is conferred by complementary base pairing between the ribozymal RNA and the target mRNA. Ribozymes can be constructed to act at specific cleavage sites in specific RNA molecules. Therefore, they have considerabl potential as agents for blocking expression of specific target genes. This can be used to investigate and validate putative molecular mechanisms of disease as well as having obvious implications for treatment.

There are several different types of ribozyme the hammerhead ribozyme is the simplest an most widely used; other types include hairpin, hepatitis δ and pRNAase ribozyme. Attempts to use ribozymes for treatment encounter many of the difficulties that affect antisense therapy or gene therapy in general Ribozymes are vulnerable to the action of nucleases and their targets may be sequestered deep within some inaccessible intracellular compartment. The main problem is to deliver a ribozyme to a precise intracellular location in quantities sufficient for it to have a biological effect. The other main problem is, as ever, to define a mRNA target that is sufficiently discriminatory: present in tumour cells, but not in normal cells.

RII Gene

A tumour suppressor gene, abnormalities of which are often found in colon cancers that exhibit **microsatellite instability** (**p. 199**). It codes for the type II receptor of TGF-β.

Risk (Danger, Dread)

The risk of an event is much the same as its **probability** (**p. 246**) — except that the outcome is, by convention, adverse. We do not talk of the risk of winning a lottery, though we might consider brightly the probability of such a welcome event. We could talk, using terms interchangeably, of the risks (or probability) of dying as a direct result of a hazardous medical procedure. Public perception of risk is not always rational — a man who smokes 20 cigarettes per day may not be prepared to travel by aeroplane because of the risk of the plane crashing. His annual risk of dying from cancer is >1 : 500; his risk of being killed in a plane crash is 1 : 20 million. Attitudes to risk vary and can be empirically demonstrated in clinical oncology — some patients are prepared to accept a high risk of operative mortality rate in seeking a surgical cure for lung cancer ('risk seeking'), others are prepared to compromise their chances of long-term cure to avoid the immediate risks of surgery ('risk averse'). The word 'dread' is useful in this context. We might disproportionately dread things that are very unlikely. Risk and danger are not equivalent. If I am late leaving home, there is a risk I might miss the train — but it is not dangerous to leave home late.

Ritumixab

Ritumixab is a genetically engineered antibody to the CD-20 antigen that is expressed on B-cells and their precursors. Nearly all B-cell non-Hodgkin's lymphomas express CD-20 and Ritumixab has been developed as an immunologically based therapy for these tumours. It is a chimeric antibody: the constant regions are human and the variable regions are murine in origin. Anti-idiotype antibodies, or the **human anti-mouse response** (**HAMA**) (**p. 147**), are unlikely to develop since the constant region is of mouse, rather than human, origin. The Fab portion of ritumixab binds to the CD-20 antigen, while the Fc portion stimulates immune cells to destroy the cells binding the antibody: both direct antibody-mediated cytotoxicity and complement-mediated lysis are important. The main toxicity is a flu-like syndrome with rigors and muscle cramps at the time of the infusion. Anaphylactic reactions may also occur.

RNA World

A theory about the origin of life in which RNA rather than DNA was the primary molecule in developing living organisms. The known forms of RNA are sufficient to support a process that could be called life: mRNA for storage; tRNA for **translation** (**p. 314**); rRNA for structures. Ribozymes would provide catalytic activity.

RNAase Protection Analysis

A technique used to quantify levels of a particular mRNA in a sample.

ROC (Receiver Operating Characteristic) Curve

A procedure used to optimise the performance of a test used in diagnosis or screening. The performance of a diagnostic test will depend on the cut-off value chosen to distinguish 'normal' from 'abnormal'. The parameters chosen may well depend on the circumstances under which the test is to be used. In screening a normal population, it may be preferable to sacrifice sensitivity to gain specificity. In assessing patients known to have a high **prevalence** (**p. 244**) of the disorder, the reverse might apply. A ROC curve is a useful means for deciding on suitable cut-off values. Sensitivity is plotted against 1− specificity (i.e. FPR) for a range of cut-off values.

Figure 1 shows ROC curves generated using raw data from a series of patients with non-seminomatous germ cell tumours (NSGCT): top left is the curve for PLAP; top right is the curve for LDH; bottom left the curve for HCG; bottom right is the curve for AFP.

Each datum point corresponds to a different threshold value used to discriminate between normal and abnormal test results. The line at 45° indicates a test with no predictive power; the further the curve bulges upwards and away from this line, the greater the predictive power of the test. This bulge can be quantified as an area-under-the curve (AUC) measurement. The data suggest that AFP and HCG are likely to be useful, but that PLAP has a limited predictive value. LDH might be helpful but the lack of a pronounced shoulder on the curve means that it might be difficult to define an optimal cut-off value.

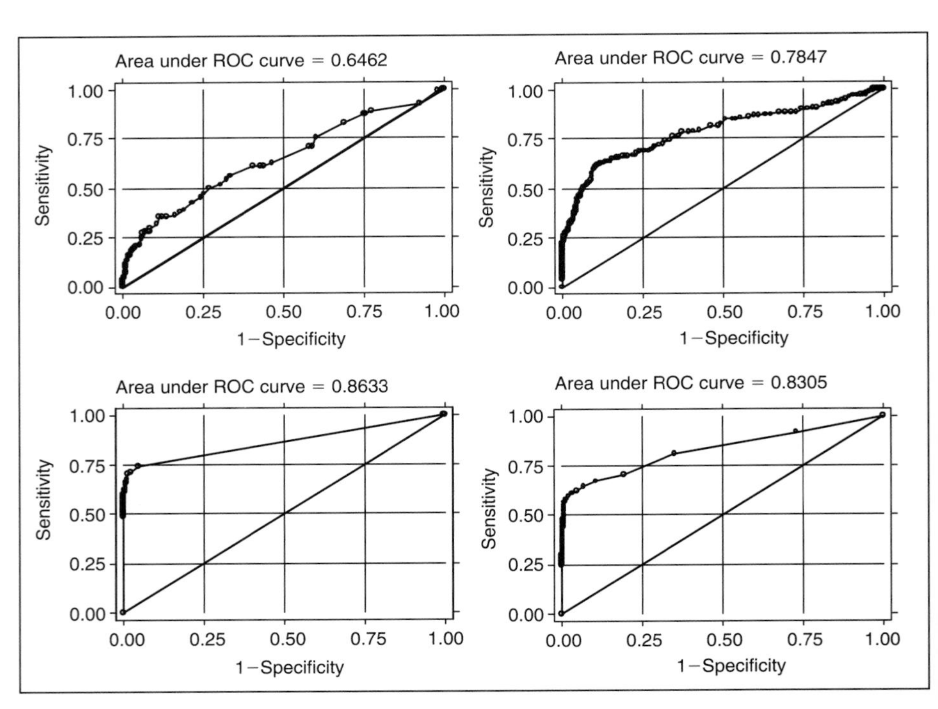

Figure 1.

:ule of Double Effect

'he rule of double effect is a doctrine
eveloped during the Middle Ages when
ıeologians were much exercised about right
ıd wrong, about the evil that men might do
ıd the eternal punishments they might
:ceive. The rule is simple: it is permitted for
ı action to cause harm if the effect is
ıticipated but unintended. Had the arrow
ıissed the apple, William Tell would not
ıve been damned for his son's murder.

In oncology, the most obvious application of the rule is in the use of opioids to treat pain in a patient with advanced cancer. The intended effect of the action is to relieve pain; the anticipated but unintended effect may be to shorten life. This rule clearly discriminates between the adverse consequences of well-meant actions and the fatal consequences of actions such as **euthanasia** (**p. 119**) where the only benefit, as such, is the victim's demise.

Salami Publication

A term used to describe that reprehensible practice whereby the same set of data is used in multiple publications. It is a means of increasing productivity without doing any more work. Not only it is dishonest, but also it is grossly misleading. Duplicate publication will vitiate the power of **meta-analysis (p. 196)** unless the duplicates can be correctly identified. This should be easy, but often is not. A recent overview of studies on the anti-emetic effects of Ondansetron showed that individual trials were being published two or even three times. Sometimes there was not a single author in common between the publications — and yet the numerical data were identical. Of 25 apparently separate trials, only 16 were unique. There were nine duplicate publications. Unsurprisingly, the conclusions of the meta-analysis were heavily biased in favour of treatment when all the duplicate studies were included. Elimination of the duplicate studies pushed the conclusions the other way: number needed to treat (NNT) (with duplicates) = 5; NNT without duplicates = 11.

Sample Size

Calculation of the appropriate sample size is important in the planning of a clinical trial. Small samples can prove nothing; large samples might prove something. In planning or interpreting a study, there is a straightforward sequence for the assessment of an appropriate sample size.

- Define a difference between the two groups, the magnitude of which you would accept as being clinically worthwhile. This will depend on factors including disease, patient population, the potential toxicities of the two treatments relative to each other, etc. For example, it would be clinically worthwhile if the 2-year survival rate of patients with **HIV (p. 151)**-related cerel lymphoma could be increased from the current 1% to 4%, provided the new treatment was safe and non-toxic. This would not be worthwhile if treatment involved craniotomy, resection of bulk disease, interstitial ^{192}Ir implant and high-dose chemotherapy. A useful way c looking at this is to use the **number needed to treat (NNT) (p. 216)**. NN' is simply the reciprocal of the **absolute rate difference (p. 1)** (%) between the two treatments: in the above example, th difference is 4 − 1%, i.e. 3%. The recipro of this is 100/3, which rounds to 33. This shows that you would need to treat 33 patients with the new, intensive treatment to produce one more survivor 2 years: 32 patients would be treated intensively but to no avail. When clinical benefit is marginal, NNT is a useful meth for putting the real issues clearly into focu
- Choose a level of statistical significance that will convince you, and others, that, should your study demonstrate a differen of this magnitude, this has not simply arisen through chance: conventionally, t is $p < 0.05$.
- Decide how sure you wish to be that yo study can demonstrate a difference of thi magnitude at the chosen level of statistic significance. The important factor to be considered here is the **power (p. 241)** o the study.

Study conclusion	True state	
	A better than B	**A no better than B**
A better than B	true positive result = power = (1 − β)	false-positive result = Type I error
A no better than B	false-negative result = Type II error (β)	correct conclusi

In most medical research power is chosen to be 80–90%, i.e. false-negative rates up to 20% are considered acceptable. Here s a worked example of **power (p. 241)** in ction. Treatment A is compared with reatment B:

Outcome	Rx A	Rx B
ive	3	1
Die	1	3

urvival after treatment A is 75% and after reatment B is 25%. *p* for this comparison is .48, i.e. it is non-significant. We cannot xclude the null hypothesis, that there is no ifference between the two treatments. It is ot unlikely that the result obtained has risen by chance. The study is continued, vith the following results:

Outcome	Rx A	Rx B
ive	302	90
ie	398	310

he survival difference is the same, 75% ersus 25%, but *p* is now highly significant: 0.0001. By increasing the size of the mple, the study has enough power to emonstrate a genuine difference with an cceptable degree of statistical significance.

Calculate, or use tables or nomograms to define, the sample size needed for your study: this calculation is based on simultaneous consideration of items 1–3.

etsky has published a method for lculating the power of a study once it s been completed: the advantage here is at the difference between the groups is own, we do not have to rely on estimates. his method is useful when we wish to clude the possibility of a Type II error ecting a published 'negative' study.

Satisfaction Measures

These measures attempt to measure the extent to which patients feel that the management of their condition has fulfilled their demands and expectations. As such, satisfaction measures open up a new set of **endpoints (p. 114)** for the evaluation of cancer treatment: endpoints that put the priorities of the consumer first. There are three main domains within which satisfaction can be assessed:

- Structure: is the system properly designed?
- Process: were the procedures carried out correctly?
- Outcomes: were the patients' aims achieved?

Mutual trust between doctor and patient and being afforded dignity and respect emerge consistently as the key elements when patients are asked to set priorities. Trust includes the ability of the patient to assume that the doctor is competent, efficient and will always act in the best interests of the individual patient.

The assessment of patient satisfaction is an entirely reasonable goal: indeed, the incorporation of patients' preferences and desires is a essential part of decision-making in oncology. Unfortunately there has been, particularly in the USA, an attempt to use satisfaction measures as a fiscal weapon. Funds are denied to institutions that cannot prove their patients are contented. Since there are few reliable instruments to measure satisfaction, this type of control is not justified by the available means.

Concepts useful when applied to short self-limiting episodes may not easily translate to more complex problems. It is easy to envisage a patient being satisfied with the management of their hernia; it is more difficult to conceive of measuring satisfaction in patients dying from advanced malignant disease. Satisfaction, in this latter context becomes a very relative thing.

Scales and Time

Biological events of considerable magnitude can be triggered by infinitesimal perturbations. A cell is very, very small — and it only takes one disordered cell for a malignant tumour to develop. In searching for causes and mechanisms in biology, there are some humbling facts that help place things into perspective:

- The retina can register the impact of a single photon.
- Six photons need to impact simultaneously on the retina for a signal to reach the cerebral cortex.
- A conformational change of 1 Å is sufficient to activate an enzyme and cause a cascade of cellular events.
- One mole of enzyme can produce 10 000 moles of product s^{-1}.
- The thickness of a cell membrane is 100 Å.
- The volume of the typical mammalian cell is 10^{-6} ml.

SCID (Severe Combined Immunodeficiency) Mouse

The SCID mouse is a useful model (a furry test tube) that can accept xenografts, e.g. human tissues or tumours. The SCID mouse repopulated with human peripheral blood lymphocytes (hu-PBL-SCID) has been widely used in studies of the pathogenesis of **HIV** (**p. 151**) infection.

Screening for Cancer

The purpose of screening for cancer is to diagnose asymptomatic disease in those who appear otherwise to be healthy. Several assumptions are involved: that in the community there are apparently healthy individuals who will unknowingly have cancer; that early tumours are more curable than more advanced tumours; that, as an extension of the above, by detecting and treating cancers early on in their clinical course the health of the community as a whole might be improved.

The concept, which at first seems simple and straightforward, turns out to be complex and complicated, by issues such as the performance of **diagnostic tests** (**p. 95**), **lead time bias** (**p. 176**) and **length time bias** (**p. 177**). In spite of the apparently compelling logic that underpins the concept of screening, it turns out to be a relatively expensive intervention in terms of life-years saved. Screening every 4 years for cancer of the cervix in young women costs ~£10 000 per life-year saved.

Some practical issues have to be addressed before a screening programme can be implemented. There has to be an intervention available to diagnose the cancer in its earliest stages ('intervention' in this context implies strategy that involves a test or combination of tests used to screen the population according to a predetermined schedule). This intervention must be: feasible, acceptable to patients, affordable and reliable. There must be evidence that earlier treatment of the cancer is effective, that it genuinely improves survival. The cancer for which screening is proposed should be sufficiently common to justify the effort and expense involved in setting up a screening programme. The overall economic benefits in terms of cost per **quality-adjusted life-year** (**QALY**) (**p. 253**) achieved through the screening programme should compare favourably with well-accepted medical interventions. Most interventions costing >£30 000 per life-year are unlikely to become routinely affordable.

'he sole criterion by which any screening rogramme for cancer should ultimately be idged is the following: is the **mortality ate (p. 207)** from a particular cancer in ıe screened population less than that in a omparable, but non-screened, population?)ther criteria such as yield, stage distribution nd comparison of survival rates between :reened and symptomatic patients have been sed to 'evaluate' screening programmes. 'hese criteria are heavily influenced by bias nd do not address the key issue: does :reening improve the health of the opulation as a whole?

'wo particular forms of **bias (p. 28)** affect ıe interpretation of screening programmes: **ad time bias (p. 176)** and **length time ias (p. 177)**. Three main factors influence ıe effectiveness of a screening strategy: :nsitivity of the test, specificity of the test nd prevalence of the disease in the screened opulation.

ome of the problems associated with the :lationship between the prevalence of the isease being sought and the behaviour of e screening test are now shown.

:reen a population of 100 000:

isease evalence	**Test A: sensitivity, 85% specificity 92%**		**Test B: sensitivity, 98% specificity 99%**	
	PPV (%)	**False-negative**	**PPV (%)**	**False-negative**
)0 per 1000	58	1500	91	200
)0 per 100 000	1.2	15	19	2

he positive predictive value (PPV) gives an dication of the probability that a patient 'hose screening test is positive actually has e disease. Bearing in mind the ychological morbidity associated with a ɔsitive test, and the expense and difficulty : subsequently proving that the 'patient' ɔes not actually have cancer, it is important that PPV be as close to 100% as possible. **PPV (p. 242)** ~50% indicates that, in terms of deciding whether someone actually has cancer, we would do as well tossing a coin as performing the test: heads they have cancer; tails they have not.

When a disease is uncommon, then even a test that is both sensitive and specific will produce mediocre results: test B has a PPV of only 9% when disease prevalence is 100 per 100 000. The test misses very few patients with cancer, only two of the 100 cases in the screened population of 100 000. A poor test performs poorly even when a disease is relatively common: test A has a PPV = 58% and, as importantly, misses 1500 of the 10 000 patients in the screened population.

Results obtained by screening females of 55–64 years of age for breast cancer and males aged 60–69 for prostate cancer: calculations are based on Scottish data for 1990:

Disease	**Test A: sensitivity 85%, specificity 92%**		**Test B: sensitivity 98%, specificity 99%**	
	PPV (%)	**False-negative**	**PPV (%)**	**False-negative**
Breast cancer	1.3	37	20	5
Prostate cancer	1.5	19	11	2

Quality control is an important issue in any screening programme. The logistics and economics of the programme are based on values for the sensitivity and specificity of the screening tests that may have been obtained under ideal circumstances. It is important to validate the procedure in the real world of everyday practice and to make sure that standards are maintained over the longer-term. These remarks particularly apply to procedures that have both a technical and an interpretative aspect to them, e.g. cervical screening or screening mammography.

Seamless Care

There should, from the point of view of the patient with cancer, be no abrupt transitions in clinical care. This is the concept of seamless care and it has been introduced to draw attention to some of the problems experienced by patients in the past. Patients should not have to cope with statements implying abandonment: 'There is nothing more I can do for you so I am going to ask Dr X to see you.' Such statements put both the patient and Dr X in an invidious position. Seamless care would imply that before the time to change the emphasis in management had occurred the patient would have met Dr X and been aware that she was an important member of the team. Seamless care is an attractive concept but it is often difficult to accomplish: not all members of a team can be available for every consultation. Nevertheless, regardless of the practical difficulties, it is a target to be aimed at, even though it might not always be achieved.

Second Malignancies

Second malignancies are one of the most unpleasant consequences of successful treatment for cancer. One of the ironies of the non-surgical treatment of cancer is that both radiation and chemotherapy are mutagenic: they have the potential to produce the very disease they are used to treat. The occurrence of second malignancies provides a classical example of a **stochastic (p. 298)** effect in oncology. As with most cancers the development of a second cancer is the result of an interaction between a genetically susceptible individual and an environmental insult, in this case treatment for cancer. One confounding issue is that given the genetic instability that is part of the malignant process, development of a second malignancy could be part of the underlying natural history of the primary tumour.

Another **confounding (p. 72)** problem is that, overall, the potential problem of second tumours may be underestimated for the simple reason that without long-term survival from the first tumour, there is little opportunity to develop a second tumour. Another major practical problem is that treatment of the second tumour may be compromised by the treatment of the original tumour.

The radiobiology of radiation-induced malignancy is complex. High doses in the centre of a **treated volume (p. 315)**, 50–65 Gy, are less likely to be carcinogenic than the doses of $\sim$5–1.5 Gy typically seen at the penumbra of a radiation treatment field. This is because the clonogenic potential of the cells in the higher dose region has been completely destroyed whereas the cells receiving lower doses have suffered a degree of damage sufficient to cause mutagenic changes in the DNA but insufficient to remove their clonogenicity. The classic mutations involved are point mutations, translocations and chromosomal rearrangements. It is also possible that radiation-induced activation of **oncogenes (p. 218)** is involved. The candidate genes include: *ras*, *fms*, *myc* and *abl*. Radiation may also induce malignancy through interference with tumour suppressor genes *Rb*, *p53* and ***DCC*** **(p. 88)**. Patients with abnormal repair **(p. 266)** genes, such as those homozygous for the A-T gene, have increased risk of second malignancy after radiation. The lifetime risk of death from malignancy for exposed adults is 8×10^{-2} Sv^{-1} for high dose rate and 4×10^{-2} Sv^{-1} for low dose rate.

The problem of second malignancies following treatment for childhood cancers is particularly important and especially distressing. Over 80% of children treated for cancer will survive: they will live long enough to run the risk of a second malignancy. These second malignancies are an important cause of lost life-years in this population. The most common second tumours are:

- Bone sarcomas (risk ×98–176, 2.8% at 20 years).
- Soft tissue sarcomas.
- Leukaemias (risk ×9–22).
- **Wilms' tumour** (**p. 328**).
- Brain tumours.

The risk of a second tumour varies according to the site of the first tumour: bone sarcomas according to original tumour site:

- **Retinoblastoma** (**p. 271**) ×999 (14% at 20 years).
- Ewing's ×649.
- Rhabdomyosarcoma ×297.
- Wilms' ×127.
- Hodgkin's ×106.

Following chemotherapy alone, the most common second malignancy is ANLL. The type of leukaemia can be related to the chemotherapeutic agents used: alkylating agents (M6 AML, with preceding myelodysplasia, 90% have translocations or deletions affecting chromosomes 5 or 7); podophyllotoxins (M4 or M5, no preceding myelodysplasia, 11q23 chromosome abnormality).

Although both radiation and chemotherapy are, in their own right, potent causes of second tumours, there is clear evidence of a supra-additive effect with the combination of drugs and radiation. It is for this reason that there is an increasing reluctance to use both modalities when one alone can suffice; hence, the declining role of radiotherapy in the management of nephroblastoma.

Secondary Prevention

Secondary prevention describes those policies of surveillance or treatment after successful treatment of a first tumour employed to prevent the development of a second malignancy. The occurrence of a second primary tumour, usually of the lung, is one of the major causes of death in patients treated for head and neck cancer. This reflects a common aetiology for both tumours: **tobacco use** (**p. 309**). The risk of developing a second primary may be reduced by treatment with retinoids (derivatives of vitamin A). Patients with curable head and neck tumours should, as a secondarily preventative measure, be strongly advised to stop smoking. By so doing, they not only will decrease their risk of second primary tumour, but also will improve their prognosis from the original tumour. Colonoscopy can be used after treatment for colorectal cancer to identify and remove any adenomas before they can develop into invasive cancers. **Tamoxifen** (**p. 304**), given after definitive treatment for breast cancer, will lower the risk of the original tumour recurring and help to prevent any new tumour developing.

Secondary Structure

A protein is not simply a long string of amino acids: the polypeptide chains are folded and convoluted in a variety of complex ways. These conformational arrangements are referred to as the higher structure of proteins. The primary structure is the amino acid sequence; the secondary structure reflects the folding of the chains under the influence of hydrogen bonds between different, non-adjacent amino acid residues. The two most frequent secondary structures are the β-pleated sheet and the α-helix. **Domains** (**p. 102**) and **monomers** (**p. 207**) are examples of **tertiary structure** (**p. 306**).

Selectins

A group of **adhesion molecules** (**p. 8**) crucially involved in interactions between blood cells and vascular endothelium. As such, selectins are concerned in inflammatory responses, tissue responses to decreased perfusion and the homing of lymphocytes.

Selective Toxicity

The principle of selective toxicity forms the basis for clinically useful cancer treatments. It aims to exploit any differences between tumour cells and normal cells in their ability to repair damage by choosing a schedule of treatment in such a way that, at the end of treatment, the normal tissues are relatively undamaged but the tumour cells, in the sense of being able to divide indefinitely, have been sterilised. The basic principle underlying the scheduling of treatment, be it with radiotherapy or chemotherapy, is that, compared with the tumour cells, normal cells will repair damage more promptly and more completely. This aim can be achieved by allowing the normal cells sufficient time to repair any damage caused by the treatment but not allowing any more time between treatments than is necessary for the repair of normal tissues. It is implicit within this approach that there will be some damage to the normal tissues (damage which will manifest itself as side-effects from treatment), but that this damage should not be permanent. It is, therefore, inevitable that treatment aimed at curing cancer will produce side-effects.

With an ideal schedule for cancer treatment, the net effect is that the normal tissues are only temporarily damaged and the tumour steadily shrinks. Unfortunately, this ideal is rarely practical. Usually, both normal tissues and the tumour are depleted. The duration of treatment will, therefore, be limited by the accumulated damage to the normal tissues. It may be necessary to stop treatment before the eradication of every last tumour cell: this would result in the eventual recurrence of the tumour. If treatment is ineffective then, even though damage to normal tissues may be kept within acceptable limits, simply maintaining the status quo is the best that could be achieved. This is fairly pointless since the patient still has the temporary toxicity without any long-term benefit. However, although these circumstances are incompatible with cure, they may be consistent with modest prolongation of life when compared with no treatment at all. The problem is that the acute, but temporary, toxicity associated with treatment is the price that has to be paid. This violates a major principle of **palliative treatment** (**p. 226**): that the treatment itself should not produce symptoms.

Of course, things could be even worse: treatment can be both toxic and ineffective, i.e. the patient suffers harm without benefit. Toxicity has been totally non-selective.

Selenium

Selenium is an important micronutrient: small quantities are essential for the production of glutathione peroxidase, a major defence against **oxidative stress** (**p. 221**).

Self-Renewal Probability

A term used in the description of the organisation of tissues and cells. If every cell division simply gives rise to one stem cell and one cell that is 'lost' (in the reproductive sense), then the self-renewal probability is 0.5. When some divisions give rise to two clonogenic cells but others do not, then self-renewal probability is between 0.5 and 1.0. If every dividing cell gives rise to two reproductively intact progeny, then self-renewal probability is 1.0.

Senescence

Cellular senescence describes the changes produced in a cell line over multiple generations. *In vitro* experiments show that or many human cell lines the total number f generations is limited to ~50. urthermore, the number of generations it is ossible to sustain *in vitro* bears an inverse elationship to the age of the person from hom the initial cells were obtained. ibroblasts from a foetus will undergo ~50 oublings *in vitro*; cells from a 40-year-old erson will yield ~40 generations; cells from a 80-year-old person will produce only 30 oublings. The molecular explanation for is may lie in the progressive shortening f the **telomere** (**p. 306**) observed, om generation to generation, in somatic ell lines. Once the limited number of enerations has been reached, the cells in ılture enter an irreversible G_0-phase. Both ancer and germ cells are effectively nmortal: there appears to be no limit on e number of possible doublings a cell line n undergo. A possible approach to the eatment of cancer would be to identify, d recruit, those mechanisms that induce nescence, that then to convert an definitely cycling population into one hose members all eventually pass reversibly into G_0.

ensitivity Analysis

his concept is applicable across a broad nge of scientific investigation. In analysing ts of data or in modelling (**p. 202**), we ften need to make assumptions, or guesses, bout the values for certain variables. This ill usually mean selecting a value from ithin a range of plausible values. This pproach is all very well if the conjecture is ccurate, but what if the guess is incorrect? ow will this affect the conclusions from the udy? Sensitivity analyses attempt to provide e answers to these questions.

he basic principle of a sensitivity analysis that we repeatedly (iteratively in the rgon) analyse the data keeping all other parameters, other than variable under consideration, constant. We allow the variable of concern to vary within plausible limits. The variation can be systematic, e.g. incremental or random, as in a Monte Carlo analysis. The effects of these variations are assessed on the outcome of the analysis. If the original conclusion is sensitive to these perturbations, then view it with caution. If the conclusion is robust, there is no need to fuss too much about the precise value chosen as the variable. Sensitivity analysis offers a means of discriminating between those critical variables whose values need to be known exactly and those for which precise information is not essential.

Examples of sensitivity analysis could include assessing the robustness of the conclusions of a **meta-analysis** (**p. 195**) by investigating the effects of **bias** (**p. 28**) on the overall conclusion: drop the most strongly positive study and see whether the overall conclusion is affected.

In the analysis of the therapeutic ratio (**p. 307**) for a novel fractionation scheme, we need to make assumptions about α/β ratios for both late effects and the tumour. There are no means of knowing what the absolute values should be for the analysis but we can define the upper and lower limits. In a sensitivity analysis using the **linear–quadratic** (**LQ**) **model** (**p. 180**), it is possible to assess the relative effects on BED of using various α/β ratios.

Sensitivity analysis is extremely useful in studies on the economics of cancer treatment or other forms of **cost–benefit analysis** (**p. 76**). It is not necessary to know costs exactly: simply assess a plausible range of values for cost by using sensitivity analysis. A similar approach to outcomes can also be used by applying a series of quality adjustments to survival times and assessing how this affects the cost per QUALY.

Sentinel Node

A sentinel node is the node within a group of regional nodes that is the first to be involved in the spread of disease. These nodes can be identified using staining with vital blue dyes (e.g. Lymphazurin) or by immunolymphoscintigraphy or other isotopic methods. If the sentinel node for a given lymph node area, groin or axilla, for example, is known, then biopsy of the sentinel node should indicate whether the whole nodal basin is involved. If the sentinel node is negative, then no further treatment may be necessary; if it is positive, then a formal lymphadenectomy might be indicated. The biopsy of sentinel nodes is being used as an aid to clinical decision-making in the management of breast cancer and malignant melanoma.

Sequence Homology

The extent to which the nucleotide sequences in separate samples of DNA are similar is described as sequence homology. The concept is crucial to the consideration of molecular evolution as well as to how different genes might code for proteins with similar functions. If sequence homology exists for DNA sequences across species, then this suggests that these sequences have been conserved, relatively intact, ever since those species diverged from a common ancestor. These sequences may have been conserved for many millions of years. This, in turn, suggests that the products of these genes might have functions vital for normal growth and development.

Sequence homologies between some **oncogenes** (**p. 218**) and **growth factors** (**p. 144**), e.g. between c-*erbB2* and EGFR, provide examples of how sequence homology might indicate the function of a gene product:

Growth factor/ receptor	Oncogene with homology
PDGF (GF)	c-*sis*
CSF-1R (rec)	c-*fms*
SLFR (rec)	c-*kit*
EGFR (rec)	c-*erbB*
IGF (GF)	c-*ros*
NGF (rec)	c-*trk*
HGF (rec)	c-*met*

GF = growth factor; rec = receptor

Sequence Tagged Sites (STS)

This concept is based on an expansion of the **polymerase chain reaction** (**PCR**) (**p. 238**) approach. A STS is a unique sequence of genomic DNA defined by a pa[ir] of PCR primers. The typical STS is 100–1000 bp, which can be used to tag, uniquely, a longer length of DNA, be it a plasmid clone (5000–20 000 bp) a **cosmid** (**p. 76**) (30 000–50 000 bp) or a **yeast artificial chromosome** (**p. 330**) (up to 10^6 bp). By looking for clones that share an identical STS, i.e. overlap, a physical map can be constructed. The STS sequence can be stored in computers and, in contrast to DNA libraries maintained *in vitro*, information can be rapidly shared without the need physically to transport DNA clone[s] between laboratories. The Human Genome Project is using STS to map the genome.

Serial Architecture

This describes how a tissue is organised in terms of its functional units. A serial pattern implies that, just as a chain is broken if a lin[k] is broken, then so if one subunit in a serial system is inactivated, then the whole system fails. An example of such a system of relevance to oncology would be the spinal cord. If a small segment of spinal cord is irradiated beyond tolerance, then the cord may effectively be transected and paraparesis results. Even if the tolerance of an organ is

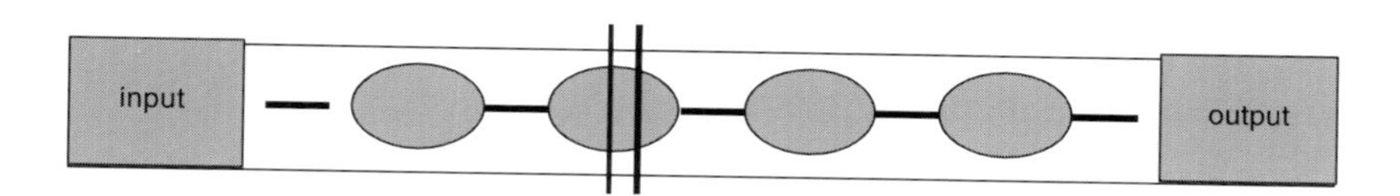

Figure 1.

high, in terms of dose, if it is organised in a serial fashion, then it is vulnerable even if only a small part of it is damaged. This is contrast to a **parallel architecture** (**p. 227**) in which there is built-in redundancy and several subunits have to be inactivated before the system fails (figure 1).

If a single subunit is inactivated, then the system as a whole fails. The important feature of this aspect of tissue organisation is that vulnerability to insult depends on two main features: the sensitivity of the individual subunits and the organisation of those subunits.

Serpins

A family of serine protease inhibitors. Their main role is in the control of the cascades concerned with blood clotting and the inflammatory response.

SF 36

Derived from the original RAND study, SF 36 represented an attempt to produce a shortened version of the comprehensive instrument developed by the RAND Corporation to measure health status. An original 20-item measure proved too brief and so the final 36-item measure was developed. It has proven **reliability** (**p. 265**) and **validity** (**p. 322**), but it is not age-specific: the questions asked of children are the same as those asked of septuagenarians. This may not be entirely sensible, although in psychometric terms there may be advantages to using a completely uniform instrument. It covers eight domains:

- Physical function.
- Physical role.
- Bodily pain.
- General perception of health.
- Vitality.
- Social function.
- Emotional health.
- Mental health.

It can be self administered, a trained interviewer is not essential.

Sickness Impact Profile

This is a complex profile-based method for measuring health status. It contains 136 items within 12 categories. There are three main clusters of category: independence, physical and psychosocial.

- Independence:
  - — sleep and rest
  - — eating
  - — work
  - — home management
  - — recreation and pastimes.
- Physical:
  - — ambulation (American for walking?)
  - — mobility
  - — body care and movement.

- Psychosocial:
  - — social interaction
  - — alertness behaviour
  - — emotional behaviour
  - — communication.

Sievert

The sievert is a unit used primarily for radiation protection. It is the dose-equivalent for a particular type of radiation and is calculated as:

$$H = DQN,$$

where H is the dose equivalent in sieverts, D is the dose (Gy), Q is the **quality factor** (**p. 253**) and N is a factor introduced to allow for differences in dose-rate or fractionation. Conventionally, 1 is substituted for N.

Signal Recognition Particle

A protein complex in the cell membrane that binds the amino acids of newly synthesised proteins. It acts as guide for the protein in the initial stages of its journey across the cell membrane.

Signal-to-Noise Ratio

A concept borrowed from wireless telegraphy. Originally the signal was the information being transmitted; noise was the extraneous sounds on the line, i.e. crackles, whistles, hisses and hums. The concept has been generalised so that the signal is the voice of truth, the noise is everything that obscures and confuses the detection of the signal. Maximisation of the signal : noise ratio, therefore, is an essential prerequisite for good **experimental design** (**p. 122**). The same framework applies to the presentation and interpretation of experimental results. It is the truth we need to hear, not extraneous cacophony.

Significance Testing

A statistical significance test will indicate no more than whether the result observed is likely to have arisen by chance. Clinical significance is an entirely separate issue. There is not necessarily agreement between the two types of significance: a large trial might show an improvement in survival from 50.5 to 51.0%: this is not clinically significant even though p might be <0.000001. Conversely, if you treat two patients with a new drug and one dies from an unexplained acute encephalopathy and this has never happened with the previous treatment, then this is clinically significant, whatever any value of p might indicate.

Statistical significance tests the null hypothesis — that there is no real difference between the two groups and that any observed difference is simply due to the play of chance. The null hypothesis epitomises the spirit of scepticism and refutation that should inform any scientific enquiry.

Sample size is important: small samples can prove nothing; large samples might prove anything. In planning or interpreting a study, there is a straightforward sequence for the assessment of an appropriate sample size:

- Define a difference between the two groups, the magnitude of which you would accept as being clinically worthwhile. This will depend on factors including disease, patient population, the potential toxicities of the two treatments relative to each other, etc. For example, it would be clinically worthwhile if the 2-year survival rate of patients with **HIV** (**p. 151**)-related cerebral lymphoma could be increased from the current 1% to 4% provided the new treatment was safe and non-toxic. This would not be worthwhile if treatment involved craniotomy, resection of bulk disease, interstitial ^{192}Ir implant and high-dose chemotherapy. A useful way of looking at this is to use the **number needed to treat** (**NNT, p. 216**). NNT is simply the reciprocal of the **absolute rate difference** (%) (**p. 1**) between the two treatments: in the above example, the difference is 4 − 1%, i.e. 3%. The reciprocal of this is 100/3, which rounds to 33. This indicates that one would need to treat 33 patients with the new intensive treatment to produce one more survivor at 2 years: 32 patients would be treated intensively but to no avail. When clinical benefit is marginal, expressed in percentages NNT is a useful method for putting the real issues clearly into focus.

Choose a level of statistical significance that will convince you, and others, that, should your study demonstrate a difference of this magnitude, this has not simply arisen through chance: conventionally, this is $p<0.05$.

Decide how sure you wish to be that your study can demonstrate a difference of this magnitude at the chosen level of statistical significance. The important factor to be considered here is the **power** (**p. 241**) of the study.

tudy onclusion	True state	
	A better than B	**A no better than B**
. better than B	true positive result = power = (1 − β)	false-positive result = Type I error
. no better than B	false-negative result = Type II error (β)	correct conclusion

ı most medical research, power is chosen as 0–90%, i.e. false-negative rates up to 20% re considered acceptable. Here is a worked xample of power in action. Treatment A is ompared with treatment B.

Outcome	Rx A	Rx B
Live	3	1
Die	1	3

Survival after treatment A is 75%; after treatment B it is 25%. p for this comparison is 0.48, i.e. it is non-significant. One cannot exclude the null hypothesis, that there is no difference between the two treatments. It is likely that the result obtained has arisen by chance. The study is continued, with the following results:

Outcome	Rx A	Rx B
Live	302	90
Die	98	310

The survival difference is the same, 75 versus 25%, but p is now highly significant: <0.0001. By increasing the size of the sample, we have given the study enough power to demonstrate a genuine difference with an acceptable degree of statistical significance.

- Calculate, or use tables or nomograms to define the sample size needed for the study: this calculation is based on simultaneous consideration of items 1–3.
- Detsky has published a method for calculating the power of a study once it has been completed: the advantage here is that one knows the difference between the groups. We do not have to rely on estimates. This method is useful when we wish to exclude the possibility of a Type II error affecting a published 'negative' study.

The ***t*-test** (**p. 303**) is often used as a test of statistical significance. The principle behind it is as follows. If we have a small sample taken from a general population, we can calculate a mean and **standard deviation** (**SD**) (**p. 296**) for the sample, but we cannot measure the mean and SD for the population as a whole.

If we increase the size of the sample, e.g. to >30, then we can be more certain that the SD observed in the sample is about the same as the SD in the population as a whole. The ***t*-test** (**p. 303**) applies a correction factor to the normal distribution to allow for the fact that, when using small samples, the SD in the sample cannot be assumed to be the same as that for the population as a whole. As sample size increases, then so the *t* distribution increasingly approximates to a Gaussian distribution. Note that it is an underlying assumption of the *t*-test that the data follow a normal distribution.

The **degrees of freedom** (**d.f.**) (**p. 91**) in a *t*-test are used to incorporate a consideration of sample size: for unpaired data, the d.f. = (number of observations − 1); for paired data, d.f. = (number of subjects − 1).

For a *t*-test to be an appropriate test of significance, the following criteria must be met:

- Is it acceptable to calculate a mean? We could not calculate a mean for T-stage or mucositis-graded on a five-point scale since these are ordinal data not measurements.
- Is distribution Gaussian (normal)? This is mainly important at small (<30) sample sizes.
- Are the samples independent of each other? Have **bias** (**p. 28**) and **confounding** (**p. 72**) variables been eliminated as far as is possible?
- Is the variance of the two samples the same? If this is not known, then use the *F*-test. The *F*-test involves calculating the ratio of the variances (larger : smaller) in the two samples and looking it up in a table of *F* distribution (d.f. sample 1 = $(n_1 - 1)$; d.f. sample 2 = $(n_2 - 1)$). If *p*, corresponding to *F*, is significant, then it is unlikely that the populations have the same variance: you should not use a *t*-test to compare them.

Another extremely useful test of statistical significance is the χ^2-test. This is based on a contingency (syn. frequency) table and is widely used in epidemiology. Its foundation is the comparison of the observed, as opposed to the expected, values within each group; using the data from both groups combined to calculate the expected values.

Parametric versus non-parametric

These terms sometimes cause confusion. Parametric tests, and the *t*-test is a good example, rely on data sets conforming to known distributions: the statistical properties of the distributions are known and so the distributions and, by extension, the data can be compared. The distributions can be defined in terms of a minimum number of parameters, typically the mean and SD; hence, the term 'parametric'. Non-parametric (syn. distribution-free) methods make no assumptions about the shapes of the distributions of the two sets of data. They work on the principle of putting the pooled data into rank order and then assessing whether the two groups of data are randomly intermingled, or whether one group tends to be at the top of the order and the other towards the bottom. The **Wilcoxon-signed ranks test** (**p. 328**) would be an example of a non-parametric test of this type. Before computers were widely available, the non-parametric methods were tedious and time-consuming and parametric methods were often preferred simply because of convenience. Nowadays, in terms of effort, there is little to choose between the two types of method. Parametric tests tend to be more powerful statistically. Through the use of a known distribution a few data points are converted into an infinite number (since a line is an infinite number of points). If there is a choice between performing a parametric test or a non-parametric one, ne usual advice is to perform the **parametric** **est (p. 228)** since it is more likely to be gnificant. The proviso is that the data in ne groups to be compared should follow istributions that are approximately normal. f we do not know how the data are istributed, then it is probably better to use non-parametric method. If the data are kewed or if the values to be compared are cores rather than measurements, then on-parametric methods must be used.

Confidence intervals versus significance tests

The heyday of the significance test is ast: some journals will not publish omparisons based solely on significance ests. **Confidence intervals** (**CI**) (**p. 70**) re now more popular, and not without eason:

- A significance test by itself says nothing about the magnitude of any observed difference.
- Statistical significance indicates only that what has been observed is unlikely to have arisen by chance.
- Concentrating on p as an outcome measure may lead to publication **bias** (**p. 28**).
- Worship of p leads down the road to multiple testing: take multiple peeks at data on work in progress. Stop the study and publish when a 'significant' p is achieved. If you take 30 peeks at a negative study, it is highly likely that one of the peeks will yield a p value <0.05.
- CI provide an easy means for assessing visually the power of a study: wide intervals indicate low power.
- The magnitude of any difference can be easily assessed using CI.
- The **central limit theorem** (**p. 61**) implies that it is entirely proper to use CI to compare non-normally distributed data.

Single Photon Emission Tomography (SPECT)

This is an imaging procedure that applies tomographic principles to conventional isotope scanning. Unlike PET scanning, it does not rely on positron-emitting isotopes. SPECT scanning can be performed using standard isotopes such as ^{99m}Tc used in imaging. The scintillation detector is rotated around the patients and the images are reconstructed on the basis of the single photon emissions. The resulting image is similar to a CT slice, but with poor spatial resolution. The advantage of SPECT is that, as with PET scanning, both function and structure can be imaged.

Small Round Cell Tumours

A term used to define a group of histologically similar, but clinically heterogeneous, tumours. Microscopically these tumours are comprised of small uniform cells with a high nuclear : cytoplasmic ratio. The important feature is that many of these tumours are treatable, even when disseminated. It is essential that, first, curable disease is recognised and, second, that the treatment given is appropriate to the particular tumour type. The differential diagnosis is extensive and surface markers are often invaluable in correctly defining the precise tumour type. Other techniques that may prove useful in characterising these tumours include electron microscopy (identification of neurosecretory granules in tumours of neuroendocrine origin, whorls and intermediate filaments in extrarenal rhabdoid tumour; myofilaments and Z-bands in rhabdomyosarcomas).

Tumour type	Markers
Peripheral neuroectodermal	neuron-specific enolase
tumour (PNET)	(NSE), HBA71
Ewing's sarcoma	neuron-specific enolase (NSE), HBA 71
Neuroendocrine carcinoma of skin (**Merkel cell carcinoma**) (**p. 195**)	neurofilament protein (NFP), epithelial membrane antigen (EMA)
Small cell carcinoma of viscera	epithelial membrane antigen (EMA),
cytokeratin	
Granulocytic sarcoma	myeloperoxidase,
lysozyme, elastase,	
	Leu-M1, CD43, CD68
'Large cell' lymphoma	leucocyte common antigen (LCA) L26 UCHL1 Ki1
Monomorphic diffuse tenosynovitis/pigmented villonodular synovitis	vimentin
Epithelioid leiomyosarcoma	vimentin
Mesenchymal chondrosarcoma	vimentin
Cellular extraskeletal myxoid chondrosarcoma	vimentin

Tumour type	Markers
Neuroblastoma (esthesioneuroblastoma)	neuron-specific enolase (NSE), neurofilament protein (NFP), S-100
Olfactomy neuroblastoma	neuron-specific enolase (NSE), S-100, neurofilament protein (NFP)
Extrarenal malignant rhabdoid tumour	cytokeratin, vimentin, epithelial membrane antigen (EMA)
Desmoplastic small round cell tumour (DSRCT)	cytokeratin, vimentin, desmin
Small cell osteosarcoma	vimentin
Primitive malignant peripheral nerve sheath tumour (primitive MPNST)	S-100 also neuron-specific enolase (NSE)
Round cell liposarcoma	Vimentin, S-100
Embryonal, alveolar and poorly differentiated rhabdomyosarcoma	ACT (α-sarcomeric actin)
Germ cell tumour	AFP (a-foetoprotein) β- HCG (human chorionic gonadotrophin) placental alkaline **phosphatase** (**p. 233**) (PLAP)

Social Class and Deprivation

The importance of these factors in oncology is increasingly apparent. There is evidence both from the USA and UK that social deprivation has a major influence on cancer survival. The reasons for this may be complex and could include:

- Inability to appreciate cardinal symptoms causing delay in diagnosis.
- Impaired access to medical care.
- Less rigorous medical care:
  - — delayed diagnosis (symptoms dismissed as trivial)
  - — incomplete staging (limited access to scans, etc.)
  - — inadequate treatment (unable to afford drugs, etc.)
  - — poor follow-up (relapse detected late).
- **Confounding** (**p. 72**) relationships: obesity (high **body mass index**) (**p. 33**), poverty and race may all be interrelated in terms of their effects on cancer incidence and survival.
- Genetic determinists might believe that the same genes that make you poor also make your cancer more aggressive.

Whatever the mechanisms involved, be they social, economic or biological, there are observable differences in cancer survival according to social deprivation. Data from the South East Thames Cancer Registry on breast cancer survival rate at 10 years gives: most affluent 59%; most impoverished 48%. This 11% difference is greater than the benefit achieved by using adjuvant chemotherapy for node-positive breast cancer in premenopausal woman. It is ironic that, as survival is improved through the introduction of better treatments, much of the improvement may be eroded by socio-economic change. The proportion of the UK population on <50% of average income was 7% in 1977; by 1995 it was 24%. In 1994, 49% of heads of household in rural Scotland had incomes <25% of the national average. As the rich get richer and the poor get poorer, socio-economic gradients in cancer survival will become steeper.

The OPCS definitions of social class are:

(1) Professional.

(2) Intermediate occupations.

(3) Non-manual, skilled.

(4) Manual, skilled.

(5) Partly skilled.

(6) Unskilled.

It is an unpleasant quirk of this system that a female spouse is given the social class of her husband's occupation and an unmarried woman is assigned to the social class of her father. A neurosurgeon married to a street cleaner will be social class I if he is male; if she is female then she will be in social class V. Perhaps deliberately, the system used for defining class will, in general, tend to obscure any inequalities between the well-off and the deprived.

To apply a little more fine structure to the problem of social deprivation, other classification systems have been proposed. The Townsend score is based on data from each household concerning: members economically active but unemployed; car ownership; overcrowding; owner-occupation. The **Carstairs index** (**p. 47**) is a similar measure but rather than owner-occupation it uses the social class of the head of the household.

Patients living in deprived circumstances can be identified using their postcodes: the postcode can be linked to the census enumeration district and a deprivation index can be calculated for that district by using the published census data. Each census enumeration district consists of ~400 households. This approach works well for urban areas but may not be so accurate in rural communities: where the serfs live at the bottom of their liege's garden. Gentrification may also be a problem, census data may be 20 years out of date and neighbourhoods change. Another difficulty, particularly in rural areas, is that because public transport is so poor, car ownership is essential. Car ownership cannot be used as a marker of affluence.

The **Calman–Hine** (**p. 39**) **report** represents an attempt to come to terms with some of the more observable inequalities in cancer survival statistics, the so-called cancer lottery.

SOCS (Suppressors of Cytokine Signalling)

Proteins involved in self-renewal processes in haemopoietic stem cells.

Somatic Mutation

A somatic mutation affects somatic, but not germline, cells. It is acquired after birth, rather than being inherited. The mutation is not transmitted through the generations and dies with the individual who bears it.

Source Skin Distance (SSD)

Distance between the source of X-ray production and the skin of the patient. When the source is very small, as in a linear accelerator, then the term 'FSD' is appropriate. When the source has finite dimensions, as in a cobalt unit, then SSD is the better term.

Southern Blotting

A technique used to demonstrate particular sequences, for which specific probes are available, within a sample of DNA fragments. The DNA is first cut into fragments using **restriction enzymes** (**p. 271**); these fragments will be of different sizes and, therefore, will migrate with different speed when subjected to an electric field. The **restriction digest** (**p. 271**) is electrophoresed on an agarose gel. Nucleic acids carry negative charge; the smaller fragments will migrate towards the anode more rapidly than the larger fragments. The fragments will be smeared across the gel, smaller fragments near the anode and larger fragments nearer the cathode. The fragments in the gel cannot be manipulated, they need to be transferred to a tractable medium. After electrophoresis the DNA on the gel is denatured and then transferred, by capillary action, to a nitrocellulose filter. The filter is exposed to UV light, which bonds the DNA to the filter. The fragments on the filter can be hybridised *in situ* with radioactive probes and the binding patterns of different DNA samples can be compared.

The technique permits identification of gene deletions, amplifications and translocations. The resolution of the technique is poor once the fragments are >20 kb in length. Other techniques, such as **pulsed-field gel electrophoresis** (**p. 251**), have been developed to deal with these larger fragments.

SP Cells

SP cells are haemopoietic stem cells that have been identified in both mouse and human. They are a distinct subpopulation in that, unlike most haemopoietic stem cells, they do not express **CD34 antigen** (**p. 52**). They can be identified by their ability to exclude Hoechst 33342, a fluorescent dye that binds to DNA. These cells, comprising <0.1% of all marrow cells, can reconstitute haemopoiesis in a lethally irradiated recipient.

Spam, Spamming

Junk email; sending junk email.

Spermatogenesis

This is the process whereby sperm is formed. It is based on a hierarchical system that has implications for the effects of radiation and cytotoxic treatment on the testis. The time to expression of damage is dictated by the time-course of normal spermatogenesis. It takes ~75 days (in man) to progress from **stem cell** (**p. 298**) to mature spermatozoon. There is a stem cell compartment, a proliferative (amplificatory) compartment, followed by a maturation compartment. The germinal epithelium is dependent on Sertoli cells for normal function (analogy with the haemopoietic micro-environment). **Apoptosis** (**p. 19**) is a physiological mechanism in the testis: a method for maintaining integrity of the genome, eliminating mistakes, and also for dealing with overproduction.

Spheroids

An *in vitro* system developed by R. Sutherland for investigating tumour cells. Spherical clumps of cells are grown in suspension, with each clump containing

100 000 cells. Histologically, the structures semble the tumour cords described in homlinson and Gray's classic paper of 1955 n the diffusion of oxygen into tumours. pheroids may, therefore, approximate more osely to *in vivo* conditions than is possible ith tumours growing in monolayer ultures. Cell types that will grow as heroids include: EMT6 mouse mammary ancer, rat 9L brain tumour, mouse RIF brosarcoma and hamster lung cancer line 79. It is possible to investigate questions of xygen diffusion and clonogenic survival in a irly standardised fashion. The advantage, hich when real life is also considered, ay be a disadvantage, is that there is no asculature, and, therefore, that nutrition and xygenation depend simply on diffusion.

pindle Poison

pindle poison is a jargon term used to efine those agents, usually drugs, that disrupt **mitosis (p. 201)**. These drugs affect the assembly, or dismantling, of the **mitotic spindle (p. 201)**. Examples include **vincristine (p. 324)**, **vinblastine (p. 324)**, the **taxanes (p. 305)** and colchicine.

Spleen Colony Assay

This technique was first described by Till and McCulloch. It represents a method of assessing clonogenic cell survival after irradiation in which the cells are irradiated *in vivo* and assessed *in vivo*, but not *in situ*. Fat pad assays in which suspensions of irradiated cells are injected into the fat pads of host animals are similar in concept. The original spleen colony assay is shown in figure 1.

The marrow cells are irradiated *in vivo* and the relative survival of stem cells after graded doses of radiation is assayed by counting the regenerating marrow colonies in the spleens of lethally irradiated recipient animals.

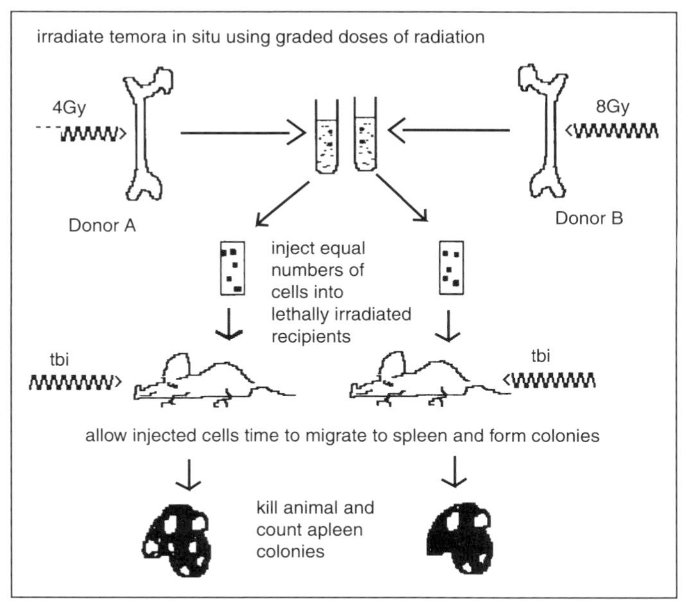

Figure 1.

Stable Disease

A term used in the evaluation of cancer treatment. The criteria that must be fulfilled for disease to be classified as stable are:

- There is a <50% reduction in the product of the diameters of assessable lesions. Assessment can be clinical or radiological.
- This reduction in size lasts at least 4 weeks.
- No new lesions develop.
- No other lesions grow in size during this period.

The presence of stable disease does not imply that a treatment is totally ineffective. Disease stability can be clinically useful in many circumstances, e.g. in metastatic cancer of the breast or prostate.

Stage Shift: How to Improve Outcome without Improving Treatment

A change in staging system, or in the techniques used to provide baseline information concerning staging, can produce 'benefits' to patients in all stages of the disease. These benefits are, however, entirely artefactual and depend simply on patients in each stage being enriched by patients with improved prognosis. The important cross-check to protect against being misled by stage shift is that the prognosis for the entire group (i.e. all stages pooled) has not been changed. The easiest way to appreciate the effects of stage shift is through a worked example.

Here (table 1) are data on the distribution of disease and on survival. The data were calculated before and after the introduction of a new test, CT chest, that could detect lung metastases which would have been missed on conventional chest X-ray.

When the new test is used, patients with very small lung metastases shift, or migrate, from Stages I to IV.

Table 1

Before				**After new staging test**			
Stage	**Distribution (%)**	**Cure (%)**	**No.**	**Stage**	**Distribution (%)**	**Cure (%)**	**No.**
I	70	90	63	I	50	94	47
II	10	80	8	II	10	80	8
III	10	80	8	III	10	80	8
IV	10	50	5	IV	30	70	21
All	100		84	All	100		84

Stage shift is also known as the **Will Roger's phenomenon** (**p. 328**) ('when the Okies moved to California they raised the intelligence level in both states'); overall survival does not improve even though the survival in each of the stages involved in the shift improves. Occult Stage IV patients are removed from Stage I, so survival in Stage I improves. Stage IV now includes patients with minimal disease and so survival in Stage IV improves. The phenomenon was probably first described by Eric Easson at the Christie Hospital.

Staging

Staging can be regarded as that series of investigations employed to map out the extent of the disease. In its simplest form, staging involves no more than the careful clinical examination of the patient and the definition of sites of disease according to a formal system of rules. The TNM (tumour nodes metastasis) system is the most widely used clinical staging system for cancer. Since each of the three categories can have several options, the result is, for any given tumour site, a large number of possible permutations of TNM staging. Various collapsed systems (such as the American Joint Committee, AJC) ave been used to get round this problem: the ·40 possible TNM categories are simplified ıto four groups (I–IV).

.evisions of the TNM system have moved om staging based purely on clinical data ıd have incorporated information from istopathology, e.g. depth of **invasion** **ɔ. 167**) of bladder wall for transitional cell ırcinomas of the bladder. For many cancers ıere are several rival staging systems, some ırely clinical, others based on a panoply of ıvestigations including imaging, tumour ıarker estimations, etc. This superfluity of ·stems indicates that there is no one ideal ·stem and the root of this problem lies in ıe fact that staging systems have to serve ·veral different purposes and that these ifferent aims may impose conflicting emands upon staging systems.

he most obvious example of such ifficulties is the 1978 TNM system for head ıd neck cancer. The staging system for the eck nodes was primarily aimed at scriminating between disease that was perable by radical neck dissection and that hich was inoperable: fixity of nodes was, ıerefore, an important factor. The system as of little use to radiotherapists who were ıuch more concerned with issues such as ze of disease and likelihood of occult sease contralaterally or in the low neck. ıbsequent revisions of the TNM system present an uneasy compromise between ıese two viewpoints.

.easons for staging cancer:

To provide prognostic information.

To facilitate rational therapy.

To provide a standard for comparison.

ɔ provide prognostic information

ıtients want to know what their chances e. These chances will depend very much on how advanced their disease is. To assess this reproducibly, so that the past can inform the future, we need to have a system that enables the current patients to be matched to patients treated in the past for whom the outcome is known. The problem is that the techniques used in the baseline assessment of patients change over time and, through the mechanism of **stage shift** (syn. Will Roger's Phenomenon, Easson's Law) (**p. 328**), patients with identical clinical stage may have very different outcomes.

To facilitate rational therapy

Staging before definitive treatment is started can prevent the use of therapy that is futile or inappropriate. Imaging the liver before surgery for rectal cancer might detect occult disease that would influence whether an abdominoperineal resection was appropriate. Similarly an abnormal chest X-ray would influence whether a radical resection should be performed for a patient with cancer of the head and neck. The approach is simple. Before starting treatment the extent of any detectable disease should be known and documented.

To provide a standard for comparison

To define optimal therapy for cancer, we need to be able to compare the results of cancer treatment. We have to compare different treatment regimens, results obtained at different centres and results achieved at present compared with those obtained in the past. In all these circumstances we need to know that like is being compared with like. It is of little interest to compare the results of screen-detected breast cancer with the results of treatment for fungating inoperable tumours. To facilitate this process of comparison, we need a shorthand for defining comparable groups of tumours.

Standard

A standard is an attribute or specification against which something can, for comparative purposes, be measured. Standards are increasingly being applied in oncology to try to even out some of the inequalities in outcomes and patterns of care observed both within and between nations. A standard can establish a set of criteria that defines an acceptable minimum; this is the usual usage. A standard set unattainably high, no matter how attractive in theory, will lack credibility. Setting standards can be regarded as an exercise in pragmatism: high enough to improve the worst; low enough to be achievable within the available resources.

Standard Deviation (SD)

SD represents an attempt to describe the variability in a set of observations in terms of a single number. Since variables and variability are the very essence of statistics, SD is an important basic concept. It is derived from the variance, a term which describes the average distance that each of a set of observations lies from the mean of the set of observations. SD is extremely useful when interpreting data that follows a Normal distribution: we can estimate accurately the chances of an observation lying in a range from mean $- 1.96 \times$ SD to mean $+ 1.96 \times$ SD. The probability is 95% with only 5% of observations lying outwith this range. It is this knowledge that enables **confidence intervals** (**CI**) (**p. 70**) to be constructed for an estimate.

Standard Error (SE)

SE is derived from standard deviation (SD) but the two are often confused. SD gives an indication of the variability within any given set of data; SE gives an indication of where the true population mean, for the data set, is likely to lie. We have to return to a consideration of the hypothetical parent population from which any observations one might have can only represent a particular subset or sample. This population will have a mean, μ, and SD, σ.

If we draw repeated samples, each of size n, from the parent population, the sample means will be Normally distributed (**central limit theorem**) (**p. 61**). In practice, we do not often take multiple samples, usually we have only one sample from the parent population. Nevertheless, using the measured SD divided by the square root of the size of our sample, gives some idea of what the parameters controlling the distribution of multiple sample means would be. We can, for the particular sample mean, calculate a 95% confidence interval (CI). This CI defines the range of values between which we can be 95% certain that the true population mean will lie.

SE is, therefore, extremely useful when there are two samples and we wish to know whether they come from the same parent population or from different populations. If the parent populations are the same, then the CI for the population means will overlap; if the parent populations are distinct, then there will be no such overlap.

Standard Gamble

A technique, originally developed by Morgenstern and von Neumann in *Games and the Theory of Economic Behaviour* that attempts to define the value an individual puts on a particular state of existence. In the jargon of **decision analysis** (**p. 88**), such values are termed 'utilities'.

To obtain an estimate of how a patient might value their life after radical surgery, in the face of diminished respiratory reserve

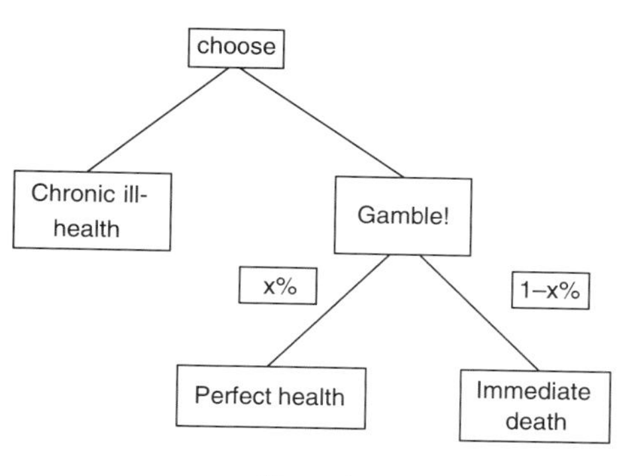

Figure 1.

for lung cancer, the procedure would be as follows. Ask the patient to imagine that they could choose to live for, say, 20 years confined to a wheelchair and dependent on oxygen or if they could take a gamble between an x% chance of 20 years of perfect health or an $(x-1)$% chance of immediate death. Schematically, the gamble is as shown in figure 1.

x is varied until the subject cannot choose between the gamble and the state of chronic ill-health: if this corresponded to a 20% chance of good health (and with it an 80% chance of immediate death), then the utility for the state of chronic ill health is defined as 20%, i.e. 0.2. Standard gambles can, in certain circumstances, be used to illustrate what we already know: that for some patients there are, for them, fates they consider to be worse than death.

Standard Normal Distribution

A transformation of the Normal distribution in which each value is treated as follows:

$$\text{New value} = \frac{(\text{untransformed value} - \text{mean})}{\text{SD}},$$

where SD is standard deviation.

These new values are conventionally termed the normal scores or standardised normal deviates. The standard normal distribution has a mean of zero and SD = 1.0.

Stathmokinetic Method

A technique used to assess cellular proliferation. After an appropriate dose of a spindle poison such as colchicine, **vincristine (p. 324)** or **vinblastine (p. 324)**, cells will accumulate in **metaphase (p. 198)** because **mitosis (p. 201)** has been blocked. The rate at which the cells accumulate gives an indication of the production rate of cells — that is the mitotic rate. The **potential doubling time (p. 107)** can then be calculated as:

$$= \log_e 2/\text{mitotic rate}.$$

The method, since it demands multiple serial samples, is not useful in clinical practice.

Statistical Launderette

This is the idea that it is possible to make a few observations, call it a study, and pass the data to a statistician who will wash, dry and press it all: turning a soggy amorphous mass into a pristine set of 'clothes' to wear to meetings — preferably in exotic locations. The only intellectual framework that supports this process is the notion that 'it will all come out in the wash'. Statistical expertise is required before, during and after a study has been performed — not just at the very end.

Status Quo Bias

A form of **bias (p. 28)** in which evidence is ignored and current attitudes are maintained, in spite of good evidence that those attitudes may be wrong. The reluctance of radiotherapists to use single fractions of treatment for palliation of bone metastases, despite several trials showing that single fractions are as effective as multiple fractions, is a neat example of status quo bias. The reasons for such bias are complex: economic motivation (if changing practice would lead to loss of income); self esteem (if being 'in charge' for longer increases the doctor's self-importance); innate suspicion of the new; and laziness — unwilling to locate or interpret evidence, etc. One of the potential problems with **peer review** (**p. 229**) is that by rejecting unusual ideas, it may perpetuate status quo bias.

Stem Cell

A stem cell is one that has the ability to reconstitute the function of a tissue or organ should that organ be damaged or insulted. Stem cells may be proliferating, as in response to injury or demand, or quiescent as in a resting tissue. Clonogenicity is a property that stem cells must possess.

Steroid Receptor Superfamily (syn. Intracellular Receptor Superfamily)

These proteins are within the cytoplasm; activation, therefore, requires that the hormone cross the cell membrane to bind to its receptor. Hormones mediated by this family of receptors include:

- Cortisol.
- Oestrogen.
- Progesterone.
- Vitamin D.
- Retinoic acid.
- Thyroid hormone.

In the inactive state, a DNA-binding **domain** (**p. 102**) is occupied by an inhibitory protein complex that, when the hormone binds to the receptor site, dissociates. This exposes a DNA-binding site that is free to bind to the DNA on the gene regulated by that particular hormone. Steroid receptors are mainly cytoplasmic; retinoid receptors are mainly nuclear.

The gene controlled by the receptor usually exhibits a biphasic response. In the initial, primary, phase, only a limited variety of proteins is produced. The secondary phase produces a wide variety of proteins, some of which are inhibitory and others that can control other pathways. This arrangement not only permits amplification of the original signal as well as feedback inhibition, but also enables one signal to trigger a complex series of events.

Stochastic

A stochastic effect is one that is not dose-dependent. There is no clear and predictable association between the quantity of **exposure** (**p. 124**) to the causative agent and the severity of the effect. Carcinogenesis induced by low doses of radiation is a classical stochastic effect. With increasing dose, the incidence of the effect increases but the severity of the individual effects is not dose-related. The laws of probability govern the effect — the higher the dose, the more probable the effect. A **non-stochastic effect** (**p. 214**) is one for which severity is directly dependent on the dose of the causative agent An example would be the dose-dependent cardiotoxicicty of **anthracyclines** (**p. 16**): there is a rapid increase in the severity of cardiomyopathy with total doses of doxorubicin >500 mg m^{-2}.

Stratification

This is the process whereby a large and possibly heterogeneous sample is divided into a series of subgroups. It provides a means for avoiding, or at least mitigating, the problem of **confounding** (**p. 72**).

Patients with poorly differentiated primary tumours of the breast have a worse prognosis than patients with well-differentiated tumours. Patients with positive axillary nodes have a worse prognosis than those with negative nodes. Patients with poorly differentiated tumours are more likely to have positive axillary nodes. In a trial of whether a new treatment was more effective against node-positive, as opposed to node-negative, patients, the degree of differentiation would be a potentially confounding variable. Stratification offers a potential solution: by subdividing the node-positive and -negative patients into groups according to the degree of differentiation and then comparing the results between strata, the effect of differentiation on prognosis can be disentangled from that of nodal status.

The problem is that an excessive number of strata can easily be defined and, ultimately, excessive stratification is much the same as no stratification at all. Stratification in a trial with four prognostic factors, each with three levels, and involving 15 institutions, with stratification for each, would produce $3 \times 3 \times 3 \times 3 \times 15$, i.e. 1215, potential categories. This is clearly ridiculous. Techniques such as minimisation have addressed this difficulty — but they only satisfy the criterion of random allocation of treatment if the patients are entered into the trial itself in a random order.

It is unwise to stratify the results of a trial retrospectively since increasing the number of strata increases the number of comparisons and leads inevitably to an increased probability of a Type 1 (false-positive) statistical error.

Sublethal Damage (SLD) Repair

SLD is cellular damage inflicted by radiation that, provided the cell is given the opportunity, can be repaired. Repair is usually complete within 4–6 h of a fraction of radiotherapy. Its extent depends on dose and it can be measured by split-dose experiments in which the increased survival after two equal fractions, compared with the same total dose given as a single fraction, gives an indication of SLD repair.

Subtoxic Dose

The subtoxic dose of a drug is one that causes consistent mild toxicity, suggesting, in a **Phase I study** (**p. 232**), that significant toxicity will be encountered at the next dose level.

Suicide Gene Therapy

In this form of gene therapy, a gene coding for an enzyme that converts an inactive prodrug into an active cytotoxic agent is introduced into the tumour genome. The **prodrug** (**p. 246**) is then administered and, in theory, only tumour cells that have incorporated the agent of their own destruction will be killed.

The approach has been tried in the treatment of glioblastoma in humans. Mouse cells, engineered to express functional retroviral components as well as the thymidine kinase (TK) sequence of HSV-1, are injected into the cavity left following surgical removal of the tumour. The retroviral elements are used to insert the mouse genomic material into dividing human cells — effectively the tumour cells, since normal neurones do not normally divide. A few days later the patient is treated with ganciclovir (GCV), a drug that, only in the presence of TK, is converted into cytotoxic nucleotides. Although normal cells also possess TK, the HSV-1 TK is ~1000 times as effective at phophorylating ganciclovir. The phosphorylated ganciclovir (MW 400 daltons) can move directly between cells, via **gap junctions** (**p. 134**) (pore size equivalent to 1000 daltons), but cannot migrate through the extracellular space. The original mouse cells are destroyed by the normal immunological response to any xenograft. Although the theory is attractive, the clinical results have been disappointing.

Other systems under investigation include cytosine deaminase/5-fluorocytosine; and *E. coli* nitroreductase/CB1954.

When suicide gene therapy is attempted, it is clear that some cells die, even though they have not actually incorporated the exogenous genetic material into their genome. This is known as the **bystander effect** (**p. 37**). The mechanism may involve the passage of a toxic metabolite, e.g. phosphorylated ganciclovir, via gap junctions, or could involve **apoptosis** (**p. 19**) of the transfected cells with subsequent ingestion by adjacent non-transfected cells of vesicles containing toxic metabolites. Immunological mechanisms may also play a role since, in some systems, the **bystander effect** (**p. 37**) is less apparent when tumours are grown and treated in the immunesuppressed mouse. In immunocompetent animals, specific anti-tumour responses have been demonstrated following **suicide gene therapy** (**p. 299**). This has implications not only just for the local effects of gene therapy, but also for systemic effects.

Superoxide Dismutase (SOD)

The human body may produce up to 2 kg of the potentially toxic **free-radical** (**p. 132**) superoxide $O^{2\cdot-}$ each year. SOD is an enzyme that converts $O^{2\cdot-}$ into hydrogen peroxide plus oxygen. The hydrogen peroxide is, in turn, metabolised by catalase or glutathione peroxidases.

Supportive Care

The concept of supportive care defines those measures implemented to mitigate the adverse effects of treatment. Implicit within the concept is the recognition that treatment for cancer not only does good, but also does harm. One of the prime considerations in supportive care is the provision of adequate information. There is much fear associated with the diagnosis and treatment of cancer: fear of the unknown is a powerful emotion. Explanation of what is likely to happen, information about what is being done and why, can do much to alleviate the fear and anxiety experienced by patients with cancer. The exploration of these areas of supportive care is part of the discipline of **psychosocial oncology** (**p. 250**).

Supportive care also has many practical aspects:

- Antibiotic prophylaxis to prevent infection.
- Haemopoietic **growth factors** (**p. 144**) to prevent neutropenic sepsis.
- Anti-emetics.
- Measures to prevent mucositis.
- Nutritional support.
- Pain control.
- Emotional support.
- Psychological support.
- Physiotherapy.
- Speech therapy.
- Rehabilitation.

No individual can provide all that is required adequately to support a patient during, and after, treatment for cancer: hence the need for a multidisciplinary approach.

Surrogate Endpoints

The classic **endpoints** (**p. 114**) (survival, recurrence) used in oncology cannot be assessed immediately after a therapeutic intervention. It may be several years before a proper assessment can be made, using these endpoints, of whether a new treatment is worthwhile. This is an unsatisfactory state of affairs. The public and its doctors want to know — have a right to know — whether a treatment works. Surrogate endpoints are intended as proxies for survival or recurrence but have the advantage that they can be assessed much sooner after treatment and thereby provide an earlier indication of the effectiveness, or lack of effectiveness, of an intervention. Local control should be regarded as a surrogate endpoint in cancer treatment, since local control does not itself guarantee the ultimate aim, i.e. survival.

Surrogate endpoints have been defined as 'a laboratory measurement or physical sign used as a substitute for a clinically meaningful endpoint that measures directly how a patient feels, functions or survives'.

A considerable amount of ingenuity goes into the development and interpretation of surrogate endpoints. Examples of clinically useful surrogate endpoints include:

- Tumour marker responses: PSA, CEA, CA 125.
- Chromosomal translocations (FISH): is an abnormal clone, e.g. with a *bcr* gene rearrangement in lymphoma, eliminated by treatment?

The proportion of patients with small tumours could be used as a surrogate endpoint in an evaluation of breast screening. The logic would be that screening provides early diagnosis: the sooner a tumour is diagnosed, the smaller it will be, the sooner a tumour is diagnosed, the sooner will treatment start, early treatment improves survival. However, the relationship between an increased proportion of small tumours and improved survival cannot be assumed, simply because it seems logical.

The problem with this approach is that to be believable, a surrogate endpoint has to be validated. Validation requires assessment against a **Gold Standard** (**p. 141**), and the Gold Standards are recurrence and survival, and these take time, and so we cannot validate a standard without waiting, and not wanting to wait was why the surrogate was wanted in the first place. We cannot simply assume that, simply because a new treatment produces a more rapid fall in tumour marker level than the conventional treatment, the new treatment will improve survival. Surrogate endpoints require validation that should be every bit as rigorous as that for any other diagnostic test. Key questions include:

- Strength and dependability of association: is there a strong, consistent and independent **correlation** (**p. 75**) between the surrogate endpoint and the clinical endpoint or outcome?
- Collateral evidence from similar studies: have other interventions, similar to that under investigation, produced changes in a surrogate marker that have correctly indicated the long-term clinical outcome?
- Applicability of previous results to study population: have any such changes been demonstrated in a population of patients similar to those in the proposed study?

Surrogate endpoints actually can mislead: the fibrates lower lipids (a surrogate endpoint), but compared with controls survival is no better in patients treated with fibrates. The fibrates are associated with decreased cardiovascular **mortality rate** (**p. 207**), but the mortality rate from other causes is increased — leading to no net benefit.

When used properly, however, surrogate **endpoints** (**p. 300**) can greatly facilitate progress: the development of estimates of viral load as surrogate endpoints in **HIV** (**p. 151**) therapy have enabled new treatments to be evaluated rapidly and effectively.

Surviving Fraction (SF)

SF expresses the results of an assay for clonogenicity. SF_2 is shorthand for the surviving fraction after a single dose of 2 Gy:

- $SF = \dfrac{\text{plating efficiency of treated cell}}{\text{plating efficiency of control cells}}.$
- $\text{Plating efficiency} = \dfrac{\text{number of colonies formed}}{\text{number of cells plated out}}.$

A colony, by definition, must contain at least 50 cells.

Systematic Review

Within the context of the Cochrane collaboration, the term 'systematic review' has a clear definition:

"A review of a clearly formulated question that uses systematic and explicit methods to identify, select and critically appraise relevant research, and to collect and analyse data from the studies that are included in the review. Statistical methods (**meta-analysis**) (**p. 195**) may or may not be used to analyse and summarise the results of the included studies."

Other so-called systematic reviews are less precisely defined. The old-fashioned review article, with selective quotation, the discarding of results not consistent with the author's (usually white and male) preconceptions, and all the other biases introduced by what is, essentially, an anecdotal approach, is sometimes described as a systematic review. It may well, in the cabaret sense, be a review – systematic it is not.

-Cell Lymphoblastic eukaemia Gene (*TCL2*)

he gene is on chromosome 11p13. It is :tivated in acute T-cell leukaemia when ırt of the T-cell receptor α-chain **locus** **. 183**) (TCRA) is translocated into its ımediate vicinity.

-Test

he *t*-test is often used as a test of statistical gnificance. It is sometimes known Student's *t*-test because it was originally ıblished under the pseudonym 'student'. he principle behind it is as follows. If we ve a small sample taken from a general pulation, we can calculate a mean d **standard deviation** (**SD**) (**p. 296**) for e sample, but the mean and SD cannot be easured for the population as a whole. If e size of the sample is increased to >30, en we can be more certain that the SD served in the sample is about the same as e SD in the population as a whole. The est applies a correction factor to the ormal distribution to allow for the fact at, when small samples are used, we nnot assume that the SD for the sample is e same as that of the whole population. As nple size increases, so the *t* distribution creasingly approximates to a Gaussian stribution. Note that it is an underlying umption of the *t*-test that the data llow a Normal distribution. The *t*-test can used for both paired and unpaired data.

ıe **degrees of freedom** (**p. 91**) (d.f.) in a est are used to incorporate a consideration sample size:

r unpaired data,

d.f. = (number of observations − 1).

r paired data,

d.f. = (number of subjects − 1).

For a *t*-test to be an appropriate test of significance, the following criteria must be met:

- Is it acceptable to calculate a mean? A mean should not be calculated for T-stage or mucositis-graded on a five-point scale since these are ordinal data, not measurements.
- Is distribution Gaussian (Normal)? This is mainly important at small (<30) sample sizes.
- Are the samples independent of each other? Have **bias** (**p. 28**) and **confounding** (**p. 72**) been, as far as is possible, eliminated?
- Is the variance of the two samples the same? If this is unknown, then the *F*-test should be used. The *F*-test involves calculating the ratio of the variances (larger : smaller) in the two samples described above and looking it up in a table of *F* distribution (d.f. sample $1 = (n_1 - 1)$; d.f. sample $2 = (n_2 - 1)$). If *p*, corresponding to *F*, is significant, then it is unlikely that the populations have the same variance: The *t*-test should not be used.

T1 (Spin-Lattice, Longitudinal Relaxation Time)

This is a term used in magnetic resonance imaging (MRI) to describe the time taken for the nuclei to resume their orientation in a magnetic field after the reorientation imposed by a brief RF pulse. In a T1-weighted image obtained using a sequence of RF pulses termed 'saturation recovery', fat appears white while water appears dark.

T2 (Spin–Spin, Transverse Relaxation Time)

A term used in magnetic resonance imaging (MRI) to describe the time taken for the synchronised, in-phase, nuclear wobbling to re-establish the out of phase precessions that existed before the application of a brief RF field. With a sequence, 'spin-echo' designed to produce a **T2-weighted** (**p. 303**) image water, e.g. cerebrospinal fluid, appears white while the grey matter of the CNS appears much darker.

Tail Moment

A means for objectively comparing the tails, representing the fragmented DNA, in a comet assay. Tail moment is calculated as:

% DNA fluorescence in the tail × distance, y.

Distance is the distance between the means of the fluorescence distributions of the tail and the head.

Tamoxifen

Tamoxifen is used as an anti-oestrogen that has some endogenous oestrogen activity. Its main clinical use is in the treatment of breast cancer, either as adjuvant or as treatment for advanced disease. It may also have a potential role in the prevention of breast cancer in women at high risk. Structurally, it is similar to diethyl-stilboestrol and clomiphene. It binds to oestrogen receptors and induces conformational changes. The actions of tamoxifen will depend on the tissue. There are at least two types of oestrogen receptor, designated α and β. Their tissue distribution varies and this no doubt explains why, in some tissues, tamoxifen has oestrogenic effects while, in many breast cancers, it acts as an anti-oestrogen. In treating cancer it is **cytostatic** (**p. 84**), not cytotoxic. Its main activity is against G_2 cells.

Tamoxifen also affects **growth factors** (**p. 144**): it inhibits insulin-like growth factor type 1 (which can stimulate breast cancer growth) and can stimulate the production of transforming **growth factor** (**p. 313**) β (TGF-β), which inhibits the growth of cancer cells. It is active after oral dosage, but the dose may need to be lowered in patients with obstruction to the biliary tracts.

Target Theory

Target theory is a term used in radiobiology to describe possible mechanisms to explain the observed relationship between dose and **radiation-induced cell killing** (**p. 256**). It attempts to unravel the biological principles underlying the shape of the cell-survival curve for irradiated cells. Many theories have been proposed: each has its implications for the predicted shape of the **cell-survival curve** (**p. 57**) and for its mathematical form.

A single ionisation along the electron track could hit a single vulnerable target (**single-target, single-hit inactivation**). The likelihood of this event will be proportional to dose. This type of damage would be linear on a conventional cell-survival plot. Alternatively there could be several targets within the cell, each of which needs to be hit before the cell is killed. The likelihood of the cell being killed will be proportional to $(\text{dose})^n$, where n is the number of targets (**multiple-target, single-hit inactivation**). This type of damage would produce a survival curve with an initial shoulder, the width of the shoulder being related to n. The smaller the n, the narrower the shoulder will be on a log–linear plot.

Neither of these target theories provides a satisfactory model for the radiation–cell-survival curve observed. The single-target, single-hit model predicts a straight line, a cell survival curve only observed for high linear energy transfer (LET) radiation. The multiple-target, single-hit model predicts a curve with an absolutely flat initial portion before the bend of the shoulder and, ultimately, the straight line.

If both mechanisms are operating simultaneously, then a hybrid curve with oth linear and bending components will be roduced. The problem is, however, that iere would still be an initial linear segment the curve and this does not correspond to xperimental observation.

'he **linear–quadratic (LQ) model (p. 180)** rovides the best pragmatic description of the bserved survival curve for irradiated iammalian cells but the underlying iechanism is far from clear. One explanation, ith its origins in target theory, would be iat there are two important components to idiation-induced cell killing. One is roportional to dose and simply represents ngle-target, single-hit inactivation. The ther, the $(dose)^2$ component, is due to iactivation that requires two separate tracks produce ionisations which are close gether both in time and space. Alternative, r supplementary, explanations of the biology ehind the LQ model deal with repair iechanisms. If there are two types of amage, that which is inevitably lethal and iat which can be lethal, but may not be if pair is allowed to occur, then this could roduce a response to irradiation that ontained elements directly proportional to ose (the inevitably lethal damage) and to $lose)^2$, the lethal, but repairable, damage.

axanes

axanes is the generic term given to a class f **cytotoxic drugs (p. 84)** originally isolated om the bark of the Pacific yew tree, *Taxus evifolia*. They were discovered in 1971 as a sult of a systematic programme of screening atural compounds for antitumour activity. he main agents in clinical use are **Paclitaxel . 226)** and **Docetaxel (p. 102)**. They have broad spectrum of clinical activity. They e **spindle poisons (p. 293)** that interact ith the tubulin subunits of microtubules. nlike the **vinca alkaloids (p. 324)**, which srupt the formation and structure of microtubules, the taxanes stabilise the microtubules. The effect is a cell frozen in **mitosis (p. 201)** and unable to dismantle the scaffolding that has been erected to permit the chromosomes to migrate to each pole.

T-Cell Antigen Receptor (TCRA)

This receptor, on the surface of T-cells, is responsible for the ability of those cells to recognise a wide variety of different antigens. Its gene is closely related to the *V* gene of the immunoglobulin heavy chain. The *TCRA* gene is on chromosome 14q11.2 and it may be translocated in acute T-cell leukaemia, the t(11;14)(p13q11) translocation, so that it is placed adjacent to the *TCL2* gene.

Telomerase

Telomerase is a ribonucleoprotein with **reverse transcriptase (p. 272)** activity. It synthesises glycine-rich sequences of DNA found in telomeres. Telomerase activity could, by preventing the shortening of telomeric DNA that occurs at each somatic cell division, circumvent the problem of cellular **senescence (p. 282)**. An important physiological role for telomerase is in germ-cells where division without telomeric shortening is essential, otherwise the human race would have died out long ago.

Cancer cells might re-activate telomerase and this might contribute to the immortality of cancer cell lines. Inhibition of telomerase, therefore, offers a tempting target for cancer treatment. The theoretical attractiveness of telomerase as a therapeutic target is beset by practical difficulties. Not all tumour cell lines have short telomeres, many have quite long telomeres and are unlikely to be telomerase-dependent. A more immediate problem is that no telomerase inhibitors actually exist — although it is possible to envisage how they might be designed. A recent approach, which works in cell culture, is to insert a mutant form of the telomerase **reverse transcriptase (p. 272)**. This inactivates the **telomerase (p. 305)**. It appears to be more effective the shorter the telomeres of cancer cells. There is, however, considerable variability in the length of telomeres in human cancer cells.

Telomere, Telomeric

A specialised area of **chromatin (p. 63)** at the end of the chromosome that is implicated in **capping (p. 46)**. The telomere consists of ~1000 short repeating sequences, i.e. TTAGGG. The function of the telomere is to protect the ends of the DNA from DNA **ligases (p. 180)**. With each cell division, in normal somatic cells a portion of the telomere is lost — this may be a means whereby cells count the number of divisions they have undergone. The progressive shortening of the telomere means that protection is, over the generations, worn away and the chromosomes become damaged. The cells lose clonogenicity and may, if damage is sufficiently severe, die by **apoptosis (p. 19)**.

Germ cells and embryonic cells are different. They can maintain their telomeres because they possess **telomerase (p. 305)**, a complex of RNA and proteins, that can prevent the progressive shortening of the telomere. Many tumours, up to 80% of human cancers, have telomerase activity and it is easy to envisage the implications this might have for the control of cell division and for the immortality of cancer cell lines.

Telophase

The last phase of **mitosis (p. 201)**: it defines the period during which the new nuclei form. It lasts ~20 min.

Teratogen, Teratogenic

A teratogen is an agent that if administered during pregnancy can produce foetal abnormalities. Ionising radiation and most **cytotoxic drugs (p. 84)** are, potentially, teratogenic.

Terminally Ill

Somewhat surprisingly, this term, often so loosely used, has at least one formal definition. Terminally ill people may be defined as those with active and progressive disease for which curative treatment is not possible or appropriate and from which death can reasonably be expected within 12 months.

Tertiary Structure

The arrangement of **motifs (p. 207)** and straight lengths of polypeptide chain that goes to form **domains (p. 102)** or subunits (**monomers, p. 207**) is referred to as the tertiary structure of a protein.

Testis: Effects of Cytotoxic Therapy

The germinal epithelium requires the support of the Sertoli cells for normal function. The time to expression of radiation-induced damage is dictated by the time-course of normal **spermatogenesis (p. 292)** (usually ~75 days from primordial cell to mature sperm). The spermatogonia are the most radiosensitive cells; it takes ~2–4 weeks after irradiation for the effect of depletion of spermatogonia to be made manifest as a falling in the sperm count. Doses of <2 Gy will produce temporary infertility and most men will recover fertility after ~12 months; doses of 5 Gy almost invariably produce permanent sterility. There is a dose-dependent prolongation of the period of infertility at intermediate doses.

Thalidomide

Thalidomide was, notoriously, introduced into medicine as a sleeping pill and, more unfortunately, as a specific treatment for the nausea associated with early pregnancy. It is teratogenic and the specific defect associated with its use during pregnancy was phocomelia. It was promptly and very publicly banned. More recently it has undergone a renaissance. It is a potent inhibitor of **tumour necrosis factor** (TNF) (**p. 317**) α. It has been used clinically for treating a wide variety of tumours: primary CNS tumours; prostate cancer; multiple myeloma. It has also been used in a number of benign conditions: erythema nodosum; Behcet's disease; Crohn's disease. It is also useful in managing wasting syndromes associated with malignant disease and/or **HIV** (**p. 151**) infection. It antagonises the action of TNF-α by increasing the rate at which the mRNA coding for the protein is degraded. It also has anti-angiogenic activity, mediated by inhibiting basic **fibroblast growth factor** (**bFGF**) (**p. 128**). As such, it is a potentially useful agent in preventing the establishment and growth of metastases. We have come full circle: its teratogenicity was probably based on inhibition of bFGF and consequent interference with the normal development of the limbs.

Other than its teratogenicity, its side-effects are relatively mild. It causes drowsiness, constipation and, with long-term use, might cause peripheral neuropathy.

Therapeutic Equivalence

Therapeutic equivalence exists when the same drug, given to different individuals, but in the same dose, produces identical therapeutic effects. **Bioequivalence** (**p. 29**) is a necessary precondition for therapeutic equivalence.

Therapeutic Ratio

All cancers could be cured if only operations were extensive enough, the doses and number of drugs high enough, or if massive doses of total body irradiation were given. We are limited in what we can safely and rationally give by the fact that surgery, drugs and radiation all damage normal tissues as well as damaging the cancer cells. The therapeutic ratio gives an indication of the extent to which any given agent is selectively effective. It can be defined as the ratio between the dose required to produce unacceptable toxicity to normal tissues and the dose required for therapeutic effect. For antibiotic therapy, the therapeutic ratio may be several thousand, 2 G day^{-1} penicillin is an adequate therapeutic dose. Direct toxic effects of penicillin do not occur until the dose >2000G day^{-1}.

In treating cancer, the therapeutic ratio may be close to unity: one reason that using arguments based on successful antibacterial therapy may be misleading when applied to the treatment of cancer. Many of the practical difficulties encountered in treating cancer stem from the low therapeutic ratios of anti-cancer treatments. Even modest changes in dose, schedule or combination may produce excessive damage to normal tissues. Sensitisation of malignant cells or protection of normal cells has to be carefully considered: there is no point in sensitising both tumour and normal cells by equal amounts. Any sensitisation should selectively affect the tumour cells. Conversely, any protection should selectively affect normal cells. Consideration of the therapeutic ratio focuses attention on these important requirements and emphasises the importance of selectivity in treating cancer. The ultimate aim is that treatment be more toxic to the tumour and less toxic to the normal tissues.

Thermal Neutrons

These are neutrons with extremely low energy, ~0.025 eV. They are used in **Boron capture therapy (p. 35)**. One of the problems with thermal neutrons is that it is difficult to collimate the beam properly and this causes problems with adequate targeting. Contamination of the beam with γ-rays also causes difficulties.

Third Space Effect

The third space effect is a pharmacological term used to describe the presence of a drug in a compartment other than that into which it might normally be expected to distribute. The classic example from oncological practice is the ability of **methotrexate (p. 199)** to remain for prolonged periods in effusions. The effusion acts as the third space and the drug slowly diffuses back into the intravascular compartment. If the phenomenon is not anticipated, it produces increased toxicity as cellular exposure to therapeutic levels of methotrexate may be prolonged. The administration of **folinic acid (p. 131)** offers a partial solution to the problem: by replenishing intracellular stores of reduced folates, the normal cells can be 'rescued' from the toxic effects of the methotrexate.

THYMITAQ (Nolatrexed)

The drug is an inhibitor of thymidylate synthase. It does not require specific transport mechanisms to enter cells; it does not underg intracellular **polyglutamation (p. 238)**. Phase I studies have established a regimen tha can safely be given as a 5-day infusion. It is now being investigated in Phase II studies.

Time to Treatment Failure

A specific form of survival analysis in which for cancer, the defining **endpoint (p. 114)** is local recurrence of disease or the development of distant metastases. It avoids the ambiguity associated with the term 'recurrence-free survival', which could be taken to mean that only local recurrences were being scored as treatment failures. Unlike progression-free survival, the term 'time to treatment failure' implies that the treatment has, for a time at least, completely controlled the disease: **complete response (p. 70)** has been achieved.

Time Trade-Off

The **standard gamble (p. 296)** is difficult to understand and to explain and so may have limited value as means of assessing patients' utilities. Torrance, therefore, introduced the time trade-off method as an alternative. This a method that attempts to measure the utility of a given health state without, as in the

Table 1

Question	Perfect health	Breathless
You can have a year of perfect health or a year with moderate breathlessness, which do you choose?	y	n
You can have a month of perfect health or a year with moderate breathlessness which do you choose?	n	y
You can have 6 months of perfect health or a year with moderate breathlessness which do you choose?	n	y
You can have 9 months of perfect health or a year with moderate breathlessness which do you choose?	y	n

standard gamble, asking patients to perform the difficult task of juggling probabilities. The subject is offered the choice of living for x years in perfect health or $x + y$ years in a specific state of compromised health, e.g. with moderate dyspnoea. y is varied until the patient cannot choose between the alternatives. The proportion of time 'traded off' in this way is an indication of their **utility** (**p. 320**) for the impaired health state.

In this example (Table 1), the subject equates 9 months of perfect health with 12 months of breathlessness. They would be prepared to trade-off 3 months of life in order not to be breathless. The calculation of utility using the time trade-off approach is:

$$\frac{(\text{life expectancy} - \text{time traded off})}{\text{life expectancy}},$$

i.e. $(12 - 3)/12 = 9/12 = 0.75$.

This is the utility value that could be used to calculate the **quality-adjusted life-year** (**QALY**) (**p. 253**) for this individual.

Tirapazamine

Tirapazamine is a bioreductive drug with activity against lung cancer. It is a benzotriazine and is broken down in hypoxic cells to produce **free radicals** (**p. 132**) that cause damage to DNA. Given this mode of action, it is possible that the combination of tirapazamine plus radiotherapy might be synergistic — particularly against tumours with a high proportion of hypoxic, but viable, cells.

Tissue Inhibitors of Metalloproteinases (TIMP)

These proteins can inhibit **metalloproteinases** (**p. 189**), enzymes produced by tumours and involved in the processes of **invasion** (**p. 167**) and **metastasis** (**p. 198**). There are four main TIMP: 1–4. Synthetic inhibitors of matrix metalloproteinases (MMP), such as marimastat, have a potential role in cancer treatment. The therapeutic results have, so far, not been encouraging.

Tissue Plasminogen Activator (tPA)

A serine protease, derived from tissue, active in initiating thrombolysis. It is used clinically in the management of acute infarction.

Tobacco Use

The most effective way to reduce the burden of cancer on a community would be to stop people from smoking cigarettes. Between 25 and 40% of deaths from cancer are related to tobacco: this corresponds to between 30 000 and 47 000 deaths per year in the UK and up to 250 000 deaths per year in the EU. As the tobacco companies face increasing difficulty pushing their products in the developed world, they are shifting their attentions to the developing world: China, India and South East Asia will face major increases in smoking-related cancers as a result of the tobacco companies' efforts to create a market for their products.

The economics of tobacco use are complex. Tobacco users are probably not, through ill health, an economic burden on the community. They pay taxes on tobacco and die prematurely. The former generates income for the government directly; the latter saves governments money because they do not have to bear the costs of care for many elderly users of tobacco. This places governments in a difficult position: do they behave with impeccable ethics and do all that they can to discourage tobacco use so that their citizens are spared the consequent miseries; or do they look to their budgets and give tacit approval to those who would encourage the use of tobacco?

Tomotherapy

The literal meaning of this term is treatment by slice, just as tomography is imaging by slice. In practice, a small beam, within which there is a miniature **multileaf collimator (MLC) (p. 210)**, is used. The beam rotates round the patient and treats a slice, the thickness of which is governed by the field length and within which the distribution of radiation intensity at depth is governed by the movements of the leaves of the MLC as it rotates around the patient. A series of slices is treated to include the whole of the target volume — this is a form of **intensity-modulated therapy (p. 161)**. The practical application is based on the multivane slit collimator. This involves vanes of lead being fired in and out of the beam path extremely rapidly as the beam scans the patient. The technique is known as MIMiC: **multivane intensity modulating collimation (p. 200)**.

Topoisomerases

The topoisomerases are enzymes that relieve the torsional stresses on DNA during **replication (p. 267)** and **transcription (p. 313)**. Were it not for the topoisomerases, the DNA would shear and tear and accurate replication would be almost impossible. There are two main classes, designated I and II. Topoisomerase I relieves the torsional strain by allowing transient single-strand breaks to form in the DNA. Topoisomerase II, also known as DNAgyrase, allows double-strand breaks to form.

Topoisomerases may also regulate gene expression. Topoisomerase I can act as a **kinase (p. 171)** and alter gene expression by changing the splicing pattern of genes. Topoisomerase II can break chromosome 11 close to the ***ATM* locus (p. 23)** (11q23) and, by so doing, may alter the function of the *AT* gene.

Inhibitors of topoisomerases are used in cancer treatment. Topoisomerase I is inhibited by **irinotecan (p. 168)**, **topotecan (p. 310)**, rubitecan and 9-AC. Topoisomerase II is inhibited by **etoposide (p. 119)**, **anthracyclines (p. 16)**, actinomycin D, amsacrine and NSC-6596871.

Topotecan

Topotecan is a **camptothecin (p. 40)** derivative. Its cytotoxic action is by virtue of its ability to inhibit topoisomerase I. It is preferentially toxic to cells in the S-phase of the **cell cycle (p. 52)**. Its main clinical uses are in the treatment of small cell cancer of the lung and, usually in combination with a **taxane (p. 305)**, in the treatment of ovarian cancer.

Total Androgen Blockade (TAB)

This is a popular, expensive and, in terms of additional benefit over monotherapy, unproven approach to the management of disseminated prostatic cancer. An anti-androgen is combined with a **leutinising-releasing hormone (LHRH) (p. 178)** agonist and the rationale is to decrease androgen production (**down-regulation (p. 107)** with LHRH agonist) and to use the anti-androgen to block the effects of any residual androgen. The approach is expensive and the literature provides a neat example of how drug company sponsorship can, through selection bias, skew the results of a **meta-analysis (p. 195)**.

Total Body Irradiation (TBI)

TBI, fractionated over long periods, has been used in clinical radiotherapy for many years, particularly in the treatment of low-grade lymphomas. A typical schedule would be 2.5 Gy in 25 fractions treating with 0.1-Gy fractions three times a week; total duration of treatment being 8.5 weeks.

›equential hemibody radiation, as used in the ›alliation of myeloma or metastatic carcinoma ›f the prostate, is a form of TBI. A period of ‑6 weeks must be allowed between each ıemibody treatment so that marrow stem ·ells can migrate from the untreated half to he irradiated half of the body. The main urrent use of TBI, however, is its use over hort periods (<7 days) in preparative egimens for patients undergoing **bone narrow transplantation** (**BMT**) (**p. 34**) or ngraftment using peripheral blood stem cells.

n the early 1970s, BMT was regarded as a isky and experimental procedure. Today, it is ·art of standard therapy for many forms of eukaemia and other haematological disorders. Not all transplant regimens include TBI. TBI s used in BMT for two main reasons: as nmunosuppression (to facilitate engraftment); nd to kill malignant cells (including those in anctuary sites, CNS, testis), which cannot be eached by cytotoxic drugs.

'he clinical indications for BMT have xpanded widely. Originally used in eukaemia, BMT is now used for aplastic naemia, lymphomas, thalassaemia, refractory olid tumours, osteopetrosis and sickle cell naemia.

'he broadest clinical experience with BMT in leukaemias and lymphomas. The process f BMT can be divided into a sequence of ages:

. Conditioning regimen: to kill malignant cells and prevent graft rejection.
. Marrow manipulation:
  — purge autologous marrow of malignant cells
  — purge allogeneic marrow of T-cells to prevent or ameliorate graft versus host disease (GVHD).
. Marrow re-infusion (transplantation).
4. Post-transplant regimen:
  — support the patient through cytopenia and minimize risks of infection and haemorrhage
  — prevent or ameliorate GVHD.

Similar procedural stages apply to peripheral blood stem cell rescue (PBSCR):

- Treat patient with **cytotoxic drugs** (**p. 84**).
- Amplify the overshoot of circulating haemopoietic precursors by timing CSF administration appropriately.
- Harvest PBSC during rebound.

Types of BMT

- Allograft:
  — from another donor
  — identical (identical twin)
  — HLA matched: from relative (35% chance of matched sibling)
  — from a panel of registered donors
  — mismatched.
- Autograft: marrow previously harvested from the patient is re-infused with or without conditioning *in vitro*. Marrow for autografting might be cryopreserved for several years.

  PBSCR is a form of autografting

TBI is but one component of an extremely complex process and it is difficult — particularly retrospectively — to tease out the precise contribution TBI, or a change therein, that may have to be made any given regimen. Policies for TBI are not uniform — a recent EBMT survey of 28 centres found 32 different techniques in use: all with different physical parameters as well as differing patterns of dose, fractionation, dose-rate and overall treatment time. The following account of the radiobiology of TBI is, therefore, grossly oversimplified but it emphasises some of the features that have emerged from the confusion.

The radiobiology of TBI is complex. LD_{50} experiments are not ethically possible in the investigation of therapeutic TBI. The $LD_{50/10}$ and $LD_{50/30}$ endpoints are applicable only to animal studies: the animal data cannot simply be extrapolated since rodent physiology and timeframes are different from those in humans.

Normal haemopoiesis is usually ablated at the doses used in TBI. Hence the need for **bone marrow transplantation (p. 34)** or infusion of autologous stem cells. Karyotypic analysis of marrow several months after TBI transplant occasionally shows that it is host marrow that is regenerated, either as sole haemopoietic cell line or as a **chimera (p. 63)** in which marrow elements from both host and donor coexist. These rare instances serve to demonstrate that normal stem cells are heterogeneous in terms of their response to the conditioning regimens used in BMT. Chromosomal abnormalities are frequently found in the haemopoietic stem cells of patients treated with chemotherapy and TBI. The α/β ratio for normal **marrow stem cells (p. 188)** has been estimated as 21 Gy — suggesting that there is a small shoulder on the survival curve and that there might be some modest sparing of these cells from fractionation.

There are abundant laboratory data to suggest that single doses of TBI are more immunosuppressive than fractionated treatments using the same dose: rat marrow graft, rat heart transplant, dog marrow, mouse marrow (α/β ratio for engraftment effect measured as 2 Gy). The radiation survival curve for human T-lymphocyte progenitors has a shoulder.

Marrow stromal cells are relatively radioresistant (D_0 between 1.4 and 4 Gy) and have a broad shoulder. Eric Hall in *Radiobiology for the Radiologist*, 3rd edn states: 'there is no instance on record of a human being having survived a dose in excess of 1000 rad (10 Gy)' (p. 368). If this were true, then TBI/BMT as currently practised would be uniformly fatal (which it is not). The contradiction is resolved through a consideration of dose-rate. Hall's statement is made in the context of radiation accidents when dose-rates are very high, >1 Gy min^{-1}.

Dose-rates used in clinical TBI are in the range 0.04–0.10 Gy min^{-1}. Acute gastrointestinal (GI) reactions do occur with therapeutic TBI: nausea, vomiting, diarrhoea. The catastrophic denudation of the GI lining seen after radiation accidents does not occur. Supportive measures including gut sterilisation and careful monitoring of fluid and electrolyte balance are sufficient to tide patients over the acute GI syndrome after therapeutic TBI.

The **late effects (p. 174)** of TBI include interstitial pneumonitis, radiation-induced cataract, infertility, endocrine deficiency (hypothyroidism, panhypopituitarism), veno-occlusive disease of the liver, growth disturbances in children, renal damage and **second malignancies (p. 280)**.

Toxic Dose Low (TDL)

A term used in preclinical evaluation of new drugs. It is the lowest dose of a drug that produces detectable adverse effects on the animals being tested: evidence of liver enzyme elevation, evidence of any myelosuppression, etc. Doubling the TDL should not kill any of the animals.

TRAFs

These are **tumour necrosis factor (p. 317)** (TNF)-associated factors that mediate the cellular response to TNF. They function as adaptor proteins that activate cytoplasmic kinases.

TRAIL (TNF-Related Apoptosis-Inducing Ligand)

There are several membrane receptors for this cytokine (R1–3, etc.). The interaction between TRAIL and TRAIL-R1 or -R2 produces **cell death (p. 55)** through **apoptosis (p. 19)**. The effect is observed predominantly in transformed cells and is less obvious in normal cells. TRAIL, therefore, has the potential to be selectively toxic to tumours and early data from mouse studies suggest that this might indeed be the case. When TRAIL interacts with the TRAIL-R3 or -R4 receptors, apoptosis is inhibited. *p53* may cause increased expression of TRAIL-R2 and could, thereby, increase TRAIL-induced apoptosis. The stimulation of apoptosis by TRAIL is mediated via **caspases (p. 47)**. The intracellular link between receptor binding and caspase activation is unknown but it may involve the death **domain (p. 102)** of an adapter protein, FADD. TRAIL is related to TNF but, unlike TNF, seems to have very little systemic toxicity. This selectivity, the mechanism for which may lie in the relative expression of receptor types on normal and malignant cells, makes it a potentially interesting agent for cancer therapy.

The **Fas ligand (p. 126)** cannot be used for cancer treatment because it causes massive hepatic damage through apoptosis of hepatocytes.

Transcription Factor

A molecule that activates the process of **transcription (p. 313)**, i.e. the formation of mRNA from DNA.

Transcription

The process whereby a complementary strand of RNA is formed on a DNA template.

Transformation Constant (syn. Decay Constant or Disintegration)

This constant, which appears in the equation used to calculate the **half-life (p. 147)** of a radioisotope, represents the fractional rate of decay of the atoms. Its symbol is λ, its units are (time)$^{-1}$ and it is defined as:

$$-(\Delta N/\Delta t)/N,$$

where $\Delta N/\Delta t$ is the rate of decay, also termed the **activity** of the sample, and N is the number of radioactive atoms present in the sample.

The half-life equation is:

$$t = \log_e 2/\lambda,$$

where t is the half-life, $\log_e 2 = 0.693$ and λ is the transformation constant.

The half-life equation is simply a special case of the general decay equation:

$$N_t = N_0\, e^{-\lambda t},$$

where N_t is the number of radioactive atoms present at time t, N_0 is the number of radioactive atoms initially present (at time zero), t is the time that has elapsed and λ is the transformation constant.

The transformation constant (λ) for each individual isotope is immutable, so, therefore, is the half-life.

Transforming Growth Factor (TGF)-β

TGF-β is a cytokine and exists in three forms: β1–3. The gene is at chromosome 19q13.1–q13.3. There are three types of cell surface receptor for TGF-β, designated I–III. The Type III receptor is the most abundant and its main role appears to be to transfer the TGF-β to the receptors, Types I and II, which actually perform the signalling function. These receptors are serine–threonine kinases and activate an intracellular signalling pathway mediated by a group of molecules known as Smad. The final common pathway is via a complex with Smad4, which moves to the nucleus and acts as a transcription factor.

TGF-β is a regulator of the **cell cycle** (**p. 52**), particularly in epithelial, endothelial and haemopoietic cells. It causes G_1 arrest, partly mediated through *p15* and partly via **cyclins** (**p. 82**) A and E. TGF-β drives normal cells away from proliferation and towards differentiation and **apoptosis** (**p. 19**). The inability to respond appropriately to TGF-β may be an early feature of malignant transformation.

After **transcription** (**p. 313**) and **translation** (**p. 314**), the function of TGF-β is extensively regulated. It is synthesised as a large inactive molecule. This contains TGF-β plus a propeptide. Before being released from the cell, TGF-β is partially split from its propeptide but remains attached to it by covalent bonds. After leaving the cell, TGF-β forms a complex with its propeptide as well as a binding protein, latent TGF-β-binding protein. Active TGF-β is only formed once this complex has been acted upon by another protein usually thrombospondin-1. It may also be activated by plasmin.

TGF-β has major effects on the extracellular matrix; it stimulates fibroblasts to produce structural proteins for the matrix as well as proteins, such as the integrins, involved in cell adhesion; it also inhibits those enzymes that degrade the matrix. It also promotes angiogenesis. TGF-β stimulates the production of fibrous tissue. Polymorphisms in receptor sensitivity may render some individuals particularly sensitive to this effect. TGF-β may be an important mediator of fibrosis after radiotherapy or of **bleomycin** (**p. 32**)-induced lung fibrosis. Some drugs inhibit the secretion or activity of TGF-β: angiotensin-converting-enzyme inhibitors, prednisolone, cyclosporin, **interferons** (**p. 163**) α and γ; and **tamoxifen** (**p. 304**).

Translation

Translation is the process whereby a polypeptide chain is formed from the template provided by mRNA.

Translational Research

Research aimed specifically at producing practical benefits from advances in basic science. The need for a specific term recognises that, for too long, there was a gap between laboratory research and the derived benefits for humanity. This gap was both cultural and temporal: a failure of pure and applied scientists to respect each others' ideals and culture; a delay in putting theoretical advances to practical use. The term is usually used specifically to describe advances in clinical science brought about by the rapid implementation of laboratory findings. The clinical use of gene therapy could be regarded as a **paradigm** (**p. 226**) for translational research.

Trapezoidal Approximation

A simple method for estimating the area under the curve (AUC) for a value plotted against time (t). It is calculated as the sum of the measurements:

$$(t_2 - t_1)(y_1 + y_2)/2,$$

where $t_2 - t_1$ is the interval between the measurements and y_1 and y_2 are the values at the two time points.

Trastuzumab

A recombinant humanised antibody to the *HER2* (c-*erbB2*) gene. It recruits natural killer (NK) cells and monocytes and causes antibody-mediated **cell death** (**p. 55**) in cells that express high levels of *HER2*. Between one-quarter and one-third of patients with breast cancer have tumours that express high levels of *HER2* and which, logically, might be treated with Trastuzumab. The prognostic significance of *HER2* expression in breast cancer is uncertain. There is some evidence from cell lines that tumours with high levels of expression may be more resistant to chemotherapy. Another problem is that expressing the receptor is not the same as expressing active receptor.

Trastuzumab has recently been licensed for the treatment of metastatic breast cancer which is refractory to other therapies. It has a number of side-effects including gastrointestinal upset, cardiomyopathy, fevers and chills. Anaphylactic reactions may occur.

Treated Volume

This term is included in the ICRU 50 definitions of terms used in external beam radiotherapy to define a volume, larger than the **planning target volume (PTV)** (**p. 235**) but smaller than the irradiated volume. Treated volume is often of a regular shape, such as box, defined by the standard beams available on the treatment machine. PTV is defined predominantly anatomically, and may, therefore, be irregularly shaped. The shape of the treatment volume conforms as closely as possible to that of the PTV but is constrained by the field shapes and beam arrangements available on the machines used for treatment. **Conformal therapy** (**p. 71**) represents, in its simplest definition, an attempt to approximate the shape of the treated volume as closely as possible to the shape of the PTV.

Troglitazone

A synthetic ligand for the **PPAR-γ** (**p. 242**) nuclear hormone receptor. Unfortunately, it causes liver failure and has, therefore, been taken off the market as a treatment for diabetes. Data from animal models on the use of **troglitazone** (**p. 315**) as a potential differentiating agent for cancer are conflicting: it may shrink some types of colon cancer but stimulate others.

Tumour-Associated Antigens (TAA)

Some tumours express antigens unique to that particular type of tumour. Some colorectal cancers, for example, express CEA and labelled antibodies directed against the CEA can be use to image the tumour. The preferential expression of tumour-specific markers has obvious potential for selective toxicity in the treatment of cancer. This is already being exploited therapeutically. An antibody (panorex) to the antigen 17 1-A expressed by colorectal cancer has been used as an **adjuvant treatment** (**p. 8**) for colorectal cancer in Phase III trials. Clear survival benefit, compared with placebo, has been shown. Antibodies to antigens, such as HMFG1, HMFG2 and AUA1 expressed by ovarian cancer, have been radiolabelled with ^{90}Yt and used successfully for intraperitoneal treatment in patients with minimal residual disease.

All of which begs the question: if tumours express antigens, why are they not rejected immunologically? The reasons are complex but include: the emergence of subclones that do not express TAA; tumours secrete factors that inhibit immune function, e.g. TGF-β and IL-10; the **fas/fas ligand** (**p. 126**) system may be exploited by tumours so that T-cells directed against the tumour are killed by **apoptosis** (**p. 19**); there may be decreased expression of mixed histocompatibility complex (MHC) antigens with consequent decrease in the efficiency of antigen presentation; there may be effects on suppressor cells; loss of signals (e.g. that mediated via B7) that are stimulatory to the immune system may occur.

Tumour Doubling Time (T_d)

A term used in the analysis of tumour growth and cell kinetics. Simply, it is a measurement of the time a tumour takes to double in volume. Since tumours contain more than just clonogenic cells, stroma and other elements have to be considered. The volume doubling time will not necessarily be the same as the time it takes the clonogens to double in number.

Tumour Flare

A phenomenon observed in the initial phase of cancer treatment in which there may be an temporary increase in growth of the tumour as a result of the therapeutic manipulation. The classic example is the growth in prostate cancer that can occur when GHRH analogues are used. This can precipitate spinal cord compression if there is an increase in the size of a metastasis adjacent to the spinal cord. The phenomenon is due to the initial **up-regulation** (**p. 320**) of leutinising hormone (LH) and follicle-stimulating hormone (FSH), with a consequent rise in serum testosterone, that precedes the sustained **down-regulation** (**p. 107**) which is the basis of the use of GHRH analogues to treat hormone-sensitive tumours. Using a directly acting anti-androgen (e.g. cyproterone acetate) during the first few weeks of treatment can prevent the problem. An analogous phenomenon can occur when **tamoxifen** (**p. 304**) is used to treat metastatic breast cancer: this may be manifest as hypercalcaemia. This is due to stimulation of tumour growth by the endogenous oestrogen-like activity of the tamoxifen.

Tumour Immunology

Hericourt and Richet attempted to raise an antiserum for cancer treatment by injecting animals with cancer cells. They did this in 1895, the year that X-rays were discovered. Immunologically based therapies for cancer are, in theory, enticing but have, in practice produced remarkably few effective treatments. There is just a hint that this is about to change: the apparent effectiveness of monoclonal antibody therapy in the adjuvant treatment of colorectal cancer may represent the first sign of a renaissance for therapies based on an immunological approach to cancer. Unfortunately, the historical record shows that this approach to treatment has, in general, promised much, but delivered little.

Experiments using induced tumours in animals have demonstrated antigenic differences between tumours and normal tissues: there is the potential for discriminating between self and non-self. Unfortunately, induced tumours in animals are very different from spontaneous tumours in humans. Encouraging experiments in animals do not simply translate into therapeutic success in humans.

There are several possible mechanisms for the relative lack of effectiveness of immunotherapy for cancer (discussed in relation to **tumour-associated antigens** (**p. 315**)). This is not to say that immunologically mediated killing of cancer cells is unimportant — simply that it is difficult to exploit therapeutically. Immunological mechanisms may be responsible for the elimination of the small number of residual clonogenic cells remaining after chemotherapy or radiotherapy. Similarly, immune surveillance may detect and eliminate circulating cancer cells and thereby decrease the likelihood of metastases developing.

The immune system has some role to play in the prevention of cancer, but its role appears to be restricted to particular tumour types.

The evidence for this statement comes from observations made on patients treated with immunosuppressive therapy. These patients have an overall incidence of malignant disease higher than would be expected. The tumours are mainly of two types: lymphomas and Kaposi's sarcoma. If the intensity of the immunosuppression is reduced, these tumours will sometimes regress. A similar observation comes from **HIV** (**p. 151**) infection where, once again in an immunocompromised population, there is an increased incidence of lymphomas and Kaposi's sarcoma.

Immunotherapy has not, until recently, had any major impact on the treatment of common tumours. A recent European study has shown that a fairly simple form of immunotherapy, simply administering a monoclonal antibody to colorectal tumours during the postoperative period, can significantly improve survival in patients with surgically resected cancer of the colon. If these early results are confirmed, then this will be the first time a clinically feasible form of immunotherapy has significantly contributed to the management of a common tumour.

More elaborate immunological therapies have had some minor successes against advanced disease. Therapy with **interleukin 2** (**IL-2**) (**p. 164**), which stimulates the production of cytotoxic T-lymphocytes, can produce responses in patients with metastatic malignant melanoma or renal cancer. In some studies IL-2 stimulates a specific subpopulation of lymphocytes, tumour-infiltrating lymphocytes (TIL), harvested from the patient's own tumour. In the longer-term, it is likely that immunologically based treatments will have more of a role to play in adjuvant therapy than in the management of more advanced disease.

Tumour Necrosis Factors (TNF)

There are two main types of TNF: α, derived mainly from monocytes and macrophages, and β, derived predominantly from lymphoid cells. They are structurally very similar. Both are capable of producing receptor-mediated **cell death** (**p. 55**) in malignant cells. Both molecules appear to share a common receptor. TNF genes are on chromosome 6p21.3 close to the *MHC HLA-B* locus.

TNF's functions as inflammatory cytokines are beneficial in small quantities and fatal in excess. Excessive production of, or sensitivity to, TNF may be responsible for tissue damage and death in patients with septic shock or malaria. It may also mediate cancer-associated cachexia. Steroids inhibit the translation of TNF-α. The ability of the drug **pentoxifylline** (**p. 231**) to reverse radiation-induced fibrosis is, in part, related its ability to prevent lipopolysaccharide induced transcription of TNF-α.

Tunel (Terminal dUTP-Biotin Nick End-Labelling)

A means of staining cells to identify **apoptosis** (**p. 19**). Apoptotic cells can be identified, and scored, on tissue sections. Apoptosis, unlike necrosis, is an ordered process and the Tunel technique recognises the ends of the DNA that have been neatly nicked as part of the cellular breakdown which occurs during apoptosis.

Turcot's Syndrome

A variant of **hereditary non-polyposis** (**p. 150**) **colorectal cancer** (**HNPCC**) in which glioblastoma and HNPCC coexist. Cerebral tumours usually present when <20 years of age.

TWiST

TWiST — time without symptoms of disease or toxicity — is an outcome measure pioneered by the Ludwig breast cancer group, which adjusts survival according to whether a patient has symptoms related to either progressive disease or treatment-related toxicity. The advantage of the approach is that it recognises that patients with progressive disease have symptoms and that treatment, although toxic in the short-term, might delay or prevent recurrence so that the overall balance is in favour of treatment rather than with withholding chemotherapy. **Q-TWiST (p. 253)** is a variant that incorporates a **quality factor (p. 253)** into the calculation of survival time.

Ubiquitin

A small protein that is, in evolutionary terms, highly conserved and which might function as a molecular **chaperone** (**p. 62**) and also be involved in the heat shock response. It is only 76 amino acids long and has a molecular weight of ~8500. It is involved in protein turnover and degradation and may be responsible for some types of **post-translational modification** (**p. 240**). The addition of ubiquitin to a protein, ubiquitination, usually marks that protein out for destruction.

Uncertainty Principle

There is some confusion here. Historically, the uncertainty principle was first formulated by Werner Heisenberg in 1927, who pointed out that we can know the position of an electron, or its momentum, but that we cannot know, at any given instant, both position and momentum. The act of measuring the one interferes with the measurement of the other. Put another way, observers influence observations.

The Oxford school of meta-analysts has used the term 'uncertainty principle' more recently. They use the term to indicate that a clinician is in a state of **equipoise** (**p. 118**) concerning whether a particular intervention is appropriate for an individual patient. In such circumstances, it is ethical for the clinician to enter the patient into a **randomised-controlled trial** (**p. 261**). The term 'clinical equipoise' is more useful in this context than 'uncertainty principle'. It avoids historical confusion and describes, both accurately and parsimoniously, the clinician's state of mind.

Uncomplicated Control

A useful concept in the analysis of **dose–response** (**p. 103**) relationships: if both the tumour control probability and the

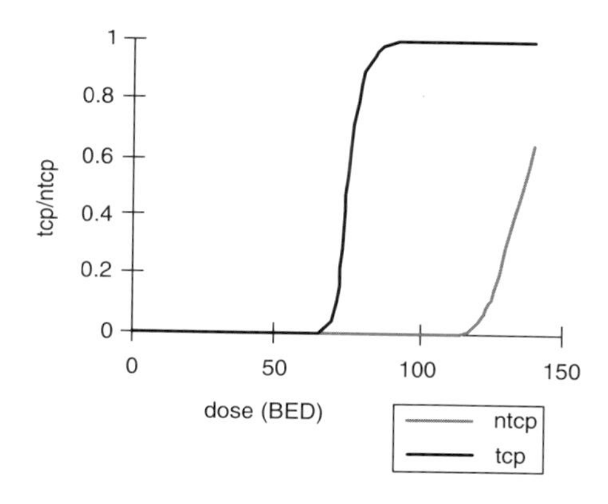

Figure 1.

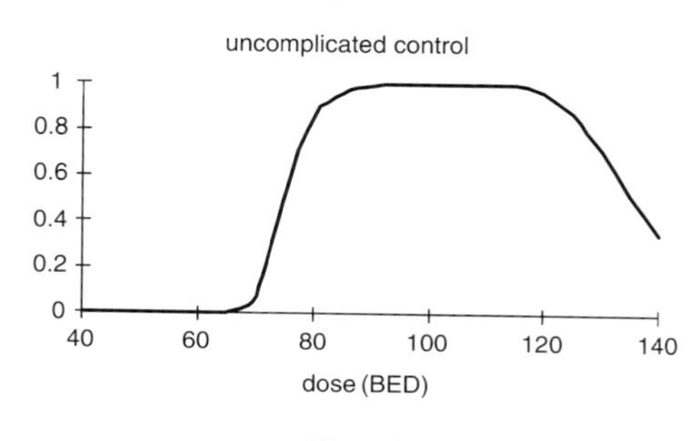

Figure 2.

probability of a complication are calculated for a range of doses, then the **probability** (**p. 246**) of uncomplicated control is simply, for each dose, the difference between the two probabilities (figure 1).

In figure 1, tcp is tumour control probability and ntcp is normal tissue complication probability. For each dose a probability can be calculated using: ntcp − cp. This is the probability of uncomplicated control (figure 2).

Provided the requisite data are available, this is an excellent method for defining the optimal therapeutic dose. The initial proviso is crucial: how do we, ethically, obtain such data using human subjects?

Unlabelled Use

Unlabelled use is when a drug is used to treat a condition for which it is not licensed.

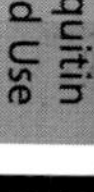

The use of the anti-helminthic agent levamisole to treat colorectal cancer was an example of unlabelled use. The term is slightly inaccurate since the bottle still has a label on it; **'off-label'** (**p. 218**) use is possibly a better term. It is related to compassionate use, but in the case of off-label use the drug is at least licensed for something.

Unplanned Gaps in Treatment

Unplanned gaps in treatment are a cause of increased overall treatment time in fractionated radiotherapy. **Repopulation** (**p. 267**) implies that such gaps may adversely affect treatment outcome and this fact is supported by abundant clinical data showing that there is a detectable loss of local tumour control when treatment is prolonged by even a few days. Unplanned gaps can be due to department-related problems (machine service days, machine breakdown, public holidays, staff sickness, transport failure) or they can be related to patients' problems (intercurrent illness, transport difficulties, rest to allow acute reactions to settle, general recalcitrance, domestic crises such as ill children or family funerals). One of the problems in identifying unplanned gaps is that overall treatment time is not explicit in most prescriptions of radiotherapy — it is only implied. Starting on a Monday, as opposed to a Thursday, may make a difference since the number of weekend gaps will increase.

There are various practical methods for compensating for unplanned gaps. The most straightforward is to keep the overall time as planned so making up for missed treatments by treating either twice per day or at weekends. An alternative is to allow the overall treatment time to lengthen but to adjust the dose using a **linear–quadratic** (**LQ**) (**p. 180**) formula with a time component. The NSD formula, or its derivatives, should no longer be used for such calculations.

Unplanned gaps affect a significant proportion of patients treated with radiotherapy. In some series, nearly 50% of patients have suffered unscheduled interruptions to treatment of up to 5 days and nearly one-quarter of patients have suffered delays of >10 days.

Up-Regulation

This rather clumsy term describes the increased expression of genes in certain circumstances. For example, in some breast cancers there is increased production of thymidine **phosphorylase** (**p. 234**), particularly in hypoxic areas: thymidine phosphorylase can, therefore, be described as being up-regulated in breast cancer. The opposite of up-regulation is down-regulation. The processes of up- and down-regulations are crucial to normal homeostasis: an excess of product can inhibit (down-regulate) the gene responsible for production. It is, as one might expect, the opposite of **down-regulation** (**p. 107**).

Utility

In its simplest sense, utility simply means usefulness: this is the sense in which philosophers have used the term. The idea was that subjects could be asked to rank objects or states in order of their utility and this would provide a sensible means for assessing their preferences. The problem is that even under conditions of certainty, human beings are inconsistent. What is a priority on Tuesday may be of no interest by Friday. This inconsistency of rankings has led to the formulation of the impossibility theorem, which simply states that because this type of assessment is impossible we should not bother with it.

Utility also has several specialised meanings. In a **decision analysis** (**p. 88**), where

onditions are uncertain, utilities are the alues applied to the possible outcomes: though even here the usage is not simple. 'he use of the term 'utility' in this context nplies an element of risk, whereas the term value' implies that the outcome is more ertain. In daily life we are prepared to acrifice value to minimise risk: this is the rinciple on which insurance premiums are ased. It is also the basis of adjuvant therapy or cancer. The overall combination of risk nd value can be defined as utility: the uantitative measure of the desirability of an utcome. Measures of utility can be used to efine health states; hence, the emergence of measures such as **quality-adjusted life-year (QALY) (p. 253)**, which are based on assessments of utility. Utility can be assessed using measures of **quality of life (p. 253)** or by other techniques such as the time trade-off or **standard gamble (p. 296)**.

Utility appears in another guise, in the principle of Utilitarianism: the philosophy of John Stuart Mill that the guiding principle of conduct should be the greatest happiness of the greatest number. This is a concept of direct relevance to the economics of healthcare and the allocation of resources.

Validity

A measurement has validity if it is both sound and powerful — in other words, it is not only just accurate and reliable, but also is adequate in terms of its explanatory and/or descriptive power. 'Chubby' is not, for scientific purposes, a particularly valid description of physique; weight is a little more valid but the **body mass index (p. 33)** is more valid still. Psychometricians have embroidered the concept of validity and, in a series of rococo variations, recognise: face validity, construct validity, criterion validity, etc.

The term is used, in a technical sense, to describe the assessment of scales and scoring systems. A variety of bewildering terms has been used to describe different aspects of validity: ultimately, however, the overall concept is fairly simple. Validity simply reflects the extent to which a test or scale fulfils the hypothesis that the method measures what it claims to measure and that, by so doing, it has a descriptive and predictive value.

Face validity assesses whether the items within a test instrument are relevant to that which is to be measured: an item concerning sleep disturbance might be relevant to a study on fatigue but would have little direct relevance to a study on alopecia. Face validity is not always completely desirable and when dealing with sensitive areas of enquiry subtler approaches may be required. The question 'Since being treated for cancer, have you turned to a life of crime?' is unlikely to produce an honest response, even though it might tell a great deal about how the experience of cancer might influence patients' attitudes to society and its laws.

Content validity assesses whether an instrument covers the whole area of relevance to the subject under consideration. A measure of **quality of life (p. 253)** would be incomplete without an assessment of a patient's ability to get out and about; a scoring system for radiation-induced skin changes would lack content validity if it simply recorded degree of erythema and did not include moist desquamation.

Criterion validity measures the extent to which a new scale produces results that agree with those produced by a standard reference measure — the so-called **Gold Standard (p. 141)**. Currently, the EORTC-QLQ C30 questionnaire might be regarded as a Gold Standard for measures of quality of life. As part of its own validation, EORTC-QLQ C30 was compared with a previous Gold Standard: **the SIP (sickness impact profile) (p. 285)**. For many outcomes in oncology there simply are no Gold Standards and we end up, rather unsatisfactorily, assessing one imperfect measure against another and, thereby, perpetuating error.

Construct validity is a difficult concept and has to do with the extent to which a scale adds to, or is consistent with, current thinking about concepts or constructs. Quality of life is an excellent example of a construct. It is not, like lung cancer, an entity for which a pathognomonic test exists. Quality of life, like intelligence, is defined largely in terms of how it is measured. It consists of a more-or-less defined grouping of domains (physical, psychological, spiritual, social, emotional) that cleave together and, in an overall sense, define the extent to which an individual's aspirations are met by their experiences. Since concepts and constructs change and evolve, construct validity is a relatively fluid notion. We are in Humpty Dumpty's world where 'words mean exactly what I want them to mean, no more, no less'

Variable

A variable is a quality or quantity that, within a series of observations, is not onstant. Consider a string quartet playing eethoven's Opus 132. The quality 'alive' is ot a variable, it is a constant since, if any ember of the quartet were not alive, the uartet would only be a trio. The quality ender' is a variable; the quartet could be omprised of varying proportions of men nd women. The quantity 'height' is also a ariable since it is not a prerequisite of a ring quartet that all members be exactly qual in height.

'ariables and variability are the very usiness of statistics. The appropriateness f particular statistical techniques will epend on the type of variable being onsidered. Classification of variables, and n understanding thereof, is an essential rst step in the practical use of statistical iethods. As a simple example, if people re being counted, such as the number aiting for radiotherapy treatment, the iedian is a more logical measure than the iean: the mean could be 83.67, which ould leave an unsolvable problem — the actional person.

'here is a taxonomy of variables:

Qualitative (categorical)

Cannot be measured numerically.

- Binary (all-or-none, quantal, existential): only two categories possible, e.g. alive or dead, pregnant or non-pregnant.
- Binary condensed: in which the **Procrustean method (p. 246)** is used to condense variables, which could have multiple categories, into only two categories, e.g. 'Are you breathless, yes or no?'
- Unordered categories: these are categories which have no natural gradient: brown eyes are neither worse nor better than blue eyes, which, in turn, are different from, but in no way superior to, green eyes, or even grey eyes, or violet eyes. Of course, matters of taste might intrude and, indeed, Elizabeth Taylor is famous for her violet eyes.
- Ordered categories: the variable can be ranked but (cf. a discrete metric scale) we cannot assume that the intervals between ranks are equal, e.g. activity classified on the MRC scale:
  - — at work or active retirement
  - — full activity but not at work
  - — out and about but activity restricted
  - — confined to home or hospital
  - — bedridden.

Quantitative

Can be measured numerically.

- Discrete and metric: scale has equal intervals but the only possible values are whole numbers (integers), e.g. number of patients waiting for radiotherapy, colony counts after irradiation of cells in a Petri dish.
- Continuous variable: can have any quantitative value with no abrupt incremental transitions. In practice, a limitation is imposed by the accuracy of the measuring instrument. Height, for example, is a continuous variable and in theory could be measured to the umpteenth decimal place. Rulers are not so accurate and one usually measures height (of patients at least) no more accurately than 1.35 m rather than 1.35291057836 m. The assumption that variables are continuous is extremely useful since many statistically useful formulae are based on it. For example, the Normal distribution, and all the tests based on it, assumes that a variable is continuous.

Vectors for Gene Therapy

Gene therapy (p. 137) requires more than just an idea and more than the production of a construct that might put the idea into practice: the delivery of the construct to the potential target is the crucial component. Vectors are delivery systems designed to achieve this goal. The lack of appropriate and reliable vector systems represents one of the main practical obstacles to successful gene therapy.

Currently available vectors can be classified as:

- Viral:
  - retrovirus (10 kb) can only deliver to dividing cells
  - adenovirus (35 kb)
  - adeno-associated virus (5 kb)
  - avipox virus (260 kb)
  - baculovirus (80–230 kb).
- Synthetic:
  - **liposomes** (**p. 182**)
  - protein–DNA complex: the protein is selected to deliver the DNA (gene) to the specific target.
- Direct/mechanical: the construct is physically placed within the target:
  - intratumoural injection
  - **gene gun** (**p. 136**).

VEGF (Vascular Endothelial Growth Factors)

This family of **growth factors** (**p. 144**) is concerned with the formation of new blood vessels. It promotes survival of endothelial cells and increases cell motility. It also tends to increase vascular permeability. Overexpression of VEGF leads to a vascular pattern characterised by increased vessel length and increased permeability.

Vesicant

A vesicant is any agent, whether inhaled, applied topically or injected intravenously, that can cause blistering of the skin or mucous membranes. Many **cytotoxic drugs** (**p. 84**) such as the anthracyclines (**p. 16**), **vinca alkaloids** (**p. 324**) and mustine are severe vesicants. Inadvertent extravasation after attempted intravenous administration can cause extensive tissue damage. Some drugs, such as **5-fluorouracil** (**5FU**) (**p. 130**), are less irritant but may cause a superficial thrombophlebitis.

Vinblastine

Vinblastine is a **vinca alkaloid** (**p. 324**). It is similar to vincristine but is less likely to cause neurotoxicity and more likely to cause myelosuppression. Its spectrum of activity is similar to that of vincristine.

Vinca Alkaloids

The vinca alkaloids are a family of cytotoxic drugs derived from the periwinkle plant, *Catharanthus rosea*. They act as **spindle poisons** (**p. 293**): disrupting the structure and function of the microtubules during the formation of the **mitotic spindle** (**p. 201**). Members include vincristine, vinblastine, and the semi-synthetic derivative, vinorelbine.

Vincristine

Vincristine is a vinca alkaloid and acts to disrupt the function of microtubules during **mitosis** (**p. 201**). As such it is a **spindle poison** (**p. 293**). It is derived, as with the other vinca alkaloids, from the periwinkle, *Catharanthus rosea*. There are two types of binding site for vincristine on microtubules: high-affinity and low-affinity. Binding to the high-affinity site interferes with the formation of microtubules, whereas binding to the low-affinity site causes disintegration of tubules already formed. The main toxicity of **vincristine** (**p. 324**) is neuropathy due to disruption of the function of neurotubules and the consequent interference with axonal transport. Autonomic neuropathy might cause diagnostic confusion: adynamic ileus caused by vincristine may be mistaken for intestinal obstruction.

Vincristine has a broad spectrum of clinical activity and, because it has little bone marrow toxicity, is particularly useful in combination therapy.

It is sometimes described as a **stathmokinetic** (**p. 297**) agent. This alludes to the fact that, if exposure is not prolonged, the block to **mitosis** (**p. 201**) may only be temporary. This phenomenon is exploited in the stathmokinetic method for assessing cellular proliferation.

Vinorelbine

Vinorelbine is a derivative of **vinblastine** (**p. 324**). It is less neurotoxic and has a broader spectrum of antitumour activity. Its mode of action is similar to that of the natural **vinca alkaloids** (**p. 324**), vincristine and vinblastine. It binds to tubulin and disrupts the formation and function of the microtubules during mitosis. It can be given orally or intravenously. Vinorelbine has activity against non-small cell lung cancer, head and neck cancer and breast cancer, as well as against lymphomas and Kaposi's sarcoma. It is moderately myelosuppressive and neurotoxic, and might cause pain at the tumour site at the time of injection.

VIP-oma

A tumour, usually benign, of the endocrine system that produces an excess of vasoactive intestinal peptide (VIP). The symptoms and signs give rise to the acronym for the associated syndrome: WDHA (watery diarrhoea hypokalaemia achlorhydria). The eponym is Werner–Morrison syndrome. VIP-omas can occur as part of the multiple endocrine neoplasia (MEN) syndromes.

Visual Analogue Scale

A technique used to obtain and quantify **responses to questions** (**p. 270**). The subject is asked to indicate their response on a line, usually 10 cm long. The ends of the line are clearly labelled, 'anchored' in the jargon, with two contrasting statements such as 'a great deal'/'not at all' or 'strongly agree'/'strongly disagree'. The distance (cm) along the line to the subjects' marks is measured and the result divided by 10 to give a score for that particular item. Calculated in this way, scores will all lie in the range 0–1.0. This technique is sometimes called **LASA** (**linear analogue self-assessment**) (**p. 173**) .

Vitalism

A form of explanation in which mysterious (and unknowable) forces are invoked to explain the difference between the living and the inanimate: the spark between the hand of God and that of Adam shown on Michelangelo's Sistine Chapel ceiling is its metaphorical apotheosis. The concept of **reductionism** (**p. 263**) is in direct opposition to such mystical notions.

Volume Doubling Time (T_d)

Time (days) taken for a tumour to double in volume. It is not the same as the time taken for the clonogenic cells in a tumour to double in number. The presence of stroma within tumours usually means that, since the stroma is clonogenically irrelevant, V_d is longer than the time it takes for the number of clonogens to double (clonogen doubling time).

There are three main factors that determine or influence V_d: **cell cycle (p. 52)** time, **growth fraction (p. 144)**, **cell loss factor (p. 56)**. V_d for human tumours varies widely: values for metastases are typically shorter than those for the corresponding primary tumours. Typical values would be:

Site	**T_d primary (days)**	**T_d metastasis (days)**
Breast	65–135	10–100
Colorectal	400–1000	80–120
Lymphoma	10–40	n/a

n/a, Not applicable.

Von Hippel Lindau Syndrome

This condition is inherited as an autosomal dominant. Clinically, there are haemangiomas affecting the retina and CNS (e.g. cerebellum) associated with renal cancer and phaeochromocytomas. The abnormality affects the *VHL* gene (3p25–p26). The gene product might be concerned in transmitting growth signals induced by changes in oxygen tension.

The development of malignant tumours falls into three main patterns:

- Renal carcinoma without phaeochromocytoma.
- Renal carcinoma plus phaeochromocytoma.
- Phaeochromocytoma alone.

Most patients die from renal cancer by the time they are 50 years of age.

WAGR Syndrome

syndrome associated with some familial ses of **Wilms' tumour (p. 328)**: Wilms', iridia, gonadoblastoma, ambiguous nitalia, genitourinary abnormalities, mihypertrophy, mental retardation. The netic abnormality is at chromosome p13, close to the catalase **locus (p. 183)**.

Web, World Wide Web

ne Web is a network of communication that n be accessed using a personal computer C) equipped with a link (modem) to an ernet service provider (ISP). The basic ncepts behind the Web include:

Open access both to providers of information and those who wish to obtain information. This implies no censorship or review — fly posting, as it were.

Linkages (p. 181) between sources so that an enquiry can move seamlessly from one area to another using single keystrokes (hyperlinks). The Web can be regarded as an enormous reference library with unlimited indexing and search capabilities.

ommercial pressures have eroded some of e more altruistic aspects of the Web and it now possible to acquire goods and services er it in exchange for payment. Many entific journals can be accessed in this way: stracts are usually free, but payment or bscription is required for the full text of icles.

arch engines provide a means of nplifying browsing. These are usually run mmercially and are funded by advertising pearing on the initial (home) Web page: g. Yahoo, Lycos, Infoseek. A vast amount health-related information is available on e Web, but, given the principle of open cess, little of it has been checked for curacy: *caveat surfor*.

Wedge

A fixed wedge is a piece of absorbent material usually of brass, steel or lead, which is triangular in cross-section and introduced into a beam of radiation to modulate its intensity. Wedges have two main uses in radiotherapy treatment planning: to compensate for the oblique incidence of a beam crossing a curved surface (e.g. neck); and to compensate for the dosage fall-off at depth from a field at right angles to the wedged field.

A **dynamic wedge (p. 110)** fulfils the same function as a physical wedge. The wedging is accomplished using software that controls the movements of the collimators so that the **dose distribution (p. 102)** is equivalent to that of a wedged field.

Wedge Angle

Used to define the steepness of the change in **isodose (p. 169)** that occurs when a wedge is introduced into a radiation field. It can be defined as the angle between the tangent to the 50% isodose and the central axis.

Weighted Average

A method for making some observations (perhaps those we have made ourselves) more important than others (perhaps those made by other people). Each value is given a weight (>0) and is then multiplied by its weight:

$$\text{Weighted average} = \frac{\text{sum (each value} \times \text{its weight)}}{\text{sum of the weights}}.$$

Western Blotting

A blotting technique, analogous to **Southern blotting (p. 292)**, used to identify protein fragments. Identification and banding usually uses antibodies to the fragments of interest.

White Literature

Information and data published in accessible journals, and which therefore can be easily identified and included in a **systematic review (p. 302)**, is sometimes referred to as the white literature.

Wilcoxon Gehan Statistic

A technique used in survival analysis that places greater emphasis on events occurring earlier during follow-up, when the numbers at risk of an event occurring are higher. It is not particularly popular, mainly because its accuracy is dependent on there being few censored results.

Wilcoxon Matched-Pairs Signed-Rank Test

A non-parametric test that, for paired data, assesses whether one set of observations is significantly different from another. This is the classic test to use for before and after measurements on a treated cohort in which each subject can serve as their own control.

The differences between the pairs of observations are ranked: under the null hypothesis (of no difference), the sum of the positive ranks should, ignoring the sign, equal the sum of the negative ranks.

Wilcoxon Rank-Sum Test (syn. Mann–Whitney Statistic)

Unlike the Wilcoxon matched pair signed ranks test, this test can be used for unmatched, unpaired, data. It is a **non-parametric (p. 227) test** based on ranking the pooled data and then assessing whether it is possible that the observed rank order could have arisen by chance or whether there is a systematic difference between the two groups: with higher ranks predominating in one group and lower rank in the other.

Will Rogers' Phenomenon

Another term used to describe **stage shift (p. 294)**: 'when the Okies moved to California they raised the intelligence level both states'. Easson, as long ago as the 1940 perhaps originally described the concept.

Wilms' Tumour (Nephroblastoma)

This eminently treatable tumour of childhood may occur in familial forms. One form, the **WAGR syndrome (p. 327** is associated with abnormalities close to the locus of the catalase gene on chromosome 11p13. There are several other candidate genes for the cause of the tumour: *Wt*1, *WT*33, *LK*15. One is a zinc finger gene at 11p13; another is at 11p15.5. Apart from the WAGR syndrome, there are several other syndromes associated with the tumour including: Denys–Drash, Frasier and Wiedemann–Beckwith.

Writhing Number

A term used to describe the supercoiling of DNA: it is simply the number of turns in th superhelix of a supercoiled molecule:

Linking number = number of double helical tur + writhing number.

Xenopus Oocyte Expression System

The clawed frog (*Xenopus laevis*) produces large oocytes. Micro-injection of exogenous mRNA into the nucleus or cytoplasm of the oocyte, or of cDNA into the nucleus, will, in appropriate circumstances, cause expression of the protein coded by the injected nucleotide sequence. The protein can then be detected and characterised. In essence, this is a system for manufacturing protein, in quantities that can be analysed, from nucleotide sequences.

Xeroderma Pigmentosa

An autosomal recessive inherited disorder characterised clinically by skin excessively sensitive to sunlight and a high incidence of skin malignancy (BCC, SCC, melanoma). The incidence of the disease (homozygotes) is ~1/100 000 live births. The basic mechanism involved is the inability of the cells to repair UV-induced **DNA damage (p. 98)**. The genetic abnormalities are heterogeneous (9q34.1; 2q21; 3p25.1; 13q32–q33, etc.). The condition is associated with a high incidence of leukaemia as well as with excessive normal tissue damage following therapeutic doses of radiation.

Y Chromosome

The smaller sex chromosome. It carries few genes and confers maleness.

Y-linked Transmission

A trait passed from father to son via the Y chromosome, in contrast to X-linked conditions such as haemophilia in which the disease is passed from mother to son. There are no clinically important Y-linked conditions: hairy ear rims are, in certain parts of India, transmitted as a Y-linked trait.

Yeast Artificial Chromosome (YAC)

A vector constructed using components from yeast chromosomes that can replicate *in vivo* and can, therefore, be used to maintain long lengths of DNA (up to 1000 kb). YACs are, therefore, extremely useful in DNA libraries.

his is identical to the Normal score and e standardised Normal deviate. It is a ansformation of the raw data in a normally stributed set of data. It is calculated as:

$$= \frac{(\text{untransformed value} - \text{mean})}{\text{SD}},$$

here SD is standard deviation.

of a Normally distributed variable has a ean of zero and SD = 1.0.

elen (Pre-Consent andomization)

procedure devised by Zelen to circumvent e problems arising from obtaining formed consent to randomised allocation experimental treatment and thereby to crease recruitment to clinical trials.

tients eligible for a trial are randomised ther to the experimental treatment or to the conventional management. Those who draw standard therapy are treated conventionally and are never informed of their participation in the study. Patients allocated to the experimental treatment are approached and asked, using the appropriate methods for obtaining informed consent, if they wish to participate in the study. The design has pragmatic attractions but, ethically, appears slightly dubious. Control patients have been unwittingly entered into a trial, never being told that they might have been allocated an experimental therapy.

Zinc Finger Proteins/Motifs

These proteins are important in the interactions between protein and DNA, particularly during **transcription** (**p. 313**). **The Wilms' tumour** (**p. 328**) suppressor gene product and the oestrogen receptor protein both contain zinc finger motifs. These proteins, since disturbing their structure and function might disrupt cell division, are an attractive target for cancer treatment.